Praise for Brian Greene's

THE HIDDEN REALITY

"If extraterrestrials landed tomorrow and demanded to know what the human mind is capable of accomplishing, we could do worse than to hand them a copy of this book."
— *The New York Times Book Review*

"The multiverse is an idea whose time has come. . . . The book serves well as an introduction . . . and will open up many people's eyes." — *The Wall Street Journal*

"Greene takes us down the rabbit hole yet again, this time setting a course for the terra incognita of parallel universes, hidden worlds, alternate realities, holographic projections, and multiverse simulations. Greene likes to drop you into the middle of the action first and then explain the backstory, but he has an elegant knack for anticipating questions and immediately dealing with any confusion or objections." — *The Daily Beast*

"An accessible and surprisingly witty handbook to parallel universes. . . . Greene is immensely gifted at finding apt and colorful everyday analogies for the arcane byways of theoretical physics." — *The Toronto Star*

"Mind-stretching. . . . [*The Hidden Reality* is] Greene's impassioned argument 'for the capacity of mathematics to reveal hidden truths about the workings of the world.'" — *The New Yorker*

"Like [Stephen] Hawking and [Roger] Penrose before him, [Greene] is an author who writes with the confidence and authority of one who . . . has seen the promised land of cosmic truth." — *Bookforum*

"The best guide available, in this universe at least."

—*Science News*

"Greene's greatest achievement is that even as you grapple with these allusive concepts, you start falling in love with these mysteries."

—*The Express Tribune*

BRIAN GREENE
THE HIDDEN REALITY

Brian Greene received his undergraduate degree from Harvard University and his doctorate from Oxford University, where he was a Rhodes Scholar. He joined the physics faculty of Cornell University in 1990, was appointed to a full professorship in 1995, and in 1996 joined Columbia University, where he is professor of physics and mathematics. He has lectured at both a general and a technical level in more than twenty-five countries and is widely regarded for a number of groundbreaking discoveries in superstring theory.

His first book, *The Elegant Universe*, was a national bestseller and a finalist for the Pulitzer Prize. His second book, *The Fabric of the Cosmos*, was also a bestseller.

He lives in Andes, New York, and New York City.

www.briangreene.org

THE
HIDDEN
REALITY

5

THE
HIDDEN
REALITY

Parallel Universes and the

Deep Laws of the Cosmos

Brian Greene

VINTAGE BOOKS

A DIVISION OF RANDOM HOUSE, INC.

NEW YORK

The Library of Congress has cataloged the Knopf edition as follows:
Greene, B. (Brian), [date]
The hidden reality: parallel universes and the deep laws of the cosmos /
By Brian Greene. — 1st ed.
p.   cm.
1. Physics — Philosophy.   2. Quantum theory.
3. General relativity (Physics).   4. Cosmology.   I. Title.
QC6.G6885 2011
530.12 — dc22   2010042710

Vintage ISBN: 978-0-307-27812-8

Book designed by Virginia Tan

www.vintagebooks.com

Printed in the United States of America
10 9 8 7 6 5 4 3 2 1

To Alec and Sophia

Contents

Preface

If there was any doubt at the turn of the twentieth century, by the turn of the twenty-first, it was a foregone conclusion: when it comes to revealing the true nature of reality, common experience is deceptive. On reflection, that's not particularly surprising. As our forebears gathered in forests and hunted on the savannas, an ability to calculate the quantum behavior of electrons or determine the cosmological implications of black holes would have provided little in the way of survival advantage. But an edge was surely offered by having a larger brain, and as our intellectual faculties grew, so, too, did our capacity to probe our surroundings more deeply. Some of our species built equipment to extend the reach of our senses; others became facile with a systematic method for detecting and expressing pattern—mathematics. With these tools, we began to peer behind everyday appearances.

What we've found has already required sweeping changes to our picture of the cosmos. Through physical insight and mathematical rigor, guided and confirmed by experimentation and observation, we've established that space, time, matter, and energy engage a behavioral repertoire unlike anything any of us have ever directly witnessed. And now, penetrating analyses of these and related discoveries are leading us to what may be the next upheaval in understanding: the possibility that our universe is not the only universe. *The Hidden Reality* explores this possibility.

In writing *The Hidden Reality*, I've presumed no expertise in physics or mathematics on the part of the reader. Instead, as in my previous books, I've used metaphor and analogy, interspersed with historical episodes, to give a broadly accessible account of some of the strangest and, should they prove correct, most revealing insights of modern physics. Many of the concepts covered require the reader to abandon comfortable modes of thought and to embrace unanticipated realms of reality. It's a journey that's all the more exciting, and understandable, for the scientific twists and turns that have blazed the trail. I've judiciously chosen from these to fill out a landscape of ideas that peak by valley stretches from the everyday to the wholly unfamiliar.

A difference in approach from my previous books is that I've not included preliminary chapters that systematically develop background material, such as special and general relativity and quantum mechanics. Instead, for the most part, I introduce elements from those subjects only on an "as needed" basis; when I find in various places that a somewhat fuller development is necessary to keep the book self-contained, I warn the more experienced reader and indicate which sections he or she may safely skip.

By contrast, the last pages of various chapters segue to a more in-depth treatment of the material, which some readers may find challenging. As we enter those sections, I offer the less experienced reader a brief summary and the option to jump ahead without loss of continuity. Nevertheless, I'd encourage everyone to read as far into these sections as interest and patience allow. While the descriptions are more involved, the material is written for a broad audience and so continues to have as its only prerequisite the will to persevere.

In this regard, the notes are different. The novice reader can skip them entirely; the more experienced reader will find in the notes clarifications or extensions that I consider important but deem too burdensome for inclusion in the chapters themselves. Many of the notes are meant for readers with some formal training in mathematics or physics.

While writing *The Hidden Reality*, I've benefited from critical comments and feedback offered by a number of friends, colleagues, and family members who read some or all of the book's chapters. I'd like to especially thank David Albert, Tracy Day,

Richard Easther, Rita Greene, Simon Judes, Daniel Kabat, David Kagan, Paul Kaiser, Raphael Kasper, Juan Maldacena, Katinka Matson, Maulik Parikh, Marcus Poessel, Michael Popowits, and Ken Vineberg. It is always a joy to work with my editor at Knopf, Marty Asher, and I thank Andrew Carlson for his expert shepherding of the book through the final stages of production. Jason Severs's wonderful illustrations greatly enhance the presentation, and I thank him for both his talent and his patience. It is also a pleasure to offer thanks to my literary agents, Katinka Matson and John Brockman.

In developing my approach to the material I cover in this book, I've benefited from a great many conversations with numerous colleagues. In addition to those already mentioned, I'd like to especially thank Raphael Bousso, Robert Brandenberger, Frederik Denef, Jacques Distler, Michael Douglas, Lam Hui, Lawrence Krauss, Janna Levin, Andrei Linde, Seth Lloyd, Barry Loewer, Saul Perlmutter, Jürgen Schmidhuber, Steve Shenker, Paul Steinhardt, Andrew Strominger, Leonard Susskind, Max Tegmark, Henry Tye, Cumrun Vafa, David Wallace, Erick Weinberg, and Shing-Tung Yau.

I started writing my first general-level science book, *The Elegant Universe*, in the summer of 1996. In the fifteen years since, I've enjoyed an unexpected and fruitful interplay between the focus of my technical research and the topics that my books cover. I thank my students and colleagues at Columbia University for creating a vibrant research environment, the Department of Energy for funding my scientific research, and also the late Pentti Kouri for his generous support of my research center at Columbia, the Institute for Strings, Cosmology, and Astroparticle Physics.

Finally, I thank Tracy, Alec, and Sophia for making this the best of all possible universes.

THE
HIDDEN
REALITY

CHAPTER 1

The Bounds of Reality

On Parallel Worlds

If, when I was growing up, my room had been adorned with only a single mirror, my childhood daydreams might have been very different. But it had two. And each morning when I opened the closet to get my clothes, the one built into its door aligned with the one on the wall, creating a seemingly endless series of reflections of anything situated between them. It was mesmerizing. I delighted in seeing image after image populating the parallel glass planes, extending back as far as the eye could discern. All the reflections seemed to move in unison—but that, I knew, was a mere limitation of human perception; at a young age I had learned of light's finite speed. So in my mind's eye, I would watch the light's round-trip journeys. The bob of my head, the sweep of my arm silently echoed between the mirrors, each reflected image nudging the next. Sometimes I would imagine an irreverent me way down the line who refused to fall into place, disrupting the steady progression and creating a new reality that informed the ones that followed. During lulls at school, I would sometimes think about the light I had shed that morning, still endlessly bouncing between the mirrors, and I'd join one of my reflected selves, entering an imaginary parallel world constructed of light and driven by fantasy.

To be sure, reflected images don't have minds of their own. But these youthful flights of fancy, with their imagined parallel realities, resonate with an increasingly prominent theme in modern science—the possibility of worlds lying beyond the one we know. This book is an exploration of such possibilities, a considered journey through the science of parallel universes.

Universe and Universes

There was a time when "universe" meant "all there is." Everything. The whole shebang. The notion of more than one universe, more than one everything, would seemingly be a contradiction in terms. Yet a range of theoretical developments has gradually qualified the interpretation of "universe." The word's meaning now depends on context. Sometimes "universe" still connotes absolutely everything. Sometimes it refers only to those parts of everything that someone such as you or I could, in principle, have access to. Sometimes it's applied to separate realms, ones that are partly or fully, temporarily or permanently, inaccessible to us; in this sense, the word relegates our universe to membership in a large, perhaps infinitely large, collection.

With its hegemony diminished, "universe" has given way to other terms that capture the wider canvas on which the totality of reality may be painted. *Parallel worlds* or *parallel universes* or *multiple universes* or *alternate universes* or the *metaverse, megaverse,* or *multiverse*—they're all synonymous and they're all among the words used to embrace not just our universe but a spectrum of others that may be out there.

You'll notice that the terms are somewhat vague. What exactly constitutes a world or a universe? What criteria distinguish realms that are distinct parts of a single universe from those classified as universes of their own? Perhaps someday our understanding of multiple universes will mature sufficiently for us to have precise answers to these questions. For now, we'll avoid wrestling with abstract definitions by adopting the approach famously applied by Justice Potter Stewart to define pornography. While the U.S. Supreme Court struggled to delineate a standard, Stewart declared, "I know it when I see it."

In the end, labeling one realm or another a parallel universe is merely a question of language. What matters, what's at the heart of the subject, is whether there exist realms that challenge convention by suggesting that what we've long thought to be *the* universe is only one component of a far grander, perhaps far stranger, and mostly hidden, reality.

Varieties of Parallel Universes

A striking fact (it's in part what propelled me to write this book) is that many of the major developments in fundamental theoretical physics—relativistic physics, quantum physics, cosmological physics, unified physics, computational physics—have led us to consider one or another variety of parallel universe. Indeed, the chapters that follow trace a narrative arc through nine variations on the multiverse theme. Each envisions our universe as part of an unexpectedly larger whole, but the complexion of that whole and the nature of the member universes differ sharply among them. In some, the parallel universes are separated from us by enormous stretches of space or time; in others, they're hovering millimeters away; in others still, the very notion of their location proves parochial, devoid of meaning. A similar range of possibility is manifest in the laws governing the parallel universes. In some, the laws are the same as in ours; in others, they appear different but have a shared heritage; in others still, the laws are of a form and structure unlike anything we've ever encountered. It's at once humbling and stirring to imagine just how expansive reality may be.

Some of the earliest scientific forays into parallel worlds were initiated in the 1950s by researchers puzzling over aspects of quantum mechanics, a theory developed to explain phenomena taking place in the microscopic realm of atoms and subatomic particles. Quantum mechanics broke the mold of the previous framework, classical mechanics, by establishing that the predictions of science are necessarily probabilistic. We can predict the odds of attaining one outcome, we can predict the odds of another, but we generally can't predict which will actually happen. This well-known departure from hundreds of years of scientific thought is surprising enough.

But there's a more confounding aspect of quantum theory that receives less attention. After decades of closely studying quantum mechanics, and after having accumulated a wealth of data confirming its probabilistic predictions, no one has been able to explain why only one of the many possible outcomes in any given situation actually happens. When we do experiments, when we examine the world, we all agree that we encounter a single definite reality. Yet, more than a century after the quantum revolution began, there is no consensus among the world's physicists as to how this basic fact is compatible with the theory's mathematical expression.

Over the years, this substantial gap in understanding has inspired many creative proposals, but the most startling was among the first. Maybe, that early suggestion went, the familiar notion that any given experiment has one and only one outcome is flawed. The mathematics underlying quantum mechanics—or at least, one perspective on the math—suggests that *all* possible outcomes happen, each inhabiting its own separate universe. If a quantum calculation predicts that a particle might be here, or it might be there, then in one universe it *is* here, and in another it *is* there. And in each such universe, there's a copy of you witnessing one or the other outcome, thinking—incorrectly—that your reality is the only reality. When you realize that quantum mechanics underlies all physical processes, from the fusing of atoms in the sun to the neural firings that constitutes the stuff of thought, the far-reaching implications of the proposal become apparent. It says that there's no such thing as a road untraveled. Yet each such road—each reality—is hidden from all others.

This tantalizing *Many Worlds* approach to quantum mechanics has attracted much interest in recent decades. But investigations have shown that it's a subtle and thorny framework (as we will discuss in Chapter 8); so, even today, after more than half a century of vetting, the proposal remains controversial. Some quantum practitioners argue that it has already been proved correct, while others claim just as assuredly that the mathematical underpinnings don't hold together.

Such scientific uncertainty notwithstanding, this early version of parallel universes resonated with themes of separate lands or alternative histories that were being explored in literature, television, and film, creative forays that continue today. (My favorites since childhood include *The Wizard of Oz*, *It's a Wonderful Life*, the *Star Trek* episode "The City on the Edge of Forever," the Borges story "The Garden of Forking Paths," and, more recently, *Sliding Doors* and *Run Lola Run*.) Collectively, these and many other works of popular culture have helped integrate the concept of parallel realities into the zeitgeist and are responsible for fueling much public fascination with the topic. But quantum mechanics is only one of numerous ways that a conception of parallel universes emerges from modern physics. In fact, it won't be the first I'll discuss.

In Chapter 2, I'll begin with a different route to parallel universes, perhaps the simplest route of all. We'll see that if space extends infinitely far—a proposition that is consistent with all observations and that is part of the cosmological model favored by many physicists and astronomers—then there must be realms out there (likely *way* out there) where copies of you and me and everything else are enjoying alternate versions of the reality we experience here. Chapter 3 will journey deeper into cosmology: the inflationary theory, an approach that posits an enormous burst of superfast spatial expansion during the universe's earliest moments, generates its own version of parallel worlds. If inflation is correct, as the most refined astronomical observations suggest, the burst that created our region of space may not have been unique. Instead, right now, inflationary expansion in distant realms may be spawning universe upon universe and may continue to do so for all eternity. What's more, each of these ballooning universes has its own infinite spatial expanse, and hence contains infinitely many of the parallel worlds encountered in Chapter 2.

In Chapter 4, our trek turns to string theory. After a brief review of the basics, I'll provide a status report on this approach to unifying all of nature's laws. With that overview, in Chapters 5 and 6 we'll explore recent developments in string theory that suggest three new kinds of parallel universes. One is string theory's *braneworld* scenario, which posits that our universe is one of potentially numerous "slabs" floating in a higher-dimensional space, much like a slice of bread within a grander cosmic loaf.[1] If we're lucky, it's an approach that may provide an observable signature at the Large Hadron Collider in Geneva, Switzerland, in the not too distant future. A second variety emerges from braneworlds that slam into one another, wiping away all they contain and initiating a new, fiery big bang–like beginning in each. As if two giant hands were clapping, this could happen over and over—branes might collide, bounce apart, attract each other gravitationally, and then collide again, a cyclic process generating universes that are parallel not in space but in time. The third scenario is the string theory "landscape," founded on the enormous number of possible shapes and sizes for the theory's required extra spatial dimensions. We'll see that, when joined with the Inflationary Multiverse, the string landscape suggests a vast collection of universes in which every possible form for the extra dimensions is realized.

In Chapter 6, we'll focus on how these considerations illuminate one of the most surprising observational results of the last century: space appears to be filled with a uniform diffuse energy, which may well be a version of Einstein's infamous cosmological constant. This observation has inspired much of the recent research on parallel universes, and it's responsible for one of the most heated debates in decades on the nature of acceptable scientific explanations. Chapter 7 extends this theme by asking, more generally, whether consideration of universes beyond our own can be rightly understood as a branch of science. Can we test these ideas? If we invoke them to solve outstanding problems, have we made progress, or have we merely swept the problems under a conveniently inaccessible cosmic rug? I've sought to lay bare the essentials of the clashing perspectives, while also emphasizing my own view that, under certain specific conditions, parallel universes fall unequivocally within the purview of science.

Quantum mechanics, with its Many Worlds version of parallel universes, is the subject of Chapter 8. I'll briefly remind you of the essential features of quantum mechanics, then focus on its most formidable problem: how to extract definite outcomes from a theory whose basic paradigm allows for mutually contradictory realities to coexist in an amorphous, but mathematically precise, probabilistic haze. I'll carefully lead you through the reasoning that, in seeking an answer, proposes anchoring quantum reality in its own profusion of parallel worlds.

Chapter 9 takes us yet further into quantum reality, leading to what I consider the strangest version of all parallel universe proposals. It's a proposal that emerged gradually over thirty years of theoretical studies on the quantum properties of black holes. The work culminated in the last decade, with a stunning result from string theory, and it suggests, remarkably, that all we experience is nothing but a holographic projection of processes taking place on some distant surface that surrounds us. You can pinch yourself, and what you feel will be real, but it mirrors a parallel process taking place in a different, distant reality.

Finally, in Chapter 10 the yet more fanciful possibility of artificial universes takes center stage. The question of whether the laws of physics give us the capacity to create new universes will be our first order of business. We'll then turn to universes created not with hardware but with software—universes that might be simulated on a superadvanced computer—and investigate whether we can be confident that we're not now living in someone's or something else's simulation. This will lead to the most unrestrained parallel universe proposal, originating in the philosophical community: that every possible universe is realized somewhere in what's surely the grandest of all multiverses. The discussion naturally unfolds into an inquiry about the role mathematics has in unraveling the mysteries of science and, ultimately, our ability, or lack thereof, to gain an ever-deeper understanding of reality.

The Cosmic Order

The subject of parallel universes is highly speculative. No experiment or observation has established that any version of the idea is realized in nature. So my point in writing this book is not to convince you that we're part of a multiverse. I'm not convinced—and, speaking generally, no one should be convinced—of anything not supported by hard data. That said, I find it both curious and compelling that numerous developments in physics, if followed sufficiently far, bump into some variation on the parallel-universe theme. It's not that physicists are standing ready, multiverse nets in their hands, seeking to snare any passing theory that might be slotted, however awkwardly, into a parallel-universe paradigm. Rather, all of the parallel-universe proposals that we will take seriously emerge unbidden from the mathematics of theories developed to explain conventional data and observations.

My intention, then, is to lay out clearly and concisely the intellectual steps and the chain of theoretical insights that have led physicists, from a range of perspectives, to consider the possibility that ours is one of many universes. I want you to get a sense of how modern scientific investigations—not untethered fantasies like the catoptric musings of my boyhood—naturally suggest this astounding possibility. I want to show you how certain otherwise confounding observations can become eminently understandable within one or another parallel-universe framework; at the same time, I'll describe the critical unresolved questions that have, as yet, kept this explanatory approach from being fully realized. My aim is that when you leave this book, your sense of what might be—your perspective on how the boundaries of reality may one day be redrawn by scientific developments now under way—will be far more rich and vivid.

Some people recoil at the notion of parallel worlds; as they see it, if we are part of a multiverse, our place and importance in the cosmos are marginalized. My take is different. I don't find merit in measuring significance by our relative abundance. Rather, what's gratifying about being human, what's exciting about being part of the scientific enterprise, is our ability to use analytical thought to bridge vast distances, journeying to outer and inner space and, if some of the ideas we'll encounter in this book prove correct, perhaps even beyond our universe. For me, it is the depth of our understanding, acquired from our lonely vantage point in the inky black stillness of a cold and forbidding cosmos, that reverberates across the expanse of reality and marks our arrival.

CHAPTER 2

Endless Doppelgängers

The Quilted Multiverse

If you were to head out into the cosmos, traveling ever farther, would you find that space goes on indefinitely, or that it abruptly ends? Or, perhaps, would you ultimately circle back to your starting point, like Sir Francis Drake when he circumnavigated the earth? Both possibilities—a cosmos that stretches infinitely far, and one that is huge but finite—are compatible with all our observations, and over the past few decades leading researchers have vigorously studied each. But for all that detailed scrutiny, if the universe is infinite there's a breathtaking conclusion that has received relatively scant attention.

In the far reaches of an infinite cosmos, there's a galaxy that looks just like the Milky Way, with a solar system that's the spitting image of ours, with a planet that's a dead ringer for earth, with a house that's indistinguishable from yours, inhabited by someone who looks just like you, who is right now reading this very book and imagining you, in a distant galaxy, just reaching the end of this sentence. And there's not just one such copy. In an infinite universe, there are infinitely many. In some, your doppelgänger is now reading this sentence, along with you. In others, he or she has skipped ahead, or feels in need of a snack and has put the book down. In others still, he or she has, well, a less than felicitous disposition and is someone you'd rather not meet in a dark alley.

And you won't. These copies would inhabit realms so distant

that light traveling since the big bang wouldn't have had time to cross the spatial expanse that separates us. But even without the capacity to observe these realms, we'll see that basic physical principles establish that if the cosmos is infinitely large, it is home to infinitely many parallel worlds—some identical to ours, some differing from ours, many bearing no resemblance to our world at all.

En route to these parallel worlds, we must first develop the essential framework of cosmology, the scientific study of the origin and evolution of the cosmos as a whole.

Let's head in.

The Father of the Big Bang

"Your mathematics is correct, but your physics is abominable." The 1927 Solvay Conference on Physics was in full swing, and this was Albert Einstein's reaction when the Belgian Georges Lemaître informed him that the equations of general relativity, which Einstein had published more than a decade earlier, entailed a dramatic rewriting of the story of creation. According to Lemaître's calculations, the universe began as a tiny speck of astounding density, a "primeval atom" as he would come to call it, which swelled over the vastness of time to become the observable cosmos.

Lemaître cut an unusual figure among the dozens of renowned physicists, in addition to Einstein, who had descended on the Hotel Metropole in Brussels for a week of intense debate on quantum theory. By 1923, he had not only completed his work for a doctorate, but he'd also finished his studies at the Saint-Rombaut seminary and been ordained a Jesuit priest. During a break in the conference, Lemaître, clerical collar in place, approached the man whose equations, he believed, were the basis for a new scientific theory of cosmic origin. Einstein knew of Lemaître's theory, having read his paper on the subject some months earlier, and could find no fault with his manipulations of general relativity's equations. In fact, this was not the first time someone had presented Einstein with this result. In 1921, the Russian mathematician and meteorologist Alexander Friedmann had come upon a variety of solutions to

Einstein's equations in which space would stretch, causing the universe to expand. Einstein balked at those solutions, at first suggesting that Friedmann's calculations were marred by errors. In this, Einstein was wrong; he later retracted the claim. But Einstein refused to be mathematics' pawn. He bucked the equations in favor of his intuition about how the cosmos *should* be, his deep-seated belief that the universe was eternal and, on the largest of scales, fixed and unchanging. The universe, Einstein admonished Lemaître, is not now expanding and never was.

Six years later, in a seminar room at Mount Wilson Observatory in California, Einstein focused intently as Lemaître laid out a more detailed version of his theory that the universe began in a primordial flash and that the galaxies were burning embers floating on a swelling sea of space. When the seminar concluded, Einstein stood up and declared Lemaître's theory to be "the most beautiful and satisfactory explanation of creation to which I have ever listened."[1] The world's most famous physicist had been persuaded to change his mind about one of the world's most challenging mysteries. While still largely unknown to the general public, Lemaître would come to be known among scientists as the father of the big bang.

General Relativity

The cosmological theories developed by Friedmann and Lemaître relied on a manuscript Einstein sent off to the German *Annalen der Physik* on the twenty-fifth of November 1915. The paper was the culmination of a nearly ten-year mathematical odyssey, and the results it presented—the general theory of relativity—would prove to be the most complete and far-reaching of Einstein's scientific achievements. With general relativity, Einstein invoked an elegant geometrical language to thoroughly refashion the understanding of gravity. If you already have a good grounding in the theory's basic features and cosmological implications, feel free to skip three sections ahead. But if you'd like a brief reminder of the highlights, stay with me.

Einstein began work on general relativity around 1907, a time when most scientists thought gravity had long since been explained by the work of Isaac Newton. As high school students around the world are routinely taught, in the late 1600s Newton came up with his so-called Universal Law of Gravity, providing the first mathematical description of this most familiar of nature's forces. His law is so accurate that NASA engineers still use it to calculate spacecraft trajectories, and astronomers still use it to predict the motion of comets, stars, even entire galaxies.[2]

Such demonstrable efficacy makes it all the more remarkable that, in the early years of the twentieth century, Einstein realized that Newton's Law of Gravity was deeply flawed. A seemingly simpleminded question revealed this starkly: How, Einstein asked, does gravity work? How, for example, does the sun reach out across 93 million miles of essentially empty space and affect the motion of the earth? There's no rope tethering them together, no chain tugging the earth as it moves, so how does gravity exert its influence?

In his *Principia*, published in 1687, Newton recognized the importance of this question but acknowledged that his own law was disturbingly silent about the answer. Newton was certain that there had to be something communicating gravity from place to place, but he was unable to identify what that something might be. In the *Principia* he gibingly left the question "to the consideration of the reader," and for more than two hundred years, those who read this challenge simply read on. That's something Einstein couldn't do.

For the better part of a decade, Einstein was consumed with finding the mechanism underlying gravity; in 1915, he proposed an answer. Although grounded in sophisticated mathematics and requiring conceptual leaps unheralded in the history of physics, Einstein's proposal had the same air of simplicity as the question it purported to address. By what process does gravity exert its influence across empty space? The emptiness of empty space seemingly left everyone empty-handed. But, actually, there is something in empty space: *space*. This led Einstein to suggest that space itself might be gravity's medium.

Here's the idea. Imagine rolling a marble across a large metal table. Because the table's surface is flat, the marble will roll in a straight line. But if a fire subsequently engulfs the table, causing it to buckle and swell, a rolling marble will follow a different trajectory because it will be guided by the table's warped and rutted surface. Einstein argued that a similar idea applies to the fabric of space. Completely empty space is much like the flat table, allowing objects to roll unimpeded along straight lines. But the presence of massive bodies affects the shape of space, somewhat as heat affects the shape of the table's surface. The sun, for example, creates a bulge in its vicinity, much like a metal bubble blistering on the hot table. And just as the table's curved surface induces the marble to travel along a curved path, so the curved shape of space around the sun guides the earth and other planets into orbit.

This brief description glides over important details. It's not just space that curves, but time as well (this is what's called spacetime curvature); earth's gravity itself facilitates the table's influence by keeping the marble pressed to its surface (Einstein contended that warps in space and time don't need a facilitator since they *are* gravity); space is three-dimensional, so when it warps it does so all around an object, not just "underneath" as the table analogy suggests. Nevertheless, the image of a warped table captures the essence of Einstein's proposal. Before Einstein, gravity was a mysterious force that one body somehow exerted across space on another. After Einstein, gravity was recognized as a distortion of the environment caused by one object and guiding the motion of others. Right now, according to these ideas, you are anchored to the floor because your body is trying to slide down an indentation in space (really, spacetime) caused by the earth.*

---

*It's easier to envision curved space than curved time, and that's why many popularizations of Einsteinian gravity focus solely on the former. However, for the gravity generated by familiar objects like the earth and sun, it is actually the curvature of time—not space—that exerts the dominant impact. For an illustration, think of two clocks, one on the ground, the other on top of the Empire State Building. Because the ground clock is closer to the earth's center, it experiences slightly stronger gravity than the clock that's high above Manhattan. General relativity shows that because of this, the rate at which time passes on each will be slightly different: the ground clock will run a tiny bit slow (billionths of a second per year) compared to the elevated clock. The temporal mismatch is an example of what we mean by time being curved or warped. General relativity then establishes that objects move toward regions where time elapses more slowly; in a sense, all objects "want" to age as slowly as possible. From an Einsteinian perspective, that explains why an object falls when you let go of it.

Einstein spent years developing this idea into a rigorous mathematical framework, and the resulting *Einstein Field Equations*, the heart of his general theory of relativity, tell us precisely how space and time will curve as a result of the presence of a given quantity of matter (more precisely, matter and energy; according to Einstein's $E = mc^2$, in which E is energy and m is mass, the two are interchangeable).[3] With equal precision, the theory then shows how such spacetime curvature will affect the motion of anything—star, planet, comet, light itself—moving through it; this allows physicists to make detailed predictions of cosmic motion.

Evidence in support of general relativity came quickly. Astronomers had long known that Mercury's orbital motion around the sun deviated slightly from what Newton's mathematics predicted. In 1915, Einstein used his new equations to recalculate Mercury's trajectory and was able to explain the discrepancy, a realization he later described to his colleague Adrian Fokker as so thrilling that for some hours it gave him heart palpitations. Then, in 1919, astronomical observations undertaken by Arthur Eddington and his collaborators showed that distant starlight passing by the sun on its way to earth follows a curved path, just the one that general relativity predicted.[4] With that confirmation—and the *New York Times* headline proclaiming LIGHTS ALL ASKEW IN THE HEAVENS, MEN OF SCIENCE MORE OR LESS AGOG—Einstein was propelled to international prominence as the world's newfound scientific genius, the heir apparent to Isaac Newton.

But the most impressive tests of general relativity were still to come. In the 1970s experiments using hydrogen maser clocks (masers are similar to lasers, but they operate in the microwave part of the spectrum) confirmed general relativity's prediction of the earth's warping of spacetime in its vicinity to about 1 part in 15,000. In 2003, the Cassini-Huygens spacecraft was used for detailed studies of the trajectories of radio waves that passed near the sun; the data collected supported the curved spacetime picture predicted by general relativity to about 1 part in 50,000. And now, befitting a theory that has truly come of age, many of us walk around with general relativity in the palm of our hand. The global positioning system you casually access from your smartphone communicates with satellites whose internal timing devices routinely take account of the spacetime curvature they experience from their orbit above earth. If the satellites failed to do so, the position readings they generate would rapidly become inaccurate. What in 1916 was a set of abstract mathematical equations that Einstein offered as a new description of space, time, and gravity is now routinely called upon by devices that fit in our pockets.

The Universe and the Teapot

Einstein breathed life into spacetime. He challenged thousands of years of intuition, built up from everyday experience, that treated space and time as an unchanging backdrop. Who would ever have imagined that spacetime can writhe and flex, providing the invisible master choreographer of motion in the cosmos? That's the revolutionary dance that Einstein envisioned and that observations have confirmed. And yet, in short order, Einstein stumbled under the weight of age-old but unfounded prejudices.

During the year after he published the general theory of relativity, Einstein applied it on the grandest of scales: the entire cosmos. You might think this a staggering task, but the art of theoretical physics lies in simplifying the horrendously complex so as to preserve essential physical features while making the theoretical analysis tractable. It's the art of knowing what to ignore. Through the so-called *cosmological principle*, Einstein established a simplifying framework that initiated the art and the science of theoretical cosmology.

The cosmological principle asserts that if the universe is examined on the largest of scales, it will appear uniform. Think of your morning tea. On microscopic scales, there is much inhomogeneity. Some H_2O molecules over here, some empty space, some polyphenol and tannin molecules over there, more empty space, and so on. But on macroscopic scales, those accessible to the naked eye, the tea is a uniform hazel. Einstein believed that the universe was like that cup of tea. The variations we observe—the earth is here, there's some empty space, then the moon, yet more empty space, followed

by Venus, Mercury, sprinkles of empty space, and then the sun—
are small-scale inhomogeneities. He suggested that on cosmological scales, these variations could be ignored because, like your tea,
they'd average out to something uniform.

In Einstein's day, evidence in support of the cosmological principle was thin at best (even the case for other galaxies was still being
made), but he was guided by a strong sense that no location in the
cosmos was special. He felt that, on average, every region of the universe should be on a par with every other and so should have essentially identical overall physical attributes. In the years since,
astronomical observations have provided substantial support for the
cosmological principle, but only if you examine space on scales at
least 100 million light-years across (which is about a thousand times
the end-to-end length of the Milky Way). If you take a box that's a
hundred million light-years on each side and plunk it down *here*,
take another such box and plunk it down way over *there* (say, a billion
light-years from *here*), and then measure the average overall properties inside each box—average number of galaxies, average amount of
matter, average temperature, and so on—you'll find it difficult to distinguish between the two. In short, if you've seen one 100-million-
light-year chunk of the cosmos, you've pretty much seen them all.

Such uniformity proves crucial to using the equations of general relativity to study the entire universe. To see why, think of a
beautiful, uniform, smooth beach and imagine that I've asked you
to describe its small-scale properties—the properties, that is, of
each and every grain of sand. You're stymied—the task is just too
big. But if I ask you to describe only the overall features of the
beach (such as the average weight of sand per cubic meter, the
average reflectivity of the beach's surface per square meter, and so
on), the task becomes eminently doable. And what makes it doable
is the beach's uniformity. Measure the average sand weight, temperature, and reflectivity over here and you're done. Doing the
same measurements over there will give essentially identical
answers. Likewise with a uniform universe. It would be a hopeless
task to describe every planet, star, and galaxy. But describing the
average properties of a uniform cosmos is monumentally easier—
and, with the advent of general relativity, achievable.

Here's how it goes. The gross overall content of a huge volume of space is characterized by how much "stuff" it contains; more precisely, the density of matter, or, more precisely still, the density of matter and energy that the volume contains. The equations of general relativity describe how this density changes over time. But without invoking the cosmological principle, these equations are hopelessly difficult to analyze. There are ten of them, and because each equation depends intricately on the others, they form a tight mathematical Gordian knot. Happily, Einstein found that when the equations are applied to a uniform universe, the math simplifies; the ten equations become redundant and, in effect, reduce to one. The cosmological principle cuts the Gordian knot by reducing the mathematical complexity of studying matter and energy spread throughout the cosmos to a single equation (you can see it in the notes).[5]

Not so happily, from Einstein's perspective, when he studied this equation he found something unexpected and, to him, unpalatable. The prevailing scientific and philosophical stance was not only that on the largest of scales the universe was uniform, but that it was also unchanging. Much like the rapid molecular motions in your tea average out to a liquid whose appearance is static, astronomical motion such as the planets orbiting the sun and the sun moving around the galaxy would average out to an overall unchanging cosmos. Einstein, who adhered to this cosmic perspective, found to his dismay that it was at odds with general relativity. The math showed that the density of matter and energy *cannot* be constant through time. Either the density grows or it diminishes, but it can't stay put.

Although the mathematical analysis behind this conclusion is sophisticated, the underlying physics is pedestrian. Picture a baseball's journey as it soars from home plate toward the center field fence. At first, the ball rockets upward; then it slows, reaches a high point, and finally heads back down. The ball doesn't lazily hover like a blimp because gravity, being an attractive force, acts in one direction, pulling the baseball toward earth's surface. A static situation, like a stalemate in a tug-of-war, requires equal and opposite forces that cancel. For a blimp, the upward push that counters downward gravity is provided by air pressure (since the blimp is filled with helium, which is lighter than air); for the ball in midair there is no counter-gravity force (air resistance does act against a ball in motion, but plays no role in a static situation), and so the ball can't remain at a fixed height.

Einstein found that the universe is more like the baseball than the blimp. Because there's no outward force to cancel the attractive pull of gravity, general relativity shows that the universe can't be static. Either the fabric of space stretches or it contracts, but its size can't remain fixed. A volume of space 100 million light-years on each side today won't be 100 million light-years on each side tomorrow. Either it will be larger, and the density of matter within it will diminish (being spread more thinly in a larger volume), or it will be smaller, and the density of matter will increase (being packed more tightly in a smaller volume).[6]

Einstein recoiled. According to the math of general relativity, the universe on the grandest of scales would be changing, because its very substrate—space itself—would be changing. The eternal and static cosmos that Einstein expected would emerge from his equations was simply not there. He had initiated the science of cosmology, but he was deeply distressed by where the math had taken him.

Taxing Gravity

It's often said that Einstein blinked—that he went back to his notebooks and in desperation mangled the beautiful equations of general relativity to make them compatible with a universe that was not only uniform but also unchanging. This is only partly true. Einstein did indeed modify his equations so they would support his conviction of a static cosmos, but the change was minimal and thoroughly sensible.

To get a feel for his mathematical move, think about filling out your tax forms. Interspersed among the lines on which you record numbers are others you leave blank. Mathematically, a blank line signifies that the entry is zero, but psychologically it connotes more.

It means you're ignoring the line because you've determined that it's not relevant to your financial situation.

If the mathematics of general relativity were arranged like a tax form, it would have three lines. One line would describe the geometry of spacetime—its warps and curves—the embodiment of gravity. Another would describe the distribution of matter across space, the source of gravity—the cause of the warps and curves. During a decade of ardent research, Einstein had worked out the mathematical description of these two features and had thus filled in these two lines with great care. But a complete accounting of general relativity requires a third line, one that is on an absolutely equal mathematical footing with the other two but whose physical meaning is more subtle. When general relativity elevated space and time into dynamic participants in the unfolding of the cosmos, they shifted from merely providing language to delineate where and when things take place to being physical entities with their own intrinsic properties. The third line on the general relativity tax form quantifies a particular intrinsic feature of spacetime relevant for gravity: *the amount of energy stitched into the very fabric of space itself.* Just as every cubic meter of water contains a certain amount of energy, summarized by the water's temperature, every cubic meter of space contains a certain amount of energy, summarized by the number on the third line. In his paper announcing the general theory of relativity, Einstein didn't consider this line. Mathematically, this is tantamount to having set its value to zero, but much as with blank lines on your tax forms, he seems to have simply ignored it.

When general relativity proved incompatible with a static universe, Einstein reengaged with the mathematics, and this time he took a harder look at the third line. He realized that there was no observational or experimental justification for setting it to zero. He also realized that it embodied some remarkable physics.

If instead of zero he entered a positive number on the third line, endowing the spatial fabric with a uniform positive energy, he found (for reasons I'll explain in the next chapter) that every region of space would push away from every other, producing something most physicists had thought impossible: *repulsive* gravity. Moreover, Einstein found that if he precisely adjusted the size of the number he put on the third line, the repulsive gravitational force produced across the cosmos would exactly balance the usual attractive gravitational force generated by the matter inhabiting space, giving rise to a static universe. Like a hovering blimp that neither rises nor falls, the universe would be unchanging.

Einstein called the entry on the third line the *cosmological member* or the *cosmological constant*; with it in place, he could rest easy. Or, he could rest easier. If the universe had a cosmological constant of the right size—that is, if space were endowed with the right amount of intrinsic energy—his theory of gravity fell in line with the prevailing belief that the universe on the largest of scales was unchanging. He couldn't explain why space would embody just the right amount of energy to ensure this balancing act, but at least he'd shown that general relativity, augmented with a cosmological constant of the right value, gave rise to the kind of cosmos he and others had expected.[7]

The Primeval Atom

It was against this backdrop that Lemaître approached Einstein at the 1927 Solvay Conference in Brussels, with his result that general relativity gave rise to a new cosmological paradigm in which space would expand. Having already wrestled with the mathematics to ensure a static universe, and having already dismissed Friedmann's similar claims, Einstein had little patience for once again considering an expanding cosmos. He thus faulted Lemaître for blindly following the mathematics and practicing the "abominable physics" of accepting an obviously absurd conclusion.

A rebuke by a revered figure is no small setback, but for Lemaître it was short-lived. In 1929, using what was then the world's largest telescope at the Mount Wilson Observatory, the American astronomer Edwin Hubble gathered convincing evidence that the distant galaxies were all rushing away from the Milky Way. The remote photons that Hubble examined had traveled to earth with a clear message: The universe is not static. It *is* expanding. Einstein's reason for introducing the cosmological constant was thus unfounded. The big bang model describing a cosmos that began enormously compressed and has been expanding ever since became widely heralded as the scientific story of creation.[8]

Lemaître and Friedmann were vindicated. Friedmann received credit for being the first to explore the expanding universe solutions, and Lemaître for independently developing them into robust cosmological scenarios. Their work was duly lauded as a triumph of mathematical insight into the workings of the cosmos. Einstein, by contrast, was left wishing he'd never meddled with the third line of the general relativity tax form. Had he not brought to bear his unjustified conviction that the universe is static, he wouldn't have introduced the cosmological constant and so might have predicted cosmic expansion more than a decade before it was observed.

Nevertheless, the cosmological constant's story was far from over.

The Models and the Data

The big bang model of cosmology includes a detail that will prove essential. The model provides not one but a handful of different cosmological scenarios; all of them involve an expanding universe, but they differ with respect to the overall shape of space—and, in particular, they differ on the question of whether the full extent of space is finite or infinite. Since the finite-versus-infinite distinction will turn out to be vital in thinking about parallel worlds, I'll lay out the possibilities.

The cosmological principle—the assumed homogeneity of the cosmos—constrains the geometry of space because most shapes are not sufficiently uniform to qualify: they bulge here, flatten out there, or twist way over there. But the cosmological principle does not imply a *unique* shape for our three dimensions of space; instead, it reduces the possibilities to a sharply culled collection of candidates. To visualize them presents a challenge even for professionals, but it is a helpful fact that the situation in *two* dimensions provides a mathematically precise analog that we can readily picture.

To this end, first consider a perfectly round cue ball. Its surface is two-dimensional (just as on earth's surface, you can denote positions on the cue ball's surface with two pieces of data—such as latitude and longitude—which is what we mean when we call a shape two-dimensional) and is completely uniform in the sense that every location looks like every other. Mathematicians call the cue ball's surface a *two-dimensional sphere* and say that it has *constant positive curvature*. Loosely speaking, "positive" means that were you to view your reflection on a spherical mirror it would bloat outward, while "constant" means that regardless of where on the sphere your reflection is, the distortion appears the same.

Next, picture a perfectly smooth tabletop. As with the cue ball, the tabletop's surface is uniform. Or nearly so. Were you an ant walking on the tabletop, the view from every point would indeed look like the view from every other, but only if you stayed far from the table's edge. Even so, complete uniformity is not hard to restore. We just need to imagine a tabletop with no edges, and there are two ways of doing so. Think of a tabletop that extends indefinitely left and right as well as back and forth. This is unusual—it's an infinitely large surface—but it realizes the goal of having no edges since there's now no place to fall off. Alternatively, imagine a tabletop that mimics an early video-game screen. When Ms. Pac-Man crosses the left edge, she reappears at the right; when she crosses the bottom edge she reappears at the top. No ordinary tabletop has this property, but this is a perfectly sensible geometrical space called a two-dimensional *torus*. I discuss this shape more fully in the notes,[9] but the only features in need of emphasis here are that, like the infinite tabletop, the video-game screen shape is uniform and it has no edges. The apparent boundaries confronting Ms. Pac-Man are fictitious; she can cross through them and remain in the game.

Mathematicians say that the infinite tabletop and the video-game screen are shapes that have *constant zero curvature*. "Zero" means that were you to examine your reflection on a mirrored tabletop or video-game screen, the image wouldn't suffer any distortion, and as before, "constant" means that regardless of where you examine your reflection, the image looks the same. The difference between the two shapes becomes apparent only from a global perspective. If you took a journey on an infinite tabletop and maintained a constant heading, you'd never return home; on a videogame screen, you could cycle around the entire shape and find yourself back at the point of departure, even though you never turned the steering wheel.

Finally—and this is a little more difficult to picture—a Pringles potato chip, if extended indefinitely, provides another completely uniform shape, one that mathematicians say has *constant negative curvature.* This means that if you view your reflection at any spot on a mirrored Pringles chip, the image will appear shrunken inward.

Fortunately, these descriptions of two-dimensional uniform shapes extend effortlessly to our real interest in the three-dimensional space of the cosmos. Positive, negative, and zero curvatures—uniform bloating outward, shrinking inward, and no distortion at all—equally well characterize uniform three-dimensional shapes. In fact, we are doubly fortunate because although three-dimensional shapes are hard to picture (when envisioning shapes, our minds invariably place them within an environment—an airplane *in* space, a planet *in* space—but when it comes to space itself, there isn't an outside environment to contain it); the uniform three-dimensional shapes are such tight mathematical analogs of their two-dimensional cousins that you lose little precision by doing what most physicists do: use the two-dimensional examples for your mental imagery.

In the table below, I've summarized the possible shapes, emphasizing that some are finite in extent (the sphere, the video-

SHAPE	TYPE OF CURVATURE	SPATIAL EXTENT
Sphere	Positive	Finite
Tabletop	Zero (or "flat")	Infinite
Video-Game Screen	Zero (or "flat")	Finite
Pringles Chip	Negative	Infinite

Table 2.1 *Possible shapes for space consistent with the assumption that every location in the universe is on a par with every other (the cosmological principle).*

game screen) while others are infinite (the endless tabletop, the endless Pringles chip). As it stands, Table 2.1 is incomplete. There are additional possibilities, with wonderful names like the *binary tetrahedral space* and the *Poincaré dodecahedral space*, that also have uniform curvature, but I've not included them because they're harder to visualize using everyday objects. By judicious slicing and paring they can be sculpted from those that I've put in the list, so Table 2.1 provides a good representative sampling. But these details are secondary to the main conclusion: *The uniformity of the cosmos articulated by the cosmological principle substantially winnows the possible shapes for the universe. Some of the possible shapes have infinite spatial extent, while others do not.*[10]

Our Universe

The expansion of space found mathematically by Friedmann and Lemaître applies verbatim to a universe that has any one of these shapes. For positive curvature, use the two-dimensional mental imagery to think of a balloon's surface expanding as it is filled with air. For zero curvature, think of a flat sheet of rubber that is being stretched uniformly in all directions. For negative curvature, mold that rubber sheet into the shape of a Pringles chip and then carry on with the stretching. If galaxies are modeled as glitter evenly sprinkled on any of these surfaces, the expansion of space results in the individual specks of glitter—the galaxies—moving apart from one another, just as Hubble's 1929 observations of distant galaxies revealed.

It's a compelling cosmological template, but if it is to be definitive and complete, we need to determine which of the uniform shapes describes our universe. We can determine the shape of a familiar object, such as a doughnut, a baseball, or a block of ice, by picking it up and turning it this way and that. The challenge is that we can't do so with the universe, and so we must determine its shape through indirect means. The equations of general relativity provide a mathematical strategy. They show that the curvature of space reduces to a single observational quantity: the density of matter (more precisely, the density of matter and energy) in space. If there is a lot of matter, gravity will cause space to curve back on itself, yielding the spherical shape. If there is little matter, space is free to flare outward in the Pringles shape. And if there is just the right amount of matter, space will have zero curvature.*

The equations of general relativity also provide a precise numerical demarcation among the three possibilities. The mathematics shows that "just the right amount of matter," the so-called critical density, weighs in today at about 2×10^{-29} grams per cubic centimeter, which is about six hydrogen atoms per cubic meter or, in more familiar terms, the equivalent of a single raindrop in every earth-sized volume.[11] Looking around, it would surely seem that the universe exceeds the critical density, but that would be a hasty conclusion. The mathematical calculation of the critical density assumes that matter is uniformly spread throughout space. So you need to envision taking the earth, the moon, the sun, and everything else and evenly dispersing the atoms they contain across the cosmos. The question then is whether each cubic meter would weigh more or less than six hydrogen atoms.

Because of its important cosmological consequences, astronomers have been trying for decades to measure the average density of matter in the universe. Their method is straightforward. With powerful telescopes, they carefully observe large volumes of space and add up the masses of the stars they can see as well as the mass of other material whose presence they can infer by studying stellar and galactic motion. Until recently, the observations indicated that the average density was on the low side, about 27 percent of the critical density—the equivalent of about two hydrogen atoms in each cubic meter—which would imply a negatively curved universe.

But then, in the late 1990s, something extraordinary hap-

---

*Given our earlier discussion of how matter curves the region in which it is immersed, you might wonder how there can be *no* curvature even though there's matter. The explanation is that a uniform presence of matter generally curves *spacetime*; in this particular case, there is zero space curvature but nonzero spacetime curvature.

pened. Through some magnificent observations and a chain of reasoning we'll explore in Chapter 6, astronomers realized that they had been leaving out an essential component of the tally: a diffuse energy that appears to be spread uniformly throughout space. The data came as a shock to most everyone. An energy suffusing space? That sounds like the cosmological constant, which, as we've seen, Einstein introduced and then famously retracted eight decades earlier. Had modern observations resurrected the cosmological constant?

We still don't know for sure. Even today, a decade after the initial observations, astronomers have yet to establish if the uniform energy is fixed or if the amount of energy in a given region of space varies over time. A cosmological constant, as its name signifies (and as its mathematical representation by a single fixed number on the general relativity tax form implies), would be unchanging. To account for the more general possibility that the energy evolves, and to also emphasize that the energy does not give off light (explaining why it had for so long evaded detection) astronomers have coined a new term: *dark energy.* "Dark" also describes well the many gaps in our understanding. No one can explain the dark energy's origin, fundamental composition, or detailed properties— issues currently under intense investigation to which we shall return in later chapters.

But, even with the numerous open questions, detailed observations using the Hubble Space Telescope and other earth-based observatories have reached consensus on the *amount* of dark energy that is now permeating space. The result differs from what Einstein long ago proposed (since he posited a value that would yield a static universe, whereas our universe is expanding). That's not surprising. Instead, what's remarkable is that the measurements have concluded that the dark energy filling space contributes approximately 73 percent of the critical density. *When added to the 27 percent of criticality astronomers had already measured, this brings the total right up to 100 percent of the critical density, just the right amount of matter and energy for a universe with zero spatial curvature.*

Current data thus favor an ever-expanding universe shaped like the three-dimensional version of the infinite tabletop or of the finite video-game screen.

Reality in an Infinite Universe

At the beginning of this chapter, I noted that we don't know whether the universe is finite or infinite. The previous sections have laid out the case that both possibilities naturally emerge from our theoretical studies, and that both possibilities are consistent with the most refined astrophysical measurements and observations. How might we one day determine observationally which possibility is right?

It's a tough question. If space is finite, then some of the light emitted by stars and galaxies might cycle around the entire cosmos multiple times before entering our telescopes. Like the repeated images generated when light bounces between parallel mirrors, cycling light would give rise to repeated images of stars or galaxies. Astronomers have looked for such multiple images but as yet haven't found any. This, in itself, doesn't prove that space is infinite, but it does suggest that if space is finite it may be so large that light hasn't had time to complete multiple laps around the cosmic racetrack. And that reveals the observational challenge. Even if the universe is finite, the larger it is the better it can masquerade as infinite.

For some cosmological questions, such as the age of the universe, the distinction between the two possibilities plays no role. Whether the cosmos is finite or infinite, at ever-earlier times, the galaxies would have been squeezed ever closer together, making the universe denser, hotter, and more extreme. We can use today's observations of the rate of expansion, together with theoretical analysis of how that rate has changed over time, to tell us how long it's been since everything we see would have been compressed into a single fantastically dense nugget, what we can call the beginning. And for either a finite or an infinite universe, state-of-the-art analyses now peg that moment at about 13.7 billion years ago.

But for other considerations, the finite-infinite distinction matters. In the finite case, for example, as we consider the cosmos at

ever-earlier times, it's accurate to picture the entirety of space continually shrinking. Although the mathematics breaks down at time
zero itself, it's correct to envision that at moments ever closer to
time zero, the universe is an ever-smaller speck. For the infinite
case, however, this description is wrong. If space is truly infinite in
size, then it always has been and always will be. When it shrinks, its
contents are squeezed ever closer together, making the density of
matter ever larger. But its overall extent remains *infinite*. After all,
shrink an infinite tabletop by a factor of 2 and what do you get? Half
of infinity, which is still infinite. Shrink by a factor of 1 million and
what do you get? Infinity still. The closer to time zero you consider
an infinite universe, the denser it becomes at every location, but its
spatial extent remains unending.

Although observations leave the finite-versus-infinite issue
undecided, I've found that when pressed, physicists and cosmologists tend to favor the proposition that the universe is infinite. Partly,
I think this view is rooted in the historical happenstance that for
many decades researchers paid little heed to the finite video-game
shape, mostly because it is more mathematically complex to analyze. Perhaps the view also reflects a common misconception that
the difference between an infinite and a huge-but-finite universe is
a cosmological distinction that's only of academic interest. After all,
if space is so large that you will only ever have access to a small portion of its entirety, should you care whether it extends for a finite or
for an infinite distance beyond what you can see?

You should. The issue of whether space is finite or infinite has a
profound impact on the very nature of reality. Which takes us to the
heart of this chapter. Let's now consider the possibility of an infinitely big cosmos and explore its implications. With minimal effort,
we'll find ourselves inhabiting one of an infinite collection of parallel worlds.

Infinite Space and the Patchwork Quilt

Let's start simply, back here on earth, far from the vast reaches of an
infinite cosmic expanse. Imagine that your friend Imelda, to satisfy

her penchant for variety in personal attire, has acquired five hundred richly embroidered dresses and a thousand pairs of designer shoes. If each day she wears one dress with one pair of shoes, at some point she will exhaust all possible combinations and duplicate an earlier outfit. It's easy to figure out when. Five hundred dresses and a thousand pairs of shoes yield 500,000 different combinations. Five hundred thousand days is about 1,400 years, so if she lived long enough Imelda would be seen in an outfit she'd already worn. If Imelda, blessed with infinite longevity, continued to cycle through every possible combination, she'd necessarily don each of her outfits an infinite number of times. An infinite number of appearances with a finite number of outfits ensures infinite repetition.

Pursuing the same theme, imagine that Randy, an expert card dealer, has shuffled a gargantuan number of decks, one by one, and neatly stacked each next to the others. Can the order of cards in every shuffled deck be different, or must they repeat? The answer depends on the number of decks. Fifty-two cards can be arranged in 80,658,175,170,943,878,571,660,636,856,403,766,975,289,505, 440,883,277,824,000,000,000,000 different ways (52 possibilities for which card will be the first, times 51 remaining possibilities for which will be the second, times 50 remaining possibilities for the next card, and so on). If the number of decks Randy shuffles exceeds the number of different possible card orderings, then some of the shuffled decks would necessarily match. If Randy were to shuffle an infinite number of decks, the orderings of the cards would repeat an infinite number of times. As with Imelda and her outfits, an infinite number of occurrences with a finite number of possible configurations ensures that outcomes are infinitely repeated.

This basic notion is of the essence for cosmology in an infinite universe. Two key steps show why.

In an infinite universe, most regions lie beyond our ability to see, even using the most powerful telescopes possible. Although light travels enormously quickly, if an object is sufficiently distant, then the light it emits—even light that may have been emitted shortly after the big bang—will simply not have had sufficient time to reach us. Since the universe is about 13.7 billion years old, you might think that anything farther away than 13.7 billion light-years would fall into this category. The reasoning behind this intuition is right on target, but the expansion of space increases the distance to objects whose light has long been traveling and has only just been received; so the maximum distance we can see is actually longer—about 41 billion light-years.[12] But the exact numbers hardly matter. The important point is that regions of the universe beyond a certain distance are regions currently beyond our observational reach. Much as ships that have sailed beyond the horizon are not visible to someone standing on shore, astronomers say that objects in space that are too far away to be seen lie beyond our *cosmic horizon*.

Similarly, the light we've been emitting can't yet have reached those distant regions, so we are beyond their cosmic horizon. And it's not that cosmic horizons solely delineate what someone can and cannot see. From Einstein's special relativity, we know that no signal, no disturbance, no information, no *anything* can travel faster than light—which means that regions of the universe so far apart that light hasn't had time to travel between them are regions that have never exchanged any influence of any kind, and so have evolved completely independently.

Using a two-dimensional analogy, we can compare the expanse of space, at a given moment of time, to a giant patchwork quilt (with circular patches) in which each patch represents a single cosmic horizon. Someone located in the center of a patch can have interacted with anything that lies in the same patch, but has had no contact with anything lying in a different patch (see Figure 2.1a), because they're too far away. Points lying near the border between two patches are closer together than their respective centers and so can have interacted, but if we consider, say, patches in every other row and every other column of the cosmic quilt, all points residing in different patches are now so far from one another that no cross-patch interactions of any kind could have taken place (see Figure 2.1b). The same idea applies in three dimensions, where the cosmic horizons—the patches in the cosmic quilt—are spherical, and the same conclusion holds: sufficiently distant patches lie beyond one another's spheres of influence and so are independent realms.

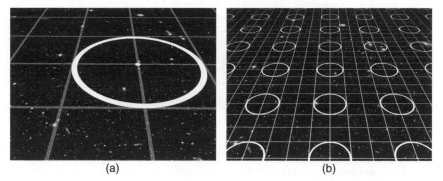

Figure 2.1 (a) *Because of light's finite speed, an observer at the center of any patch (called the observer's cosmic horizon) can have interacted only with things lying in that same patch.* (b) *Sufficiently distant cosmic horizons are too far apart to have had any interactions, and so have evolved completely independently of one another.*

If space is large but finite, we can divide it into a large but finite number of such independent patches. If space is infinite, then there are an *infinite* number of independent patches. It's this latter possibility that's of particular allure, and the second part of the argument tells why. As we will now see, in any given patch the particles of matter (more precisely, matter and all forms of energy) can be arranged in only a finite number of different configurations. Using the reasoning rehearsed with Imelda and Randy, this means that conditions in the infinity of far-flung patches—in regions of space like the one we inhabit but distributed through a limitless cosmos—*necessarily repeat.*

Finite Possibilities

Imagine it's a hot summer night and there's an annoying fly buzzing around your bedroom. You've tried the swatter, you've tried the nasty spray. Nothing's worked. In desperation, you try reason. "This is a big bedroom," you tell the fly. "There are so many other places you could be. There's no reason to keep buzzing around my ear." "Really?" the fly slyly counters. "How many places are there?"

In a classical universe, the answer is "Infinitely many." As you tell the fly, he (or, more precisely, his center of mass) could move 3 meters to the left, or 2.5 meters to the right, or 2.236 meters up, or 1.195829 meters down, or . . . you get the idea. Since the fly's position can vary continuously, there are infinitely many places it can be. In fact, as you explain all this to the fly, you realize that not only does position present the fly with infinite variety, so does velocity. At one moment the fly can be here, heading to the right at a kilometer per hour. Or it might be heading to the left at half a kilometer per hour, or heading up at a quarter of a kilometer per hour, or heading down at .349283 kilometers per hour, and so on. Although the fly's speed is constrained by a number of factors (including the limited energy it possesses, since the faster it flies, the more energy it needs to expend), it can vary continuously and hence provides another source of infinite variety.

The fly isn't convinced. "I'm with you when you talk about moving a centimeter, or half a centimeter, or even a quarter of a centimeter," the fly responds. "But when you speak of locations that differ by a ten-thousandth or a hundred-thousandth of a centimeter, or even less, you've lost me. To an egghead, those might be different locations, but it flies in the face of experience to say that *here* and a billionth of a centimeter to the left of *here* are really different. I can't sense such a tiny change in location and so I don't count them as different places. Same goes for speed. I can tell the difference between going a kilometer per hour and going at half that rate. But the difference between .25 kilometers per hour and .249999999 kilometers per hour? Please. Only a wise fly would claim to be able to tell the difference. Fact is, none of us can. So as far as I'm concerned, those are the same speeds. There's far less variety available than you're describing."

The fly has raised an important point. In principle, he can occupy an infinite variety of positions and attain an infinite variety of speeds. But in any practical sense, there is a limit to how fine the differences in location and speed can be before they go completely unnoticed. This is true even if the fly employs the best of equipment. There is always a limit on how small an increment in position or speed can be and yet still register. And regardless of how fine those minimal increments are, if they're not zero, they radically reduce the range of possible experience.

For instance, if the smallest increments that can be detected are a hundredth of a centimeter, then each centimeter offers not an infinite number of detectably different locations, but only a hundred. Each cubic centimeter would thus provide $100^3 = 1,000,000$ different locations, and your average bedroom would offer about 100 trillion. Whether the fly would find this array of options sufficiently impressive to keep away from your ear is difficult to say. The conclusion, though, is that *anything but measurements with perfect resolution reduces the number of possibilities from infinite to finite.*

You might counter that the inability to distinguish between tiny spatial separations or differences in speed reflects nothing more than a technological limitation. With progress, the precision of equipment always improves, so the number of discernibly distinct positions and speeds available to a well-funded fly will also always increase. Here I must invoke some basic quantum theory. According to quantum mechanics, there's a precise sense in which there *is* a fundamental limit on how accurate particular measurements can be, and this limit can't ever be surpassed, regardless of technological progress—ever. The limit arises from a central feature of quantum mechanics, the *uncertainty principle.*

The uncertainty principle establishes that regardless of what equipment you use or what techniques you employ, if you increase the resolution of your measurement of one property, there is an unavoidable cost: you necessarily reduce how accurately you can measure a complementary property. As a prime example, the uncertainty principle shows that the more accurately you measure an object's position, the less accurately you can measure its speed, and vice versa.

From the perspective of classical physics, the physics that informs much of our intuition about how the world works, this limitation is completely foreign. But as a rough analogy, think about photographing that impish fly. If your shutter speed is high, you'll get a sharp image that records the fly's location at the moment you snapped the picture. But because the photo is crisp, the fly appears motionless; the image gives no information about the fly's speed. If you set your shutter speed low, the resulting blurry image will convey something of the fly's motion, but because of that blurriness it also provides an imprecise measurement of the fly's location. You can't take a photo that gives sharp information about position and speed simultaneously.

Using the mathematics of quantum mechanics, Werner Heisenberg provided a precise limit on how imprecise the combined measurements of position and speed necessarily are. This inescapable imprecision is what quantum physicists mean by uncertainty. For our purpose, there's a particularly useful way of framing his result. Much as a sharper photograph requires that you use a higher shutter speed, Heisenberg's math shows that a sharper measurement of an object's position requires that you use a higher energy probe. Turn on your bedside lamp, and the resulting probe—diffuse, low-energy light—allows you to make out the general shape of the fly's legs and eyes; illuminate him with higher energy photons, like x-rays (keeping the photon bursts short to avoid cooking him), and the finer resolution reveals the minuscule muscles that flap the fly's wings. But perfect resolution, according to Heisenberg, requires a probe with infinite energy. That's unattainable.

And so, the essential conclusion is at hand. Classical physics makes clear that perfect resolution is unattainable in practice. Quantum physics goes further and establishes that perfect resolution is unattainable in principle. If you imagine both the speed and the position of an object—be it a fly or an electron—changing by sufficiently small amounts, then according to quantum mechanics, you are imagining something meaningless. Changes that are too small to be measured, even in principle, are not changes at all.[13]

By the same reasoning we used in our pre-quantum analysis of the fly, the limit on resolution reduces from infinite to finite the number of distinct possibilities for an object's position and speed. And since the limited resolution entailed by quantum mechanics is entwined in the very fibers of physical law, this reduction to finite possibilities is unavoidable and unassailable.

Cosmic Repetition

So much for flies in bedrooms. Now consider a larger region of space. Consider a region the size of today's cosmic horizon, a sphere with a radius of 41 billion light-years. A region, that is, which is the size of a single patch in the cosmic quilt. And consider filling it not with a single fly but with particles of matter and radiation. Here's the question: How many different arrangements of the particles are possible?

Well, as with a box of Legos, the more pieces you have—the more matter and radiation you cram into the region—the greater the number of possible arrangements. But you can't cram pieces in indefinitely. Particles carry energy, so more particles means more energy. If a region of space contains too much energy, it will collapse under its own weight and form a black hole.* And if after a black hole forms you try to cram yet more matter and energy into the region, the black hole's boundary (its *event horizon*) will grow larger, encompassing more space. There is thus a limit to how much matter and energy can exist within a region of space of a given size. For a region of space as large as today's cosmic horizon, the limits involved are huge (about 10^{56} grams). But the size of the limit is not central. What's central is that there *is* a limit.

Finite energy within a cosmic horizon entails a finite number of particles, be they electrons, protons, neutrons, neutrinos, muons, photons, or any of the other known or as yet unidentified species in the particle bestiary. Finite energy within a cosmic horizon also entails that each of these particles, like the annoying fly in your bedroom, has a finite number of distinct possible locations and speeds. Collectively, a finite number of particles, each of which can have finitely many distinct positions and velocities, means that

---

*I will discuss black holes more fully in later chapters. Here we'll stick to the familiar notion, by now well ingrained in popular culture, of a spatial region—think of it as a ball in space—whose gravitational pull is so strong that nothing crossing its edge can escape. The bigger the black hole's mass, the larger its size, so when anything falls in, not only does the black hole's mass increase but its size does too.

within any cosmic horizon only a finite number of different particle arrangements are available. (In the more refined language of
quantum theory proper, which we'll encounter in Chapter 8, we
don't speak of particle positions and velocities per se, but rather of
the *quantum state* of these particles. From this perspective, we
would say there are only a finite number of observably distinct
quantum states for the particles in the cosmic patch.) Indeed, a
short calculation—described in the notes, if you're curious about
the details—reveals that the number of distinct possible particle
configurations within a cosmic horizon is about $10^{10^{122}}$ (a 1 followed
by 10^{122} zeros). This is a huge but decidedly finite number.[14]

The limited number of different clothes combinations ensures
that with enough outings, Imelda's attire will necessarily repeat.
The limited number of different card orderings ensures that with
enough decks, Randy's shuffles will necessarily repeat. By the same
reasoning, the limited number of particle arrangements ensures
that with enough patches in the cosmic quilt—enough independent cosmic horizons—*the particle arrangements, when compared
from patch to patch, must somewhere repeat.* Even if you were able
to play cosmic designer and tried to arrange each patch to be different from the ones you'd examined before, with a big enough
expanse you'd eventually run out of distinct designs and would be
forced to repeat a previous arrangement.

In an infinitely big universe, the repetition is yet more extreme.
There are infinitely many patches in an infinite expanse of space;
so, with only finitely many different particle arrangements, the
arrangements of particles within patches must be duplicated an
infinite number of times.

That's the result we've been after.

Nothing but Physics

In interpreting the implications of this statement, I should declare
my bias. I believe that a physical system is completely determined by
the arrangement of its particles. Tell me how the particles making
up the earth, the sun, the galaxy, and everything else are arranged,

and you've fully articulated reality. This reductionist view is common among physicists, but there are certainly people who think otherwise. Especially when it comes to life, some believe that an essential nonphysical aspect (spirit, soul, life force, chi, and so on) is required to animate the physical. Although I remain open to this possibility, I've never encountered any evidence to support it. The position that makes the most sense to me is that one's physical and mental characteristics are nothing but a manifestation of how the particles in one's body are arranged. Specify the particle arrangement and you've specified everything.[15]

Adhering to this perspective, we conclude that if the particle arrangement with which we're familiar were duplicated in another patch—another cosmic horizon—that patch would look and feel like ours in every way. This means that if the universe is infinite in extent, you are not alone in whatever reaction you are now having to this view of reality. There are many perfect copies of you out there in the cosmos, feeling exactly the same way. And there's no way to say which is *really* you. All versions are physically and hence mentally identical.

We can even estimate the distance to the nearest copy. If the particle arrangements are randomly distributed from patch to patch (an assumption that's compatible with the refined cosmological theory we will encounter in the next chapter), then we can expect that the conditions in our patch will be duplicated as frequently as those in any other. In every collection of $10^{10^{122}}$ cosmic patches, we thus expect there to be, on average, one patch that looks just like ours. That is, in every region of space that's roughly $10^{10^{122}}$ meters across, there should be a cosmic patch that replicates ours—one that contains you, the earth, the galaxy, and everything else that inhabits our cosmic horizon.

If you lower your sights and don't seek an exact replica of our entire cosmic horizon, but would be satisfied with an exact copy of a region a few light-years in radius and centered on our sun, the order is more easily filled: on average, in every region that's about $10^{10^{100}}$ meters across, you should find one such copy. Still easier to find are approximate copies. After all, there is only one way to duplicate a region exactly, but many ways to *almost* duplicate it.

Were you to visit these inexact copies, you'd find some that are barely distinguishable from ours, while in others the differences would range from obvious to exhilarating to shocking. Every decision you've ever made is tantamount to a particular particle arrangement. If you turned left, your particles went one way; if you turned right, your particles went the other. If you said yes, the particles in your brain, lips, and vocal cords proceeded through one pattern; if you said no, they proceeded through a different pattern. And so every possible action, every choice you've made and every option you've discarded, will be played out in one patch or another. In some, your worst fears about yourself, your family, and life on earth have been realized. In others, your wildest dreams have come to pass. In others still, the differences arising from the close but distinct particle arrangements have combined to yield an unrecognizable environment. And in most patches, the particle complexion would not include the highly specialized arrangements we recognize as living organisms, so the patches would be lifeless, or at least devoid of life as we know it.

Over time, the size of the cosmic patches laid out in Figure 2.1b will increase; with more time, light can travel farther and so each of the cosmic horizons will grow larger. Ultimately, the cosmic horizons will overlap. And when they do, the regions can no longer be considered as separate and isolated; the parallel universes will no longer be parallel—they will have merged. Nevertheless, the result we've found will continue to hold. Just lay out a new grid of cosmic patches with patch size set by the distance light can have traveled since the big bang through this later moment. The patches will be bigger, so to fill out a pattern like that in Figure 2.1b their centers will need to be farther apart, but with infinite space at our disposal, there's ample room to accommodate this adjustment.[16]

And so we've come to a conclusion that's both general and provocative. Reality in an infinite cosmos is not what most of us would expect. At any moment in time, the expanse of space contains an infinite number of separate realms—constituents of what I'll call the *Quilted Multiverse*—with our observable universe, all we see in the vast night sky, being but one member. Canvassing this infinite collection of separate realms, we find that particle arrangements necessarily repeat infinitely many times. The reality that holds in any given universe, including ours, is thus replicated in an infinite number of other universes across the Quilted Multiverse.[17]

What to Make of This?

It's possible that the conclusion we've reached strikes you as so outlandish that you're inclined to turn the discussion on its head. You might argue that the bizarre nature of where we've gotten—infinite copies of you and everyone and everything—is evidence of the faulty nature of one or more of the assumptions that led us here.

Might the assumption that the entire cosmos is inhabited by particles be wrong? Perhaps beyond our cosmic horizon is a vast realm containing nothing but empty space. It's possible, but the theoretical contortions required to accommodate such a picture render it thoroughly unconvincing. The most refined cosmological theories, to be encountered shortly, don't lead us anywhere near this possibility.

Might the very laws of physics change beyond our cosmic horizon, corrupting our ability to perform any reliable theoretical analyses of those distant realms? Again, it's possible. But as we will see in the next chapter, recent developments yield a compelling argument that although the laws can vary, that variation doesn't invalidate our conclusions regarding the Quilted Multiverse.

Might the universe's spatial expanse be finite? Sure. Definitely possible. If space were finite yet large enough, there could still be some interesting patches way out there. But a smallish finite universe could easily fail to have adequate space to accommodate substantial numbers of distinct patches, let alone any that are duplicates of our own. A finite universe poses the most convincing way to upend the Quilted Multiverse.

But in the last few decades, physicists working to push the big bang theory back to time zero—in search of a deeper understanding of the origin and nature of Lemaître's primeval atom—have developed an approach called *inflationary cosmology*. In the inflationary framework, the argument in support of an infinitely large cosmos not only garners strong observational and theoretical support but, as we will see in the next chapter, becomes an almost inevitable conclusion.

What's more, inflation brings to the fore another, even more exotic, variety of parallel worlds.

CHAPTER 3

Eternity and Infinity

The Inflationary Multiverse

A pioneering group of physicists in the mid-1900s realized that if you were to shut off the sun, remove the other stars from the Milky Way, and even sweep away the more distant galaxies, space would not be black. To the human eye it would appear black, but if you could see radiation in the microwave part of the spectrum, then every which way you turned you'd see a uniform glow. Its origin? *The* origin. Remarkably, these physicists discovered a pervasive sea of microwave radiation filling space that is a present-day relic of the universe's creation. The story of this breakthrough recounts a phenomenal achievement of the big bang theory, but in time it also revealed one of the theory's fundamental shortcomings and thus set the stage for the next major breakthrough in cosmology after the pioneering works of Friedmann and Lemaître: the *inflationary theory*.

Inflationary cosmology modifies the big bang theory by inserting an intense burst of enormously fast expansion during the universe's earliest moments. This modification, as we will see, proves essential to explaining some otherwise perplexing features of the relic radiation. But more than that, inflationary cosmology is a key chapter in our story because scientists have gradually realized over the last few decades that the most convincing versions of the theory yield a vast collection of parallel universes, radically transforming the complexion of reality.

Relics of a Hot Beginning

George Gamow, a hulking six-foot-three Russian physicist known for important contributions to quantum and nuclear physics in the early twentieth century, was as quick-witted and fun-loving as he was hard-living (in 1932, he and his wife tried to defect from the Soviet Union by paddling across the Black Sea in a kayak stocked with a healthy assortment of chocolate and brandy; when bad weather sent the two scurrying back to shore, Gamow was able to fast-talk the authorities with a tale of the unfortunately failed scientific experiments he'd been undertaking at sea). In the 1940s, after having successfully slipped past the iron curtain (on dry land, with less chocolate) and settled in at George Washington University, Gamow turned his attention to cosmology. With critical assistance from his phenomenally talented graduate student Ralph Alpher, Gamow's research resulted in a far more detailed and vivid picture of the universe's earliest moments than had been revealed by the earlier work of Friedmann (who had been Gamow's teacher back in Leningrad) and Lemaître. With a little modern updating, Gamow and Alpher's picture looks like this.

Just after its birth, the stupendously hot and dense universe experienced a frenzy of activity. Space rapidly expanded and cooled, allowing a particle stew to congeal from the primordial plasma. For the first three minutes, the rapidly falling temperature remained sufficiently high for the universe to act like a cosmic nuclear furnace, synthesizing the simplest atomic nuclei: hydrogen, helium, and trace amounts of lithium. But with the passing of just a few more minutes, the temperature dropped to about 10^8 Kelvin (K), roughly 10,000 times the surface temperature of the sun. Although immensely high by everyday standards, this temperature was too low to support further nuclear processes, and so from this time on the particle commotion largely abated. For eons that followed, not much happened except that space kept expanding and the particle bath kept cooling.

Then, some 370,000 years later, when the universe had cooled to about 3000 K, half the sun's surface temperature, the cosmic monotony was interrupted by a pivotal turn of events. To that point, space had been filled with a plasma of particles carrying electric charge, mostly protons and electrons. Because electrically charged particles have the unique ability to jostle photons—particles of light—the primordial plasma would have appeared opaque; the photons, incessantly buffeted by electrons and protons, would have provided a diffuse glow similar to a car's high beams cloaked by a dense fog. But when the temperature dropped below 3000 K, the rapidly moving electrons and nuclei slowed sufficiently to amalgamate into atoms; electrons were captured by the atomic nuclei and drawn into orbit. This was a key transformation. Because protons and electrons have equal but opposite charges, their atomic unions are electrically neutral. And since a plasma of electrically neutral composites allows photons to slip through like a hot knife through butter, the formation of atoms allowed the cosmic fog to clear and the luminous echo of the big bang to be released. The primordial photons have been streaming through space ever since.

Well, with one important caveat. Although no longer knocked to and fro by electrically charged particles, the photons have been subject to one other important influence. As space expands, things dilute and cool, including photons. But unlike particles of matter, photons don't slow down when they cool; being particles of light, they always travel at light speed. Instead, when photons cool their vibrational frequencies decrease, which means they change color. Violet photons will shift to blue, then to green, to yellow, to red, and then into the infrared (like those visible with night goggles), the microwave (like those that heat food by bouncing around your microwave oven), and finally into the domain of radio frequencies.

As Gamow first realized and as Alpher and his collaborator Robert Herman worked out with greater fidelity, all this means that if the big bang theory is correct, then space everywhere should now be filled with *remnant photons from the creation event*, streaming every which way, whose vibrational frequencies are determined by how much the universe has expanded and cooled during the billions of years since they were released. Detailed mathematical calculations showed that the photons should have cooled close to absolute zero, placing their frequencies in the microwave part of the spectrum. For this reason, they are called the *cosmic microwave background radiation*.

I recently reread the papers of Gamow, Alpher, and Herman

that in the late 1940s announced and explained these conclusions. They are marvels of theoretical physics. The technical analyses involved require hardly more than a grounding in undergraduate physics, and yet the results are profound. The authors concluded that we are all immersed in a bath of photons, a cosmic heirloom bequeathed to us by the universe's fiery birth.

With that buildup, you may find it surprising that the papers were ignored. This was mostly because they were written during an era dominated by quantum and nuclear physics. Cosmology had yet to make its mark as a quantitative science, so the physics culture was less receptive to what seemed like fringe theoretical studies. To some degree, the papers also languished because of Gamow's unusually playful style (he once modified the authorship of a paper he was writing with Alpher to include his friend the future Nobel laureate Hans Bethe, just to make the paper's byline—Alpher, Bethe, Gamow—sound like the first three letters of the Greek alphabet), which resulted in some physicists taking him less seriously than he deserved. Try as they might, Gamow, Alpher, and Herman could not interest anyone in their results, let alone persuade astronomers to devote the significant effort required to attempt to detect the relic radiation they predicted. The papers were quickly forgotten.

In the early 1960s, unaware of the earlier work, the Princeton physicists Robert Dicke and Jim Peebles went down a similar path and also realized that the big bang's legacy should be the presence of a ubiquitous background radiation filling space.[1] Unlike the members of Gamow's team, however, Dicke was a renowned experimentalist and so didn't need to persuade anyone to seek the radiation observationally. He could do it himself. Together with his students David Wilkinson and Peter Roll, Dicke devised an experimental scheme to capture some of the big bang's vestigial photons. But before the Princeton researchers could put their plan to the test, they received one of the most famous telephone calls in the history of science.

While Dicke and Peebles had been calculating, the physicists Arno Penzias and Robert Wilson at Bell Labs, less than thirty miles from Princeton, had been struggling with a radio communications antenna (coincidentally, it was based on a design Dicke had come up with in the 1940s). No matter what adjustments they made, the antenna hissed with a steady, unavoidable background noise. Penzias and Wilson were convinced that something was wrong with their equipment. But then came a serendipitous chain of conversations. It began with a talk Peebles gave in February 1965 at Johns Hopkins University, which was attended by the Carnegie Institution radio astronomer Kenneth Turner, who mentioned the results he heard Peebles present to his MIT colleague Bernard Burke, who happened to be in touch with Penzias at Bell Labs. Hearing of the Princeton research, the Bell Labs team realized that their antenna was hissing for good reason: *it was picking up the cosmic microwave background radiation.* Penzias and Wilson called Dicke, who quickly confirmed that they had unintentionally tapped into the reverberation of the big bang.

The two groups agreed to publish their papers simultaneously in the prestigious *Astrophysical Journal.* The Princeton group discussed their theory of the background radiation's cosmological origin, while the Bell Labs team reported, in the most conservative of language and with no mention of cosmology, the detection of uniform microwave radiation permeating space. Neither paper mentioned the earlier work of Gamow, Alpher, and Herman. For their discovery, Penzias and Wilson were awarded the 1978 Nobel Prize in physics.

Gamow, Alpher, and Herman were deeply dismayed, and in the years that followed struggled mightily to have their work recognized. Only gradually and belatedly has the physics community saluted their primary role in this monumental discovery.

The Uncanny Uniformity of Ancient Photons

During the decades since it was first observed, the cosmic microwave background radiation has become a crucial tool in cosmological investigations. The reason is clear. In a great many fields, researchers would give their eyeteeth to have an unfettered, direct glimpse of the past. Instead, they generally have to piece together a

view of remote conditions on the basis of evidence from rem-
nants—weathered fossils, decaying parchments, or mummified
remains. Cosmology is the one field in which we can actually witness history. The pinpoints of starlight we can see with the naked
eye are streams of photons that have been traveling toward us for a
few years or a few thousand. The light from more distant objects,
captured by powerful telescopes, has been traveling toward us
far longer, sometimes for billions of years. When you look at such
ancient light, you are seeing—literally—ancient times. Those primeval comings and goings transpired far away, but the apparent
large-scale uniformity of the universe argues strongly that what was
happening there was also, on average, happening here. In looking
up, we are looking back.

The cosmic microwave photons allow us to make the most of
this opportunity. No matter how technology may improve, the
microwave photons are the oldest we can hope to see, because their
elder brethren were trapped by the foggy conditions that prevailed
during earlier epochs. When we examine the cosmic microwave
background photons, we are glimpsing how things were nearly
14 billion years ago.

Calculations show that today there are about 400 million of
these cosmic microwave photons racing through every cubic meter
of space. Although our eyes can't see them, an old-fashioned television set can. About 1 percent of the snow on a television that's been
disconnected from the cable signal and tuned to a station that's
ceased broadcasting is due to reception of the big bang's photons.
It's a curious thought. The very same airwaves that carry reruns
of *All in the Family* and *The Honeymooners* are infused with some
of the universe's oldest fossils, photons communicating a drama
that played out when the cosmos was but a few hundred thousand
years old.

The big bang model's correct prediction that space would be
filled with microwave background radiation was a triumph. During
a mere three hundred years of scientific thought and technological
progress, our species went from peering through rudimentary telescopes and dropping balls from leaning towers to grasping physical
processes at work just after the universe was born. Nevertheless, fur-

ther investigation of the data raised a pointed challenge. Ever more refined measurements of the radiation's temperature, made not with television sets but with some of the most precise astronomical equipment ever built, showed that the radiation is thoroughly—uncannily—uniform across space. Regardless of where you point your detector, the temperature of the radiation is 2.725 degrees above absolute zero. The puzzle is to explain how such fantastic uniformity came to be.

Given the ideas presented in Chapter 2 (and my comment four paragraphs ago), I can imagine your saying, "Well, that's just the cosmological principle at work: no location in the universe is special when compared with any other, so the temperature at each should be the same." Fair enough. But remember that the cosmological principle was a simplifying *assumption* that physicists, including Einstein, invoked to make the mathematical analysis of the universe's evolution tractable. Since the microwave background radiation is indeed uniform throughout space, it provides convincing observational evidence for the cosmological principle, and it strengthens our confidence in conclusions the principle helped reveal. But the radiation's astounding uniformity shines a glaring spotlight on the cosmological principle itself. Reasonable though the cosmological principle may sound, what mechanism established the cosmos-wide uniformity that observations confirm?

Faster Than the Speed of Light

We've all had the mildly unsettling sensation of shaking someone's hand and finding it steamy hot (not so bad) or clammy cold (definitely worse). But were you to hold on to that hand, you'd find that the modest temperature differential would quickly subside. When objects are in contact, heat migrates from the hotter to the colder, until their temperatures are equal. You experience this all the time. It's why coffee left on your desk eventually comes to room temperature.

Similar reasoning would seem to explain the uniformity of the microwave background radiation. As with holding hands and standing coffee, the uniformity presumably reflects the familiar reversion of an environment to an overall common temperature. The sole novelty of the process is that the reversion is supposed to have taken place over cosmic distances.

In the big bang theory, however, the explanation fails. For places or things to reach a common temperature, an essential condition is mutual contact. It may be direct, as with shaking hands, or, minimally, through an exchange of information so that conditions at distinct locations can become correlated. Only through such mutual influence can a shared, communal environment be achieved. A thermos is designed to prevent such interactions, thwarting the drive to uniformity and preserving temperature differences.

This simple observation highlights the problem with the naïve explanation of the cosmic temperature uniformity. Locations in space that are very far apart—say, one point way off to your right, so deep in the night sky that the first light it ever emitted has only just reached you, and a second, similar point way off to your left—have never interacted. Although you can see both, light from one still has an enormous distance to cover before it reaches the other. Thus, hypothetical observers situated at the distant left and right locations have yet to see each other, and since the speed of light sets the upper limit for how fast anything can travel, they've yet to interact in any way. To use the language of the previous chapter, they are beyond each other's cosmic horizon.

This description makes the mystery manifest. You'd be floored if inhabitants of these distant locations spoke the same language and had libraries filled with the same books. With no contact, how could a common heritage have been established? You should be equally floored to learn that without any apparent contact, these widely separated regions share a common temperature, one that matches to an accuracy of better than four decimal places.

Years ago, when I first learned of this puzzle, I *was* floored. But on further thought, I became puzzled by the puzzle. How could two objects that were once close together—as we believe all things in the observable universe were at the time of the big bang—have separated so quickly that light emitted by one wouldn't have time to reach the other? Light sets the cosmic speed limit, so how could the objects achieve a spatial separation greater than what light would have had time to traverse? The answer highlights a point that's often not adequately stressed. The speed limit set by light refers solely to the motion of objects *through space*. But galaxies recede from one another not because they are traveling through space—galaxies don't have jet engines—but rather because space itself is swelling and the galaxies are being dragged along by the overall flow.[2] And the thing is, relativity places no limit on how fast space can swell, so there is no limit on how fast galaxies that are being pushed apart by the swell recede from one another. The rate of recession between any two galaxies can exceed any speed, including the speed of light.

Indeed, the mathematics of general relativity shows that in the universe's earliest moments, space would have swelled so fast that regions would have been propelled apart at greater than light speed. As a result, they would have been unable to exert any influence on one another. The difficulty then is to explain how nearly identical temperatures were established in independent cosmic domains, a puzzle cosmologists have named the *horizon problem.*

Broadening Horizons

In 1979, Alan Guth (then working at the Stanford Linear Accelerator Center) came up with an idea that, with subsequent critical refinements made by Andrei Linde (then carrying out research at the Lebedev Physical Institute in Moscow), and by Paul Steinhardt and Andreas Albrecht (a professor-student duo who were then working at the University of Pennsylvania), is widely believed to solve the horizon problem. The solution, *inflationary cosmology,* relies on some subtle features of Einstein's general relativity that I'll describe in a moment, but its broad outline can be readily summarized.

The horizon problem afflicts the standard big bang theory because regions of space separate too quickly for thermal equality to be established. The inflationary theory resolves the problem by slowing the speed with which the regions were separating very early on, providing them ample time to come to the same temperature. The theory then proposes that after the completion of these "cosmic handshakes" there came a brief burst of enormously fast and ever-quickening expansion — called *inflationary expansion* — which more than compensated for the sluggish start, rapidly driving the regions to vastly distant positions in the sky. The uniform conditions we observe no longer pose a mystery, since a common temperature was established before the regions were rapidly driven apart.[3] In broad strokes, that's the essence of the inflationary proposal.*

Bear in mind, however, that physicists don't dictate how the universe expands. As far as we can tell from our most refined observations, Einstein's equations of general relativity do. The viability of the inflationary scenario thus depends on whether its proposed modification to the standard big bang expansion can emerge from Einstein's mathematics. At first glance, this is far from obvious.

For example, I'm pretty sure that if you were to bring Newton up to date by giving him a five-minute primer on general relativity, explaining the outlines of warped space and the expanding universe, he'd find your subsequent description of the inflationary proposal preposterous. Newton would sternly maintain that regardless of fancy math and newfangled Einsteinian language, gravity is still an attractive force. And so, he would emphasize with a pound on the table, gravity acts to pull objects together, slowing any cosmic divergence. Expansion that starts out dawdling, then sharply quickens for a brief period, might solve the horizon problem, but it's a fiction. Newton would declare that just as gravitational attraction implies that the speed of a batted baseball diminishes as the ball moves upward, it similarly implies that the cosmic expansion must slow over time. Sure, if the expansion drops all the way to zero and then turns into cosmic contraction, the implosion can speed up

---

*Equivalently, superfast accelerated expansion means that today's distant regions would have been much closer together in the early universe than is suggested by the traditional big bang theory — ensuring that a common temperature could be established before the burst separated them.

over time, much as the ball's speed can increase when it starts its downward journey. But the speed of the outward spatial expansion can't increase.

Newton's making a mistake, but you can't blame him. The burden lies with the cursory summary you gave him of general relativity. Don't get me wrong. It's understandable that, given only five minutes (one of which was spent explaining baseball), you focused on curved spacetime as the source of gravity. Newton himself had called attention to the fact that there was no known mechanism for transmitting gravity, and he always viewed that as a yawning hole in his own theory. Naturally, you wanted to show him Einstein's resolution. But Einstein's theory of gravity did much more than merely fill a gap in Newtonian physics. Gravity in general relativity differs in its essence from gravity in Newton's physics, and in the present context, there is one feature that cries out for emphasis.

In Newton's theory, gravity arises solely from an object's mass. The bigger the mass, the bigger the object's gravitational pull. In Einstein's theory, gravity arises from an object's mass (and energy) *but also from its pressure.* Weigh a sealed bag of potato chips. Weigh it again, but this time squeeze the bag so that the air inside is under higher pressure. According to Newton, the weight will be the same, because there's been no change in mass. According to Einstein, the squeezed bag will weigh slightly more, because although the mass is the same there's been an increase in pressure.[4] In everyday circumstances we're not aware of it, because for ordinary objects the effect is fantastically tiny. Even so, general relativity, and the experiments that have shown it to be correct, makes it perfectly clear that pressure contributes to gravity.

This deviation from Newton's theory is critical. Air pressure, whether the air is in a bag of potato chips, an inflated balloon, or the room where you're now reading, is positive, meaning that the air pushes outward. In general relativity, positive pressure, like positive mass, contributes positively to gravity, resulting in increased weight. But whereas mass is always positive, there are situations in which pressure can be negative. Think of a stretched rubber band. Rather than pushing outward, the rubber band's straining molecules pull inward, exerting what physicists call *negative pressure* (or, equivalently, *tension*). And much as general relativity shows that positive pressure gives rise to attractive gravity, it shows that negative pressure gives rise to the opposite: *repulsive* gravity.

Repulsive gravity?

This would blow Newton's mind. For him, gravity was only attractive. But your mind should remain intact: you've already encountered this strange clause in general relativity's contract with gravity. Remember Einstein's cosmological constant, discussed in the previous chapter? I declared there that by infusing space with a uniform energy, a cosmological constant generates repulsive gravity. But in that earlier encounter, I didn't explain why this happens. Now I can. A cosmological constant not only endows the spatial fabric with a uniform energy determined by the constant's value (the number on the third line of the apocryphal relativity tax form), but it also fills space with a uniform negative pressure (we will see why in a moment). And, as above, when it comes to the gravitational force each produces, negative pressure does the opposite of positive mass and positive pressure. It yields repulsive gravity.*

In Einstein's hands, repulsive gravity was used for a single erroneous purpose. He proposed finely adjusting the amount of negative pressure that permeates space to ensure that the repulsive gravity produced would exactly counter the attractive gravity exerted by the universe's more familiar material contents, yielding a static universe. As we've seen, he subsequently renounced this move. Six decades later, the developers of the inflationary theory proposed a kind of repulsive gravity that differed from Einstein's version much as the finale of Mahler's Eighth differs from the drone of a tuning fork. Rather than a moderate and steady outward

---

*You might think that negative pressure would pull inward and thus be at odds with repulsive—outward-pushing—gravity. Actually, *uniform* pressure, regardless of its sign, doesn't push or pull at all. Your eardrums pop only when there is nonuniform pressure, lower on one side than the other. The repulsive push I'm describing here is the *gravitational force generated by the presence of the uniform negative pressure*. This is a difficult but essential point. Again, whereas the presence of positive mass or positive pressure generates attractive gravity, the presence of negative pressure generates the less familiar repulsive gravity.

push that would stabilize the universe, the inflationary theory envisions a gargantuan surge of repulsive gravity that's astoundingly short and thunderingly intense. Regions of space had ample time before the burst to come to the same temperature, but then, riding the surge, covered the great distances necessary to reach their observed positions in the sky.

At this point, Newton would surely shoot you another disapproving look. Ever the skeptic, he would find another problem with your explanation. After catching up on the more intricate details of general relativity by racing through one of the standard textbooks, he would accept the strange fact that gravity can—in principle—be repulsive. But, he'd ask, what's all this talk of negative pressure permeating space? It's one thing to use the inward pull of a stretched rubber band as an example of negative pressure. It's another to argue that billions of years ago, just around the time of the big bang, space was momentarily permeated by an enormous and uniform negative pressure. What thing, or process, or entity has the capacity to supply such a fleeting but pervasive negative pressure?

The genius of inflation's pioneers was to provide an answer. They showed that the negative pressure required for an antigravity burst naturally emerges from a novel mechanism involving ingredients known as *quantum fields*. For our story, the details are crucial because the manner in which inflationary expansion comes about is central to the version of parallel universes it yields.

Quantum Fields

In Newton's day, physics concerned itself with the motion of objects you can see—stones, cannonballs, planets—and the equations he developed closely reflected this focus. Newton's laws of motion are a mathematical embodiment of how such tangible bodies move when they're pushed, pulled, or shot through the air. For more than a century, this was a wonderfully fruitful approach. But in the early 1800s, the English scientist Michael Faraday initiated a transformation in thinking with the elusive but demonstrably powerful concept of the *field*.

Take a strong refrigerator magnet and place it an inch above a paper clip. You know what happens. The clip jumps up and sticks to the magnet's surface. This demonstration is so commonplace, so thoroughly familiar, that it's easy to overlook how bizarre it is. Without touching the paper clip, the magnet can make it move. How is this possible? How can an influence be exerted in the absence of any contact with the clip itself? These and a multitude of related considerations led Faraday to postulate that though the magnet proper does not touch the paper clip, the magnet produces something that *does*. That something is what Faraday called a *magnetic field*.

We can't see the fields produced by magnets; we can't hear them; none of our senses are attuned to them. But that reflects physiological limitations, nothing more. As a flame generates heat, so a magnet generates a magnetic field. Lying beyond the physical boundary of the solid magnet, the magnet's field is a "mist" or "essence" that fills space and does the magnet's bidding.

Magnetic fields are but one kind of field. Charged particles give rise to another: electric fields, such as those responsible for the shock you sometimes receive when you reach for a metal doorknob in a room with wall-to-wall wool carpeting. Unexpectedly, Faraday's experiments showed that electric and magnetic fields are intimately related: he found that a changing electric field generates a magnetic field, and vice versa. In the late 1800s, James Clerk Maxwell put mathematical might behind these insights, describing electric and magnetic fields in terms of numbers assigned to each point in space; the numbers' values reflect the field's ability, at that location, to exert influence. Places in space where the magnetic field's numerical values are large, for instance an MRI's cavity, are places where metal objects will feel a strong push or pull. Places in space where the electric field's numerical values are large, for instance the inside of a thundercloud, are places where powerful electrical discharges such as lightning may occur.

Maxwell discovered equations, which now bear his name, that govern how the strength of electric and magnetic fields varies from point to point in space and moment to moment in time. These very same equations govern the sea of rippling electric and magnetic fields, so-called *electromagnetic waves*, within which we're all immersed. Turn on a cell phone, a radio, or a wireless computer, and the signals received represent a tiny portion of the thicket of electromagnetic transmissions silently rushing by and through you every second. Most stunning of all, Maxwell's equations revealed that visible light itself is an electromagnetic wave, one whose rippling patterns our eyes *have* evolved to see.

In the second half of the twentieth century, physicists united the field concept with their burgeoning understanding of the microworld encapsulated by quantum mechanics. The result, *quantum field theory*, provides a mathematical framework for our most refined theories of matter and nature's forces. Using it, physicists have established that in addition to electric and magnetic fields, there exists a whole panoply of others with names like *strong* and *weak nuclear fields* and *electron, quark,* and *neutrino fields*. One field that to date remains wholly hypothetical, the *inflaton field*, provides a theoretical basis for inflationary cosmology.*

Quantum Fields and Inflation

Fields carry energy. Qualitatively, we know this because fields accomplish tasks that require energy, such as causing objects (like paper clips) to move. Quantitatively, the equations of quantum field theory show us how, given the numerical value of a field at a particular location, to calculate the amount of energy it contains. Typically, the larger the value, the larger the energy. A field's value can vary from place to place, but should it be constant, taking the same value everywhere, it would fill space with the same energy at every point. Guth's critical insight was that such uniform field configurations fill space not only with uniform energy but also with uniform negative pressure. And with that, *he found a physical mechanism to generate repulsive gravity.*

---

*The rapid expansion of space is called inflation, but following the historical pattern of invoking names that end in "on" (electron, proton, neutron, muon, etc.), when physicists refer to the field driving inflation, they drop the second "i." Hence, inflaton field.

To see why a uniform field yields negative pressure, think first about a more ordinary situation that involves positive pressure: the opening of a bottle of Dom Pérignon. As you slowly remove the cork, you can feel the positive pressure of the champagne's carbon dioxide pushing outward, driving the cork from the bottle and into your hand. A fact you can directly verify is that this outward exertion drains a little energy from the champagne. You know those vapor tendrils you see near the bottle's neck when the cork is out? They form because the energy expended by the champagne in pushing against the cork results in a drop in temperature, which, much as with your breath on a wintry day, causes surrounding water vapor to condense.

Now imagine replacing the champagne with something less festive but more pedagogical—a field whose value is uniform throughout the bottle. When you remove the cork this time, your experience will be very different. As you slide the cork outward, you make a little extra volume inside the bottle available for the field to permeate. Since a uniform field contributes the same energy at every location, the larger the volume the field fills, the *greater* the total energy the bottle contains. Which means that, unlike with champagne, the act of removing the cork adds energy to the bottle.

How could that be? Where would the energy come from? Well, think about what happens if the bottle's contents, rather than pushing the cork outward, *pull the cork inward.* This would require you to pull on the cork to remove it, an exertion of effort that in turn would transfer energy from your muscles to the contents of the bottle. To explain the increase in the bottle's energy we thus conclude that, unlike champagne, which pushes outward, a uniform field sucks inward. That's what we mean by a uniform field's resulting in a negative—not positive—pressure.

Although there's no sommelier uncorking the cosmos, the same conclusion holds: if there's a field—the hypothetical inflaton field—that has a uniform value throughout a region of space, it will fill that region not only with energy but also with negative pressure. And, as is now familiar, such negative pressure yields repulsive gravity, which drives an ever-quickening expansion of space. When Guth slotted into Einstein's equations the likely numerical values for the inflaton's energy and pressure consonant with the extreme environment of the early universe, the mathematics revealed that the resulting repulsive gravity would be stupendous. It would easily be many orders of magnitude stronger than the repulsive force Einstein envisioned years earlier when he dallied with the cosmological constant, and would propel a spectacular spatial stretching. That alone was exciting. But Guth realized there was an indispensable bonus.

The same reasoning that explains why a uniform field has negative pressure applies as well to a cosmological constant. (If the bottle contains empty space endowed with a cosmological constant, then when you slowly remove the cork the extra space you make available within the bottle contributes extra energy. The only source for this extra energy is your muscles, which therefore must have strained against an inward, negative pressure supplied by the cosmological constant.) And, as with a uniform field, a cosmological constant's uniform negative pressure also yields repulsive gravity. But the vital point here is not the similarities, per se, but the manner in which a cosmological constant and a uniform field differ.

A cosmological constant is just that—a constant, a fixed number inserted on the third line of general relativity's tax form that would generate the same repulsive gravity today as it would have billions of years ago. By contrast, the value of a field can change, and generally will. When you turn on your microwave oven, you change the electromagnetic field filling its interior; when the technician flips the switch on an MRI machine, he or she changes the electromagnetic field threading the cavity. Guth realized that an inflaton field filling space could behave similarly—turning on for a burst and then turning off—which would allow repulsive gravity to operate during only a brief window of time. That's essential. Observations establish that if the blistering growth of space happened at all, it must have happened billions of years ago and then sharply dropped off to the statelier-paced expansion evidenced by detailed astronomical measurements. So an all-important feature of the inflationary proposal is that the era of powerful repulsive gravity be transient.

The mechanism for turning on and then shutting off the inflationary burst relies on physics that Guth initially developed but that Linde, and Albrecht and Steinhardt, refined substantially. To get a feel for their proposal, think of a ball—better still, think of nearly round Eric Cartman—perched precariously on one of South Park's snow-covered mountains. A physicist would say that because of his position, Cartman embodies energy. More precisely, he embodies *potential energy*, meaning that he has pent-up energy that's ready to be tapped, most easily by his tumbling downward, which would transform the potential energy into the energy of motion (*kinetic energy*). Experience attests, and the laws of physics make precise, that this is typical. A system harboring potential energy will exploit any opportunity to release that energy. In short, things fall.

The energy carried by a field's nonzero value is also potential energy: it, too, can be tapped, resulting in an incisive analogy with Cartman. Just as the increase in Cartman's potential energy as he climbs the mountain is determined by the shape of the slope—in flatter regions his potential energy varies minimally as he walks, because he gets hardly any higher, while in steeper regions his potential energy rises sharply—the potential energy of a field is described by an analogous shape, called its *potential energy curve*. Such a curve, as in Figure 3.1, determines how a field's potential energy varies with its value.

Following inflation's pioneers, let's then imagine that in the earliest moments of the cosmos, space is uniformly filled with an inflaton field, whose value places it high up on its potential energy curve. Imagine further, these physicists urge us, that the potential energy curve flattens out into a gentle plateau (as in Figure 3.1), allowing the inflaton to linger near the top. Under these hypothesized conditions, what will happen?

Two things, both critical. While the inflaton is on the plateau, it fills space with a large potential energy and negative pressure, driving a burst of inflationary expansion. But, just as Cartman releases his potential energy by rolling down the slope, so the inflaton releases its potential energy by its value, throughout space, rolling to lower numbers. And as its value decreases, the energy and negative pressure it harbors dissipate, bringing an end to the period of

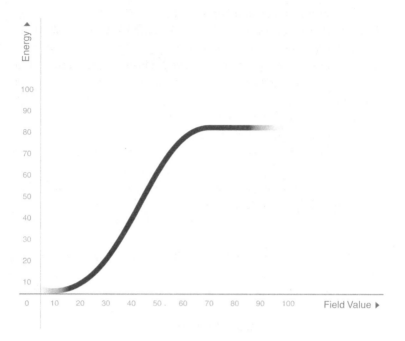

Figure 3.1 *The energy contained in an inflaton field (vertical axis) for given values of the field (horizontal axis).*

blistering expansion. Just as important, the energy released by the inflaton field isn't lost—instead, like a cooling vat of steam condensing into water droplets, the inflaton's energy condenses into a uniform bath of particles that fill space. This two-step process—brief but rapid expansion, followed by energy conversion to particles—results in a huge, uniform spatial expanse that's filled with the raw material of familiar structures like stars and galaxies.

Precise details depend on factors that neither theory nor observation has as yet determined (the initial value of the inflaton field, the exact shape of the potential energy slope, and so on)[5] but in typical versions the mathematical calculations show that the inflaton's energy would roll down the slope in a tiny fraction of a second, on the order of 10^{-35} seconds. And yet, during that brief span, space would expand by a colossal factor, perhaps 10^{30} if not more. These numbers are so extreme that they defy analogy. They imply that a region of space the size of a pea would be stretched larger than the observable universe in a time interval so short that the blink of an eye would overestimate it by a factor larger than a million billion billion billion.

However difficult it is to envision such a scale, what's essential is that the region of space that spawned the observable universe was so small that it would easily have come to a uniform temperature before it was stretched into our grand cosmic expanse by the rapid burst. The inflationary expansion, and billions of years of subsequent cosmological evolution, resulted in this temperature cooling substantially, but the uniformity set in place early on dictates a uniform result today. This resolves the mystery of how the universe's uniform conditions came to be. In inflation, a uniform temperature across space is inevitable.[6]

Eternal Inflation

During the nearly three decades since its discovery, inflation has become a fixture of cosmological investigation. But to have an accurate picture of the research panorama, you should be aware that inflation is a cosmological framework, but it is not a specific theory. Researchers have shown that there are many ways to skin an inflationary cat, differing in details such as the number of inflaton fields supplying the negative pressure, the particular potential energy curves to which each field is subject, and so on. Fortunately, the sundry realizations of inflation have some implications in common, so we can draw conclusions even in the absence of a definitive version.

Among these, one first fully realized by Andrei Linde of Stanford University and also Alexander Vilenkin of Tufts University (and subsequently developed by Linde), is of great importance.[7] In fact, it's the very reason I've spent the first half of this chapter explaining the inflationary framework.

In many versions of the inflationary theory, the burst of spatial expansion is not a onetime event. Instead, the process by which our region of the universe formed—rapid stretching of space, followed by a transition to a more ordinary, slower expansion, together with the production of particles—may happen over and over again at various far-flung locations throughout the cosmos. From a bird's-eye view, the cosmos would appear riddled with innumerable widely separated regions, each being the aftermath of a portion of space transitioning out of the inflationary burst. Our realm, what we have always thought of as *the* universe, would then be but one of these numerous regions, floating within a vastly larger spatial expanse. If intelligent life exists in the other regions, those beings would just as surely have thought their universe to be *the* universe, too. And so inflationary cosmology steers us headlong into our second variation on the theme of parallel universes.

To grasp how this *Inflationary Multiverse* comes about, we need to engage two complications that my Cartman analogy glossed over.

First, the image of Cartman perched high on a mountaintop offered an analogy to an inflaton field harboring significant potential energy and negative pressure, poised to roll to lower values. But whereas Cartman is perched on a single mountaintop, the inflaton field has a value at *each* point in space. The theory posits that the inflaton field starts off with the same value at each location within an initial region. And so we'd achieve a more faithful rendering of the science if we imagine something a little odd: numerous Cartman clones perched on numerous, closely packed, identical mountaintops throughout a spatial expanse.

Second, we've so far barely touched on the *quantum* aspect of quantum field theory. The inflaton field, like everything else in our quantum universe, is subject to quantum uncertainty. This means that its value will undergo random quantum jitters, momentarily rising a little here and dropping a little there. In everyday situations, quantum jitters are too small to notice. But calculations show that the larger the energy an inflaton has, the greater the fluctuations it will experience from quantum uncertainty. And since the inflaton's energy content during the inflationary burst was extremely high, the jitters in the early universe were big and dominant.[8]

We should thus not only picture a platoon of Cartmans perched high on identical mountaintops; we should also imagine that they are all subject to a random series of tremors—strong here, weak there, very strong way over there. With this setup, we can now determine what will happen. Different Cartman clones will stay perched on their mountaintops for different durations. In some locations, a strong tremor knocks most Cartmans down their slopes; in other locations, a mild tremor coaxes only a few to tumble down; in others still, some Cartmans may have started to roll down until a strong tremor knocked them back *up*. After a while, the terrain will be divided into a random assortment of domains—much as the United States is divided into states—in some of which no Cartmans are left on mountaintops, while in others many Cartmans remain securely perched.

The random nature of quantum jitters yields a similar conclusion for the inflaton field. The field begins high up on its potential energy slope at every point in a region of space. The quantum jitters then act like tremors. Because of this, as illustrated in Figure 3.2, the expanse of space rapidly divides into domains: in some, quantum jitters cause the field to topple down the slope, while in others it remains high.

So far, so good. But now stay with me closely; here's where cosmology and Cartmans differ. A field that's perched high up on its energy curve affects its environment far more significantly than a similarly perched Cartman does. From our familiar refrain—a field's uniform energy and negative pressure generate repulsive gravity—we recognize that the region the field permeates expands at a fantastic rate. This means that the inflaton field's evolution across space is driven by two opposing processes. Quantum jitters, by tending to knock the field off its perch, *decrease* the amount of space suffused with high field energy. Inflationary expansion, by rapidly enlarging those domains in which the field remains perched, *increases* the volume of space suffused with high field energy.

Which process wins?

In the vast majority of proposed versions of inflationary cosmology, the increase occurs at least as quickly as the decrease. The reason is that an inflaton field that can be knocked off its perch too quickly typically generates too little inflationary expansion to solve the horizon problem; in cosmologically successful versions of inflation, the increase thus wins over the decrease, ensuring that the

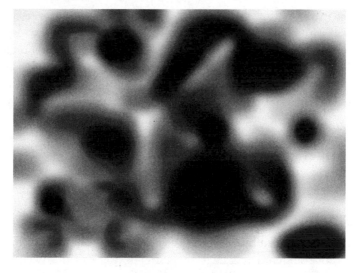

Figure 3.2 *Various domains in which the inflaton field has dropped down the slope (darker gray) or remains high (lighter gray).*

total volume of space in which the field's energy is high increases over time. Recognizing that such field configurations yield yet further inflationary expansion, we see that once inflation begins it never ends.

It's like the spread of a viral pandemic. To eradicate the threat, you need to wipe out the virus faster than it can reproduce. The inflationary virus "reproduces"—a high field value generates rapid spatial expansion and thus infuses a yet larger domain with that same high field value—and it does so faster than the competing process eliminates it. The inflationary virus effectively resists eradication.[9]

Swiss Cheese and the Cosmos

Collectively, these insights show that inflationary cosmology leads to a vastly new picture of reality's expanse, one that can be grasped most easily with a simple visual aid. Think of the universe as a gigantic block of Swiss cheese, with the cheesy parts being regions where the inflaton field's value is high and the holes being regions where it's low. That is, the holes are regions, like ours, that have transitioned out of the superfast expansion and, in the process, converted the inflaton field's energy into a bath of particles, which over time may coalesce into galaxies, stars, and planets. In this language, we've found that the cosmic cheese acquires more and more holes because quantum processes knock the inflaton's value downward at a random assortment of locations. At the same time, the cheesy parts stretch ever larger because they're subject to inflationary expansion driven by the high inflaton field value they harbor. Taken together, the two processes yield an ever-expanding block of cosmic cheese riddled with an ever-growing number of holes. In the more standard language of cosmology, each hole is called a *bubble universe* (or a *pocket universe*).[10] Each is an opening tucked within the superfast stretching cosmic expanse (Figure 3.3).

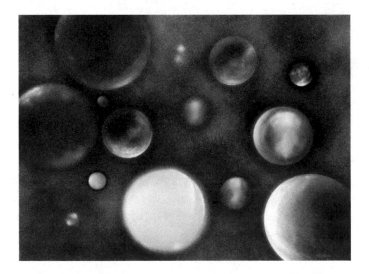

Figure 3.3 *The Inflationary Multiverse arises when bubble universes continually form within an ever-expanding spatial environment permeated by a high-valued inflaton field.*

Don't let the descriptive but diminutive-sounding "bubble universe" fool you. Our universe is gigantic. That it may be a single region embedded within an even larger cosmic structure—a single bubble in an enormous block of cosmic cheese—speaks to the fantastic expanse, in the inflationary paradigm, of the cosmos as a whole. And this goes for the other bubbles too. Each would be as much a universe—a real, gigantic, dynamic expanse—as ours.

There are versions of the inflationary theory in which inflation is not eternal. By fiddling with details such as the number of inflaton fields and their potential energy curves, clever theorists can arrange things so that the inflaton would, in due course, be knocked off its perch everywhere. But these proposals are the exception rather than the rule. Garden-variety inflationary models yield a gargantuan number of bubble universes carved into an eternally expanding spatial expanse. And so, if the inflationary theory is on the mark, and if, as many theoretical investigations conclude, its physically relevant realization is eternal, the existence of an Inflationary Multiverse would be an inevitable consequence.

Changing Perspectives

Back in the 1980s, when Vilenkin realized the eternal nature of inflationary expansion and the parallel universes to which it would give rise, he excitedly visited Alan Guth at MIT to tell him about it. Midway through the explanation, Guth's head drooped forward. He'd fallen asleep. This was not necessarily a bad sign; Guth is famous for nodding off during physics seminars—he's caught a few winks during talks I've given—then opening his eyes midway through to ask the most insightful of questions. But the broader physics community was no more enthusiastic than Guth was, so Vilenkin shelved the idea and moved on to other projects.

Sentiment today is very different. When Vilenkin was first thinking about the Inflationary Multiverse, the evidence in direct support of the inflationary theory itself was thin. So, to the few who paid any attention at all, ideas about inflationary expansion yielding a vast collection of parallel universes seemed like speculation piled upon speculation. But in the years since, the observational case for inflation has grown much stronger, once again thanks largely to precise measurements of the microwave background radiation.

Even though the observed uniformity of the microwave background radiation was one of the prime motivations for developing the inflationary theory, early proponents realized that rapid spatial expansion would not render the radiation *perfectly* uniform. Instead, they argued that quantum mechanical jitters stretched large by the inflationary expansion would overlay the uniformity with minuscule temperature variations, like tiny ripples on the surface of an otherwise smooth pond. This has proved to be a spectacular and enormously influential insight.* Here's how it goes.

Quantum uncertainty would have caused the value of the inflaton field to jitter. Indeed, if the inflationary theory is correct, the burst of inflationary expansion stopped here because a large and lucky quantum fluctuation, nearly 14 billion years ago, knocked the inflaton off its perch in our vicinity. Yet there's more to the story. As the inflaton's value rolled down its slope headlong toward the point of bringing inflation in our bubble universe to a close, its value would still have been subject to quantum jitters. The jitters, in turn, would have made the inflaton's value a little higher here and a little lower there, like the wavy surface of an unfurled sheet as it descends to your mattress. This would have produced slight variations in the energy the inflaton harbored across space. Normally, such quantum variations are so tiny and happen over such minuscule scales that they are irrelevant over cosmological distances. But inflationary expansion is anything but normal.

The expansion of space is so rapid, even during the transition out of the inflationary phase, that the microscopic would have been stretched to the macroscopic. And much as a tiny message scribbled on a deflated balloon becomes easier to read when air stretches the balloon's surface, so the influence of quantum jitters becomes visible when inflationary expansion stretches the cosmic fabric. More

---

*Among those who played a leading role in this work were Viatcheslav Mukhanov, Gennady Chibisov, Stephen Hawking, Alexei Starobinsky, Alan Guth, So-Young Pi, James Bardeen, Paul Steinhardt, and Michael Turner.

particularly, minute energy differences caused by quantum jitters are stretched into temperature variations that become imprinted in the cosmic microwave background radiation. Calculations show that the temperature differences wouldn't exactly be huge, but could be as large as a thousandth of a degree. If the temperature is 2.725 K in one region, the stretched-out quantum jitters would result in its being a touch colder, say 2.7245 K, or a touch hotter, 2.7255 K, at nearby regions.

Painstakingly precise astronomical observations have sought these temperature variations. They've found them. Just as the theory predicted, they measure about a thousandth of a degree (see Figure 3.4). More impressive still, the tiny temperature differences fit a pattern on the sky that is explained spot-on by the theoretical calculations. Figure 3.5 compares theoretical predictions of how the temperature should vary as a function of the distance between two regions (measured by the angle between their respective lines of sight when viewed from earth) with the actual measurements. The agreement is stunning.

Figure 3.4 *The enormous spatial expansion in inflationary cosmology stretches quantum fluctuations from the microscopic to the macroscopic, resulting in observable temperature variations in the cosmic microwave background radiation (the darker splotches are slightly colder than the lighter ones).*

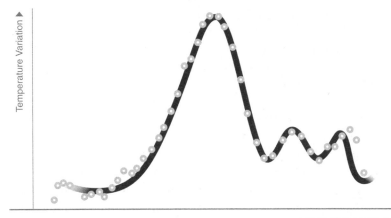

Figure 3.5 *The pattern of temperature differences in the cosmic microwave background radiation. Temperature variation is the vertical axis; the separation between two locations (measured by the angle between their respective lines of sight when viewed from earth — larger angles to the left, smaller angles to the right) is the horizontal axis.*[11] *The theoretical curve is solid; the observational data are given by the circles.*

The 2006 Nobel Prize in Physics was awarded to George Smoot and John Mather, who led more than a thousand researchers on the Cosmic Background Explorer team in the early 1990s to the first detection of these temperature differences. During the past decade, every new and more accurate measurement, yielding data such as those in Figure 3.5, has resulted in yet more precise verification of the predicted temperature variations.

These works have capped a thrilling story of discovery that began with the insights of Einstein, Friedmann, and Lemaître, was pushed sharply forward by the calculations of Gamow, Alpher, and Herman, was reinvigorated by the ideas of Dicke and Peebles, was shown relevant by the observations of Penzias and Wilson, and has now culminated in the handiwork of armies of astronomers, physicists, and engineers whose combined efforts have measured a fantastically minute cosmic signature that was set in place billions of years ago.

On a more qualitative level, we should all be thankful for the blotches in Figure 3.4. At the close of inflation in our bubble universe, regions with slightly more energy (equivalently, via $E = mc^2$, regions with slightly more mass) exerted a slightly stronger gravitational pull, attracting more particles from their surroundings and thus growing larger. The larger aggregate, in turn, exerted an even stronger gravitational pull, thus attracting yet more matter and growing larger still. In time, this snowball effect resulted in the formation of clumps of matter and energy that, over billions of years, evolved into galaxies and the stars within them. In this way, inflationary theory establishes a remarkable link between the largest and smallest structures in the cosmos. The very existence of galaxies, stars, planets, and life itself derives from microscopic quantum uncertainty amplified by inflationary expansion.

Inflation's theoretical underpinnings may be rather tentative: the inflaton, after all, is a hypothetical field whose existence has yet to be demonstrated; its potential energy curve is posited by researchers, not revealed by observation; the inflaton must somehow start at the top of its energy curve across a region of space; and so on. Despite all that, and even if some details of the theory are not quite right, the agreement between theory and observation has convinced many that the inflationary scheme taps into a deep truth about cosmic evolution. And since a great many versions of inflation are eternal, yielding an ever-growing number of bubble universes, theory and observation combine to make an indirect yet compelling case for this second version of parallel worlds.

Experiencing the Inflationary Multiverse

In a Quilted Multiverse, there's no sharp divide between one parallel universe and another. All are part of a single spatial expanse whose overall qualitative features are similar from region to region. The surprise lies in the details. Most of us wouldn't expect worlds to repeat; most of us wouldn't expect, every so often, to encounter versions of ourselves, our friends, our families. But if we could journey sufficiently far, that's what we would find.

In an Inflationary Multiverse, the member universes are sharply divided. Each is a hole in the cosmic cheese, separated from the others by domains in which the inflaton's value remains high. Since such intervening regions are still undergoing inflationary expansion, the bubble universes are rapidly driven apart, with a speed of recession proportional to the amount of swelling space between them. The farther apart they are, the greater the expansion's speed; the ultimate result is that distant bubbles move apart faster than the speed of light. Even with unlimited longevity and technology, there's no way to cross such a divide. There's no way to even send a signal.

All the same, we can still contemplate an imaginary voyage to one or more of the other bubble universes. On such a journey, what would you find? Well, because each bubble universe results from the same process—the inflaton is knocked from its perch, yielding a region that drops out of the inflationary expansion—they are all governed by the same physical theory and so are all subject to the same set of physical laws. But, much as the behavior of identical twins can differ profoundly as a result of environmental differences, identical laws can manifest themselves in profoundly different ways in different environments.

Imagine, for example, that one of the other bubble universes looks much like ours, dotted by galaxies containing stars and planets, but with one essential difference. Permeating the universe is a magnetic field, thousands of times stronger than that created in our most advanced MRI machines, and one that can't be switched off by a technician. Such a powerful field would affect the way a great many things behave. Not only would objects containing iron have a nasty habit of flying off in the direction of the field, but even basic properties of particles, atoms, and molecules would shift. A sufficiently strong magnetic field would so disrupt cellular function that life as we know it couldn't take hold.

Yet just as the physical laws operating inside an MRI are the very same laws that operate outside, so the fundamental physical laws operating in this magnetic universe would be the same as ours.

The discrepancies in experimental results and observable features would be due solely to an aspect of the environment: the strong magnetic field. Talented scientists in the magnetic universe would in time tease out this environmental factor and home in on the same mathematical laws we've discovered.

Over the past forty years, researchers have built a case for a similar scenario right here in our own universe. The most lauded theory of fundamental physics, the *Standard Model of particle physics*, posits that we are immersed in an exotic mist called the *Higgs field* (named after the English physicist Peter Higgs, who with important contributions from Robert Brout, François Englert, Gerald Guralnik, Carl Hagen, and Tom Kibble pioneered this idea in the 1960s). Both Higgs fields and magnetic fields are invisible and hence can fill space without directly revealing their presence. However, according to modern particle theory, a Higgs field camouflages itself far more fully. As particles move through a uniform, space-filling Higgs field, they don't speed up, they don't slow down, they are not coaxed to follow particular trajectories, as some would in the presence of a strong magnetic field. Instead, the theory claims, they're influenced in ways more subtle and profound.

As fundamental particles burrow through a Higgs field, *they acquire and maintain the mass that experiments have revealed them to possess.* According to this idea, when you push against an electron or quark in an effort to change its speed, the resistance you feel comes from the particle's "rubbing" against the molasses-like Higgs field. It's this resistance that we call the particle's mass. Were you to remove the Higgs field from some region, particles passing through would suddenly become massless. Were you to double the value of the Higgs field in another region, particles passing through would suddenly have twice their usual mass.*

Such human-induced changes are hypothetical, because the

---

*I stress *fundamental* particles, like electrons and quarks, because for composite particles, like protons and neutrons (each made from 3 quarks), much of the mass arises from interactions between the constituents (the energy carried by gluons of the strong nuclear force, which bind the quarks inside protons and neutrons, contributes most of the mass of these composite particles).

energy required to substantially modify a Higgs field's value in even a small region of space is enormously beyond what we can muster. (The changes are also hypothetical because the existence of the Higgs fields is still up in the air. Theorists eagerly anticipate highly energetic collisions between protons at the Large Hadron Collider chipping off small chunks of the Higgs field—Higgs particles—that may be detected in the coming years.) But in many versions of inflationary cosmology, *a Higgs field would naturally have different values in different bubble universes.*

A Higgs field, much like an inflaton field, has a curve that records the amount of energy it contains for various values it can assume. An essential difference from the inflaton field's energy curve, though, is that the Higgs typically settles not at the value 0 (as in Figure 3.1), but rather rolls to one of the troughs illustrated in Figure 3.6a. Picture, then, an early stage in each of two bubble universes, ours and another. In both, the hot, tempestuous frenzy causes the value of the Higgs field to undulate wildly. As each universe expands and cools, the Higgs field calms and its value rolls

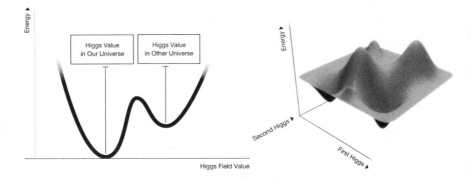

Figure 3.6 (a) *A potential energy curve for a Higgs field that has two troughs. The familiar features of our universe are associated with the field settling down in the left trough; in another universe, however, the field can settle down in the right trough, yielding different physical features.* **(b)** *A sample potential energy curve for a theory with two Higgs fields.*

toward one of the troughs in Figure 3.6a. In our universe, the Higgs field's value settles down in, say, the left trough, giving rise to the particle properties familiar from experimental observation. But in the other universe, the Higgs' motion may result in its value settling down in the right trough. If it did, that universe would have properties substantially different from ours. Although the underlying laws in both universes would be the same, the masses and various other properties of particles would not.

Even a modest difference in particle properties would have weighty consequences. If the electron mass in another bubble universe were a few times larger than it is here, electrons and protons would tend to merge, forming neutrons and thus preventing the widespread production of hydrogen. The fundamental forces—the electromagnetic force, the nuclear forces, and (we believe) gravity—are also communicated by particles. Change the particle properties and you drastically change the properties of the forces. The heavier a particle, for example, the more sluggish its motion and so the shorter the distance over which the corresponding force is transmitted. The formation and stability of atoms in our bubble universe rely on the properties of the electromagnetic and nuclear forces. If you substantially modify those forces, atoms will fall apart or, more likely, not coalesce in the first place. An appreciable change to the properties of particles would thus disrupt the very processes that give our universe its familiar features.

Figure 3.6a illustrates only the simplest case, in which there is a single species of Higgs field. But theoretical physicists have explored more complicated scenarios involving multiple Higgs fields (we will shortly see that such possibilities naturally emerge from string theory), which translate into an even richer set of distinct bubble universes. An example with two Higgs fields is illustrated in Figure 3.6b. As before, the various troughs represent Higgs field values that one or another bubble universe could settle into.

Permeated by such unfamiliar values of various Higgs fields, these universes would differ from ours considerably, as schematically illustrated in Figure 3.7. This would make a journey through the Inflationary Multiverse a perilous undertaking. Many of the other universes would not be places you'd want high on your itiner-

Figure 3.7 *Because fields can settle down to different values in different bubbles, the universes in the Inflationary Multiverse can have different physical features, even though the universes are all governed by the same fundamental physical laws.*

ary, because the conditions would be incompatible with the biological processes essential to survival, giving new meaning to the saying that there's no place like home. In the Inflationary Multiverse, our universe could well be an island oasis in a gigantic but largely inhospitable cosmic archipelago.

Universes in a Nutshell

Because of their fundamental differences, the Quilted and Inflationary Multiverses might appear unrelated. The quilted variety emerges if the extent of space is infinite; the inflationary variety emerges from eternal inflationary expansion. Yet, there is a deep and wonderfully satisfying connection between them, one that brings the discussion in the previous two chapters full circle. The parallel universes arising from inflation generate their quilted cousins. The process has to do with time.

Of the many strange things Einstein's work revealed, the fluidity of time is the hardest to grasp. Whereas everyday experience convinces us that there is an objective concept of time's passage, relativity shows this to be an artifact of life at slow speeds and weak gravity. Move near light speed, or immerse yourself in a powerful gravitational field, and the familiar, universal conception of time will evaporate. If you're rushing past me, things I insist happened at the same moment will appear to you to have occurred at different moments. If you're hanging out near the edge of a black hole, an hour's passage on your watch will be monumentally longer on mine. This isn't evidence of a magician's trickery or a hypnotist's deception. The passage of time depends on the particulars— trajectory followed and gravity experienced—of the measurer.[12]

When applied to the entire universe, or to our bubble in an inflationary setting, this immediately raises a question: How does such malleable, custom-made time comport with the notion of an absolute cosmological time? We freely speak of the "age" of our universe, but given that galaxies are moving rapidly relative to one another, at speeds dictated by their various separations, doesn't the relativity of time's passage create a nightmarish accounting problem for any would-be cosmic timekeeper? More pointedly, when we speak of our universe being "14 billion years old," are we using a particular clock to measure that duration?

We are. And a careful consideration of such cosmic time reveals a direct link between parallel universes of the inflationary and quilted varieties.

Every method we use to measure time's passage involves an examination of change that occurs to some particular physical system. Using a common wall clock, we examine the change in position of its hands. Using the sun, we examine the change in its position in the sky. Using carbon 14, we examine the percentage of an original sample that's undergone radioactive decay to nitrogen. Historical precedent and general convenience have led us to use the rotation and revolution of the earth as physical referents, giving rise to our standard notions of "day" and "year." But when we're thinking on cosmic scales, there is another, more useful, method for keeping time.

We've seen that inflationary expansion yields vast regions whose properties on average are homogeneous. Measure the temperature, pressure, and average density of matter in two large but separate regions within a bubble universe, and the results will agree. The results can change over time, but the large-scale uniformity ensures that, on average, the change *here* is the same as the change *there*. As an important case in point, the mass density in our bubble universe has steadily decreased over our multibillion-year history, thanks to the relentless expansion of space, but because the change has occurred uniformly, our bubble's large-scale homogeneity has not been disrupted.

This proves important because just as the steadily decreasing amount of carbon 14 in organic matter provides a means of measuring time's passage on earth, so the steadily decreasing mass density provides a means of measuring time's passage across space. And because the change has happened uniformly, mass density as a marker of time's passage provides our bubble universe with a global standard. If everyone diligently calibrates their watches to the average mass density (and recalibrates after trips to black holes, or periods of travel at near light speed), the synchronicity of our timepieces across our bubble universe will be maintained. When we speak of the age of the universe—the age of our bubble, that is—it is on such cosmically calibrated watches that we imagine time's passage being measured; it is only with respect to them that cosmic time is a sensible concept.

In the earliest era of our bubble universe, the same reasoning would have applied with one change of detail. Ordinary matter had yet to form, so we can't speak of the average mass density in space. Instead, the inflaton field carried our universe's storehouse of energy—energy that would shortly be converted into familiar particles—so we need to envisage setting our clocks by the density of the inflaton field's energy.

Now, the inflaton's energy is determined by its value, as summarized by its energy curve. To determine what time it is at a given location in our bubble, we therefore need to determine the value of the inflaton at that location. Then, just as two trees are the same age if they have the same number of tree rings, and just as two samples of glacial sediment are the same age if they have the same percentage of radioactive carbon, *two locations in space are passing through the same moment in time when they have the same value of the inflaton field.* That's how we set and synchronize clocks in our bubble universe.

The reason I've brought all this up is that when applied to the cosmic Swiss cheese of the Inflationary Multiverse, these observations yield a strikingly counterintuitive implication. Much as Hamlet famously declares, "I could be bounded in a nutshell, and count myself a king of infinite space," each of the bubble universes appears to have *finite* spatial extent when examined from the outside, but *infinite* spatial extent when examined from the inside. And that's a marvelous realization. Infinite spatial extent is just what we need for quilted parallel universes. So we can meld the Quilted Multiverse into the inflationary story.

The extreme disparity between the outsider's and insider's perspectives arises because they have vastly different conceptions of time. Although the point is far from obvious, we'll now see that *what appears as endless time to an outsider appears as endless space, at each moment of time, to an insider.*[13]

Space in a Bubble Universe

To grasp how this comes about, imagine that Trixie, floating within a rapidly expanding inflaton-filled region of space, is observing the formation of a nearby bubble universe. Focusing her inflaton-meter on the growing bubble, she is able to directly track its changing inflaton field value. Although the region—the hole in the cosmic cheese—is three-dimensional, it's simpler to examine the field along a one-dimensional cross section across its diameter, and as Trixie does so she records the data in Figure 3.8a. Each higher row shows the inflaton's value at a successive moment in time, from Trixie's perspective. And as is apparent from the figure, Trixie sees the bubble universe—represented in the figure by the lighter locations where the inflaton's value has dropped—grow ever larger.

70	60	50	40	30	20	10	0	10	20	30	40	50	60	70
80	70	60	50	40	30	20	10	20	30	40	50	60	70	80
90	80	70	60	50	40	30	20	30	40	50	60	70	80	90
100	90	80	70	60	50	40	30	40	50	60	70	80	90	100
100	100	90	80	70	60	50	40	50	60	70	80	90	100	100
100	100	100	90	80	70	60	50	60	70	80	90	100	100	100
100	100	100	100	90	80	70	60	70	80	90	100	100	100	100
100	100	100	100	100	90	80	70	80	90	100	100	100	100	100
100	100	100	100	100	100	90	80	90	100	100	100	100	100	100
100	100	100	100	100	100	100	90	100	100	100	100	100	100	100
100	100	100	100	100	100	100	100	100	100	100	100	100	100	100

Figure 3.8a *Each row chronicles the inflaton's value at one moment of time from an outsider's perspective. Higher rows correspond to later moments. The columns denote positions across space. A bubble is a region of space that stops inflating because of a drop in the inflaton's value. The lighter entries denote the value of the inflaton field within the bubble. From the perspective of the outside observer, the bubble grows ever larger.*

Now imagine that Norton is also examining this very same bubble universe, but from the inside; he's hard at work making detailed astronomical observations with his own inflaton-meter. Norton, unlike Trixie, adheres to a notion of time that's calibrated by the value of the inflaton. This is key to the conclusion we're chasing, so I need you to buy into it fully. Imagine, if you will, that everyone in the bubble universe wears a watch that measures and displays the inflaton's value. When Norton throws a dinner party, he instructs the guests to show up at his house when the inflaton's value is 60. Since everyone's watch is calibrated to the same, uniform standard—the inflaton field's value—the party goes off without a hitch. Everyone shows up at the same moment because everyone is attuned to the same concept of synchronicity.

With this understanding, it's a simple matter for Norton to work out the size of the bubble universe at any given moment of his time. In fact, it's child's play: all Norton has to do is paint by numbers. By connecting all points that have the same numerical value for the inflaton field, Norton can delineate all locations within the bubble at a single moment of time. His time. Insider's time.

Norton's drawing in Figure 3.8b says it all. Each curve, connecting points with the same inflaton-field value, represents all of space at a given moment of time. As the figure makes clear, each curve extends indefinitely far, which means that the size of the bubble universe, according to its inhabitants, is *infinite*. This reflects that endless outsider time, experienced by Trixie as the endless number of rows in Figure 3.8, appears as endless space, at each moment of time, according to an insider like Norton.

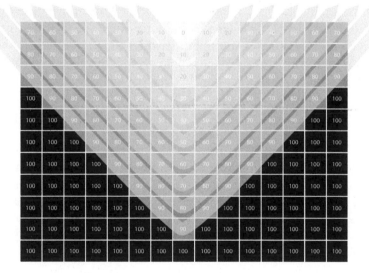

Figure 3.8b *The same information as in Figure 3.8a is organized differently by someone within the bubble. Inflaton values that agree correspond to identical moments, so the curves drawn sweep through all those points in space that exist at the same moment in time. Smaller inflaton values correspond to later moments. Note that the curves could be extended infinitely far, so from an insider's perspective, space is infinite.*

That's a powerful insight. In Chapter 2, we found that the Quilted Multiverse was contingent upon space being infinitely large, something that, as we discussed there, might or might not be the case. Now we see that each bubble within the Inflationary Multiverse is spatially finite from the outside but spatially infinite from the inside. If the Inflationary Multiverse is real, then the inhabitants of a bubble—us—would thus be members not only of the Inflationary Multiverse but of the Quilted Multiverse, too.[14]

When I first learned of the Quilted and Inflationary Multiverses, it was the inflationary variety that struck me as more plausible. Inflationary cosmology resolves a number of long-standing puzzles while yielding predictions that match up well with observations. And by the reasoning we've recounted, inflation is naturally a process that never ends; it produces bubble universes upon bubble universes, of which we inhabit but one. The Quilted Multiverse, on the other hand, by having its full force when space is not just large but truly infinite (you might have repetition in a large universe, but you are guaranteed repetition in an infinite one), seemed avoidable: it might be the case, after all, that the universe has finite size. But we now see that eternal inflation's bubble universes, when properly analyzed from the viewpoint of their inhabitants, *are* spatially infinite. Inflationary parallel universes beget quilted ones.

The best available cosmological theory for explaining the best available cosmological data leads us to think of ourselves as occupying one of a vast inflationary system of parallel universes, each of which harbors its own vast collection of quilted parallel universes. Cutting-edge research yields a cosmos in which there are not only parallel universes but parallel parallel universes. It suggests that reality is not only expansive but abundantly expansive.

CHAPTER 4

Unifying Nature's Laws

On the Road to String Theory

From the big bang to inflation, modern cosmology traces its roots to a single scientific nexus: Einstein's general theory of relativity. With his new theory of gravity, Einstein upended the accepted conception of a rigid and immutable space and time; science now had to embrace a dynamic cosmos. Contributions of this magnitude are rare. Yet, Einstein dreamed of scaling even greater heights. With the mathematical arsenal and geometric intuition he'd amassed by the 1920s, he set out to develop a *unified field theory*.

By this, Einstein meant a framework that would stitch all of nature's forces into a single, coherent mathematical tapestry. Rather than have one set of laws for these physical phenomena and a different set for those, Einstein wanted to fuse all the laws into a seamless whole. History has judged Einstein's decades of intense work toward unification as having had little lasting impact—the dream was noble, the timing was early—but others have taken up the mantle and made substantial strides, the most refined proposal being *string theory*.

My previous books *The Elegant Universe* and *The Fabric of the Cosmos* covered the history and essential features of string theory. In the years since they appeared, the theory's general health and status have faced a spate of public questioning. Which is completely reasonable. For all its progress, string theory has yet to make definitive predictions whose experimental investigation could prove the theory right or wrong. As the next three multiverse varieties we will encounter (in Chapters 5 and 6) emerge from a string theoretic perspective, it's important to address the current state of the theory as well as the prospects for making contact with experimental and observational data. Such is the charge of this chapter.

A Brief History of Unification

At the time Einstein pursued the goal of unification, the known forces were gravity, described by his own general relativity, and electromagnetism, described by Maxwell's equations. Einstein envisioned melding the two into a single mathematical sentence that would articulate the workings of all nature's forces. Einstein had high hopes for this unified theory. He considered Maxwell's nineteenth-century work on unification an archetypal contribution to human thought—and rightly so. Before Maxwell, the electricity flowing through a wire, the force generated by a child's magnet, and the light streaming to earth from the sun were viewed as three separate, unrelated phenomena. Maxwell revealed that, in actuality, they formed an intertwined scientific trinity. Electric currents *produce* magnetic fields; magnets moving in the vicinity of a wire *produce* electric currents; and wavelike disturbances rippling through electric and magnetic fields *produce* light. Einstein anticipated that his own work would carry forward Maxwell's program of consolidation by making the next and possibly final move toward a fully unified description of nature's laws—a description that would unite electromagnetism and gravity.

This wasn't a modest goal, and Einstein didn't take it lightly. He had an unparalleled capacity for single-minded devotion to problems he'd set for himself, and during the last thirty years of his life the problem of unification became his prime obsession. His personal secretary and gatekeeper, Helen Dukas, was with Einstein at the Princeton Hospital during his penultimate day, April 17, 1955. She recounts how Einstein, bedridden but feeling a little stronger, asked for the pages of equations on which he had been endlessly manipulating mathematical symbols in the fading hope that the unified field theory would materialize. Einstein didn't rise with the morning sun. His final scribblings shed no further light on unification.[1]

Few of Einstein's contemporaries shared his passion for unification. From the mid-1920s through the mid-1960s, physicists, guided by quantum mechanics, were unlocking the secrets of the atom and learning how to harness its hidden powers. The lure of prying apart matter's constituents was immediate and powerful. While many agreed that unification was a laudable goal, it was of only passing interest in an age when theorists and experimenters were working hand in glove to reveal the laws of the microscopic realm. With Einstein's passing, work on unification ground to a halt.

His failure was compounded when subsequent research showed that his quest for unity had been too narrowly focused. Not only had Einstein downplayed the role of quantum physics (he believed the unified theory would supersede quantum mechanics and so it needn't be incorporated from the outset), but his work failed to take account of two additional forces revealed by experiments: the *strong nuclear force* and the *weak nuclear force*. The former provides a powerful glue that holds together the nuclei of atoms, while the latter is responsible for, among other things, radioactive decay. Unification would need to combine not two forces but four; Einstein's dream seemed all the more remote.

During the late 1960s and 1970s, the tide turned. Physicists realized that the methods of quantum field theory, which had been successfully applied to the electromagnetic force, also provided descriptions of the weak and strong nuclear forces. All three of the nongravitational forces could thus be described using the same mathematical language. Moreover, detailed study of these quantum field theories—most notably in the Nobel Prize–winning work of Sheldon Glashow, Steven Weinberg, and Abdus Salam, as well as in the subsequent insights of Glashow and his Harvard colleague Howard Georgi—revealed relationships suggesting a potential unity among the electromagnetic, weak, and strong nuclear forces. Following Einstein's nearly half-century-old lead, theoreticians argued that these three apparently distinct forces might actually be manifestations of a single monolithic force of nature.[2]

These were impressive advances toward unification, but set against the encouraging backdrop was a pesky problem. When scientists applied the methods of quantum field theory to nature's fourth force, gravity, the math just wouldn't work. Calculations involving quantum mechanics and Einstein's general relativistic description of the gravitational field yielded jarring results that amounted to mathematical gibberish. However successful general relativity and quantum mechanics had been in their native domains, the large and the small, the nonsensical output from the attempt to unite them spoke to a deep fissure in the understanding of nature's laws. If the laws you have prove mutually incompatible, then—clearly—the laws you have are not the right laws. Unification had been an aesthetic goal; now it was transformed into a logical imperative.

The mid-1980s witnessed the next pivotal development. That's when a new approach, *superstring theory*, captured the attention of the world's physicists. It ameliorated the hostility between general relativity and quantum mechanics, and so provided hope that gravity could be brought within a unified quantum mechanical fold. The era of superstring unification was born. Research proceeded at an intense pace, and thousands of journal pages were quickly filled with calculations that fleshed out aspects of the approach and laid the groundwork for its systematic formulation. An impressive and intricate mathematical structure emerged, but much about superstring theory (*string theory*, for short) remained mysterious.[3]

Then, beginning in the mid-1990s, theorists intent on unraveling those mysteries unexpectedly thrust string theory squarely into the multiverse narrative. Researchers had long known that the mathematical methods being used to analyze string theory invoked a variety of approximations and so were ripe for refinement. When some of those refinements were developed, researchers realized that the math suggested plainly that our universe might belong to a multiverse. In fact, the mathematics of string theory suggested not just one but a number of different kinds of multiverses of which we might be a part.

To fully grasp these compelling and contentious developments, and to assess their role in our ongoing search for the deep laws of the cosmos, we need to take a step back and first evaluate the state of string theory.

Quantum Fields Redux

Let's begin by taking a closer look at the traditional, highly success-ful framework of quantum field theory. This will prepare us to dis-cuss string unification as well as highlight pivotal connections between these two approaches for formulating nature's laws.

Classical physics, as we saw in Chapter 3, describes a field as a kind of mist that permeates a region of space and can carry distur-bances in the form of ripples and waves. Were Maxwell to describe the light that's now illuminating this text, for example, he'd wax enthusiastic about electromagnetic waves, produced by the sun or by a nearby lightbulb, undulating across space on their way to the printed page. He'd describe the waves' movement mathematically, using numbers to delineate the field's strength and direction at each point in space. An undulating field corresponds to undulating numbers: the field's numerical value at any given location cycles up and down and up again.

When quantum mechanics is brought to bear on the concept of a field, the result is quantum field theory, which is characterized by two essential new features. We've already encountered both, but they're worth a refresher. First, quantum uncertainty causes the value of a field at each point in space to jitter randomly—think of the fluctuating inflaton field from inflationary cosmology. Second, quantum mechanics establishes that, somewhat as water is com-posed of H_2O molecules, a field is composed of infinitesimally small particles known as the field's *quanta*. For the electromagnetic field, the quanta are photons, and so a quantum theorist would modify Maxwell's classical description of your lightbulb by saying that the bulb emits a steady stream comprising 100 billion billion photons each second.

Decades of research have established that these features of quantum mechanics as applied to fields are completely general. Every field is subject to quantum jitters. And every field is associ-ated with a species of particle. Electrons are quanta of the electron field. Quarks are quanta of the quark field. For a (very) rough men-tal image, physicists sometimes think of particles as knots or dense nuggets of their associated field. This visualization notwithstanding, the mathematics of quantum field theory describes these particles as dots or points that have no spatial extent and no internal structure.[4]

Our confidence in quantum field theory comes from one essential fact: there is not a single experimental result that counters its predictions. To the contrary, data confirm that the equations of quantum field theory describe the behavior of particles with astounding accuracy. The most impressive example comes from the quantum field theory of the electromagnetic force, *quantum electrodynamics*. Using it, physicists have undertaken detailed calculations of the electron's magnetic properties. The calculations are not easy, and the most refined versions have taken decades to complete. But they've been worth the effort. The results match measurements to a precision of *ten* decimal places, an almost unimaginable agreement between theory and experiment.

With such success, you might anticipate that quantum field theory would provide the mathematical framework for understanding all of nature's forces. An illustrious coterie of physicists shared this very expectation. By the late 1970s, the hard work of many of these visionaries had established that, indeed, the weak and strong nuclear forces fit squarely within the rubric of quantum field theory. Both forces are accurately described in terms of fields—the weak and the strong fields—that evolve and interact according to the mathematical rules of quantum field theory.

But, as I indicated in the historical overview, many of these same physicists quickly realized that the story for nature's remaining force, gravity, was far subtler. Whenever the equations of general relativity commingled with those of quantum theory, the mathematics balked. Use the combined equations to calculate the quantum probability of some physical process—such as the chance of two electrons ricocheting off each other, given both their electromagnetic repulsion and their gravitational attraction—and you'd typically get the answer *infinity*. While some things in the universe can be infinite, such as the extent of space and the quantity of matter that may fill it, probabilities are not among them. By definition, the value of a probability must be between 0 and 1 (or, in terms of percentages, between 0 and 100). An infinite probability does not mean that something is very likely to happen, or is certain to happen; rather, it's meaningless, like speaking of the thirteenth egg in an even dozen. An infinite probability sends a clear mathematical message: the combined equations are nonsense.

Physicists traced the failure to the jitters of quantum uncertainty. Mathematical techniques had been developed for analyzing the jitters of the strong, weak, and electromagnetic fields, but when the same methods were applied to the gravitational field—a field that governs the curvature of spacetime itself—they proved ineffective. This left the mathematics saturated with inconsistencies such as infinite probabilities.

To get a feel for why, imagine you're the landlord of an old house in San Francisco. If you have tenants who throw raucous parties, it might take effort to deal with the situation, but you don't worry that the festivities will compromise the building's structural integrity. However, if there's an earthquake, you're facing something far more serious. The fluctuations of the three nongravitational forces—fields that are tenants within the house of spacetime—are like the building's incessant partyers. It took a generation of theoretical physicists to grapple with their raucous jitters, but by the 1970s they'd developed mathematical methods capable of describing the quantum properties of the nongravitational forces. The fluctuations of the gravitational field, however, are qualitatively different. They're more like an earthquake. Because the gravitational field is woven within the very fabric of spacetime, its quantum jitters shake the entire structure through and through. When used to analyze such pervasive quantum jitters, the mathematical methods collapsed.[5]

For years, physicists turned a blind eye to this problem because it surfaces only under the most extreme conditions. Gravity makes its mark when things are very massive, quantum mechanics when things are very small. And rare is the realm that is both small and massive, so that to describe it you must invoke both quantum mechanics and general relativity. Yet, there are such realms. When gravity and quantum mechanics are together brought to bear on either the big bang or black holes, realms that *do* involve extremes of enormous mass squeezed to small size, the math falls apart at a critical point in the analyses, leaving us with unanswered questions regarding how the universe began and how, at the crushing center of a black hole, it might end.

Moreover—and this is the truly daunting part—beyond the specific examples of black holes and the big bang, you can calculate how massive and how small a physical system needs to be for both gravity and quantum mechanics to play a significant role. The result is about 10^{19} times the mass of a single proton, the so-called *Planck mass*, squeezed into a fantastically small volume of about 10^{-99} cubic centimeters (roughly a sphere with a radius of 10^{-33} centimeters, the so-called *Planck length* graphically illustrated in Figure 4.1).[6] The dominion of quantum gravity is thus more than a million billion times beyond the scales we can probe even with the world's most powerful accelerators. This vast expanse of uncharted territory could easily be rife with new fields and their associated particles—and who knows what else. To unify gravity and quantum mechanics requires trekking from here to there, grasping the known and the unknown across an enormous expanse that, for the most part, is experimentally inaccessible. That's a hugely ambitious task, and many scientists concluded that it was beyond reach.

You can thus imagine the surprise and skepticism when, in the mid-1980s, rumors started racing through the physics community that there had been a major theoretical breakthrough toward unification with an approach called string theory.

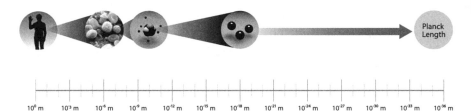

Figure 4.1 *The Planck length, where gravity and quantum mechanics confront each other, is some 100 billion billion times smaller than any domain that's been explored experimentally. Reading across the chart, each of the equally spaced tick marks represents a decrease in size by a factor of 1,000; this allows the chart to fit on a page but visually downplays the huge range of scales. For a better feel, note that if an atom were magnified to be as large as the observable universe, the same magnification would make the Planck length the size of an average tree.*

String Theory

Although string theory has an intimidating reputation, its basic idea is easy to grasp. We've seen that the standard view, prior to string theory, envisions nature's fundamental ingredients as point particles—dots with no internal structure—governed by the equations of quantum field theory. With each distinct species of particle is associated a distinct species of field. String theory challenges this picture by suggesting that the particles are not dots. Instead, the theory proposes that they're tiny, stringlike, vibrating filaments, as in Figure 4.2. Look closely enough at any particle previously deemed elementary and the theory claims you'll find a minuscule vibrating string. Look deep inside an electron, and you'd find a string; look deep inside a quark, and you'd find a string.

With even more precise observation, the theory argues, you'd notice that the strings within different kinds of particles are identical, the leitmotif of string unification, but vibrate in different patterns. An electron is less massive than a quark, which according to string theory means that the electron's string vibrates less energetically than the quark's string (reflecting again the equivalence of energy and mass embodied in $E = mc^2$). The electron also has an electric charge whose magnitude exceeds that of a quark, and this difference translates into other, finer differences between the string vibrational patterns associated with each. Much as different vibrational patterns of strings on a guitar produce different musical

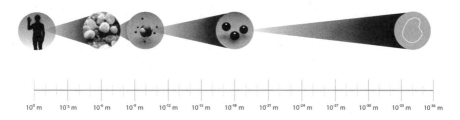

10^0 m 10^{-3} m 10^{-6} m 10^{-9} m 10^{-12} m 10^{-15} m 10^{-18} m 10^{-21} m 10^{-24} m 10^{-27} m 10^{-30} m 10^{-33} m 10^{-36} m

Figure 4.2 *String theory's proposal for the nature of physics at the Planck scale envisions that the fundamental constituents of matter are string-like filaments. Because of the limited resolving power of our equipment, the strings appear as dots.*

notes, different vibrational patterns of the filaments in string theory produce different particle properties.

In fact, the theory encourages us to think of a vibrating string not merely as dictating the properties of its host particle but rather as *being* the particle. Because of the string's infinitesimal size, on the order of the Planck length—10^{-33} centimeters—even today's most refined experiments cannot resolve the string's extended structure. The Large Hadron Collider, which slams particles together with energies just beyond 10 trillion times that embodied by a single proton at rest, can probe to scales of about 10^{-19} centimeters; that's a millionth of a billionth the width of a strand of hair, but still orders of magnitude too *large* to resolve phenomena at the Planck length. And so, just as earth would look dotlike if viewed from Pluto, strings would appear dotlike when studied even with the most advanced particle accelerator in the world. Nevertheless, according to string theory, particles *are* strings.

In a nutshell, that's string theory.

Strings, Dots, and Quantum Gravity

String theory has many other essential features, and the developments it has undergone since it was first proposed have greatly enriched the bare-bones description I've so far given. In the rest of this chapter (as well as Chapters 5, 6, and 9), we will encounter some of the most pivotal advances, but I want to stress here three overarching points.

First, when a physicist proposes a model of nature using quantum field theory, he or she needs to choose the particular fields the theory will contain. The choice is guided by experimental constraints (each known particle species dictates the inclusion of an associated quantum field) as well as theoretical concerns (hypothetical particles and their associated fields, like the inflaton and Higgs fields, are invoked to address open problems or puzzling issues). The *Standard Model* is the prime example. Considered the crowning achievement of twentieth-century particle physics because of its capacity to accurately describe the wealth of data collected by particle accelerators worldwide, the Standard Model is a quantum field theory containing *fifty-seven* distinct quantum fields (the fields corresponding to the electron, the neutrino, the photon, and the various kinds of quarks—the up-quark, the down-quark, the charm-quark, and so on). Undeniably, the Standard Model is tremendously successful, but many physicists feel that a truly fundamental understanding would not require such an ungainly assortment of ingredients.

An exciting feature of string theory is that the particles emerge from the theory itself: a distinct species of particle arises from each distinct string vibrational pattern. And since the vibrational pattern determines the properties of the corresponding particle, if you understood the theory well enough to delineate all vibrational patterns, you'd be able to explain *all* properties of *all* particles. The potential and the promise, then, is that string theory will transcend quantum field theory by deriving all particle properties mathematically. Not only would this unify everything under the umbrella of vibrating strings, it would also establish that future "surprises"—such as the discovery of currently unknown particle species—are built into string theory from the outset and so would be accessible, in principle, to sufficiently industrious calculation. String theory doesn't build piecemeal toward an ever more complete description of nature. It seeks a complete description from the get-go.

The second point is that among the string's possible vibrations, there is one with just the right properties to be the quantum particle of the gravitational field. Even though pre–string theoretic attempts to merge gravity and quantum mechanics were unsuccessful, research did reveal the properties that any hypothesized particle associated with the quantum gravitational field—dubbed the *graviton*—would necessarily possess. The studies concluded that the graviton must be massless and chargeless, and must have the quantum mechanical property known as *spin-2*. (Very roughly, the graviton should spin like a top, twice as fast as the spin of a photon.)[7] Wonderfully, early string theorists—John Schwarz, Joël Scherk, and, independently, Tamiaki Yoneya—found that right there on the list of the string's vibrational patterns was one whose properties matched those of the graviton. Precisely. When convincing arguments were put forward in the mid-1980s that string theory was a mathematically consistent quantum mechanical theory (largely due to the work of Schwarz and his collaborator Michael Green), the presence of gravitons implied that *string theory provided a long-sought quantum theory of gravity.* This is the most important accomplishment on string theory's résumé and the reason it quickly soared to worldwide scientific prominence.*[8]

Third, however radical a proposal string theory may be, it recapitulates a revered pattern in the history of physics. Successful new theories usually do not render their predecessors obsolete. Instead, successful theories typically embrace their predecessors, while greatly extending the range of physical phenomena that can be accurately described. Special relativity extends understanding into the realm of high speeds; general relativity extends understanding further still, to the realm of large masses (the domain of strong gravitational fields); quantum mechanics and quantum field theory extend understanding into the realm of short distances. The concepts these theories invoke and the features they reveal are unlike anything previously envisioned. Yet, apply these theories in the familiar domains of everyday speeds, sizes, and masses and they reduce to the descriptions developed prior to the twentieth century—Newton's classical mechanics and the classical fields of Faraday, Maxwell, and others.

String theory is potentially the next and final step in this progression. In a *single* framework, it handles the domains claimed by relativity and the quantum. Moreover, and this is worth sitting up straight to hear, string theory does so in a manner that fully embraces all the discoveries that preceded it. A theory based on vibrating filaments might not seem to have much in common with general relativity's curved spacetime picture of gravity. Nevertheless, apply string the-

---

*If you'd like to know how string theory surmounts the problems that blocked earlier attempts to join gravity and quantum mechanics, see *The Elegant Universe*, Chapter 6; for a sketch, see note 8. For an even briefer summary, note that whereas a point particle exists at a single location, a string, because it has length, is slightly spread out. This spreading, in turn, dilutes the raucous short-distance quantum jitters that stymied previous attempts. By the late 1980s, there was strong evidence that string theory successfully melds general relativity and quantum mechanics; more recent developments (Chapter 9) make the case overwhelming.

ory's mathematics to a situation where gravity matters but quantum mechanics doesn't (to a massive object, like the sun, whose size is large) and out pop Einstein's equations. Vibrating filaments and point particles are also quite different. But apply string theory's mathematics to a situation where quantum mechanics matters but gravity doesn't (to small collections of strings that are not vibrating quickly, moving fast, or stretched long; they have low energy—equivalently, low mass—so gravity plays virtually no role) and the math of string theory morphs into the math of quantum field theory.

This is graphically summarized in Figure 4.3, which shows the logical connections between the major theories physicists have

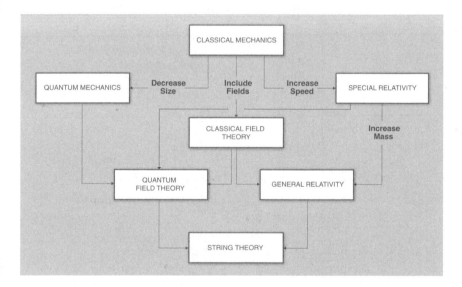

Figure 4.3 *A graphical representation of the relationships between the major theoretical developments in physics. Historically, successful new theories have extended the domain of understanding (to faster speeds, larger masses, shorter distances) while reducing to previous theories when applied in less extreme physical circumstances. String theory fits this pattern of progress: it extends the domain of understanding while, in suitable settings, reducing to general relativity and quantum field theory.*

developed since the time of Newton. String theory could have required a sharp break from the past. It could have stepped clear off the chart provided in the figure. Remarkably, it doesn't. String theory is sufficiently revolutionary to transcend the barriers that hemmed in twentieth-century physics. Yet, the theory is sufficiently conservative to allow the past three hundred years of discovery to snuggly fit within its mathematics.

The Dimensions of Space

Now for something stranger. The passage from dots to filaments is only part of the new framework introduced by string theory. In the early days of string theory research, physicists encountered pernicious mathematical flaws (called *quantum anomalies*), entailing unacceptable processes like the spontaneous creation or destruction of energy. Typically, when problems like this afflict a proposed theory, physicists respond swiftly and sharply. They discard the theory. Indeed, many in the 1970s thought this the best course of action regarding strings. But the few researchers who stayed the course came upon an alternative way of proceeding.

In a dazzling development, they discovered that the problematic features were entwined with the number of dimensions of space. Their calculations revealed that were the universe to have more than the three dimensions of everyday experience—more than the familiar left/right, back/forth, and up/down—then string theory's equations could be purged of their problematic features. Specifically, in a universe with nine dimensions of space and one of time, for a total of ten spacetime dimensions, the equations of string theory become trouble-free.

I'd love to explain in purely nontechnical terms how this comes about, but I can't, and I've never encountered anyone who can. I made an attempt in *The Elegant Universe*, but that treatment only describes, in general terms, how the number of dimensions affects aspects of string vibrations, and doesn't explain where the specific number ten comes from. So, in one slightly technical line, here's the mathematical skinny. There's an equation in string theory that has a contribution of the form $(D - 10)$ times (*Trouble*), where D represents the number of spacetime dimensions and *Trouble* is a mathematical expression resulting in troublesome physical phenomena, such as the violation of energy conservation mentioned above. As to why the equation takes this precise form, I can't offer any intuitive, nontechnical explanation. But if you do the calculation, that's where the math leads. Now, the simple but key observation is that if the number of spacetime dimensions is ten, not the four we expect, the contribution becomes 0 times *Trouble*. And since 0 times anything is 0, in a universe with ten spacetime dimensions the trouble gets wiped away. That's how the math plays out. Really. And that's why string theorists argue for a universe with more than four spacetime dimensions.

Even so, no matter how open you may be to following the trail blazed by mathematics, if you've never encountered the idea of extra dimensions, the possibility may nevertheless sound nutty. Dimensions of space don't go missing like car keys or one member of your favorite pair of socks. If there were more to the universe than length, width, and height, surely someone would've noticed. Well, not necessarily. Even as far back as the early decades of the twentieth century, a prescient series of papers by the German mathematician Theodor Kaluza and by the Swedish physicist Oskar Klein suggested that there might be dimensions that are proficient at evading detection. Their work envisioned that unlike the familiar spatial dimensions that extend over great, possibly infinite, distances, there might be additional dimensions that are tiny and curled up, making them difficult to see.

To picture this, think of a common drinking straw. But for the purpose at hand, make it decidedly uncommon by imagining it as thin as usual but as tall as the Empire State Building. The surface of the tall straw (like that of any straw) has two dimensions. The long vertical dimension is one; the short circular dimension, which curls around the straw, is the other. Now imagine viewing the tall straw from across the Hudson River, as in Figure 4.4a. Because the straw is so thin, it looks like a vertical line stretching from ground to sky. At this distance, you don't have the visual acuity to see the straw's tiny circular dimension, even though it exists at every point along the straw's long vertical extent. This leads you to think, incorrectly, that the straw's surface is one-dimensional, not two.[9]

For another visualization, think of a huge carpet blanketing Utah's salt flats. From an airplane, the carpet looks like a flat surface with two dimensions that extend north/south and east/west. But after you parachute down and view the carpet up close, you realize that its surface is composed of a tight pile: tiny cotton loops attached to each point on the flat carpet backing. The carpet has two large, easy-to-see dimensions (north/south and east/west), but also one small dimension (the circular loops) that is harder to detect (Figure 4.4b).

The Kaluza-Klein proposal suggested that a similar distinction, between dimensions that are big and easily seen, and others that are tiny and thus more difficult to reveal, might apply to the fabric of space itself. The reason we are all aware of the familiar three dimensions of space would be that their extent, like the vertical dimension of the straw and the north/south and east/west dimensions of the carpet, is huge (possibly infinite). However, if an extra dimension of space were curled up like the circular part of the straw or carpet, but to an extraordinarily small size—millions or even billions of times smaller than a single atom—it could be as ubiquitous as the familiar unfurled dimensions and yet remain beyond our ability to detect even with today's most powerful magnifying equipment. The dimension would indeed go missing. Such was the beginning of *Kaluza-Klein theory*, the proposition that our universe has spatial dimensions beyond the three of everyday experience (Figure 4.5).

This line of thought establishes that the suggestion of "extra" spatial dimensions, however unfamiliar, is not absurd. That's a good start, but it invites an essential question: Why, back in the 1920s, would someone invoke such an exotic idea? Kaluza's motivation came from an insight he had shortly after Einstein published the general theory of relativity. He found that with a single stroke of the pen—literally—he could modify Einstein's equations to make them apply to a universe with one additional dimension of space. And when he analyzed those modified equations, the results were so thrilling that, as his son has recounted, Kaluza discarded his normally reserved demeanor, pounded his desk with both hands, shot

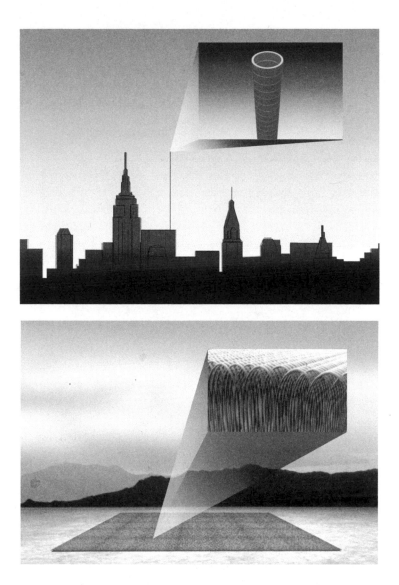

Figure 4.4 (a) *The surface of a tall straw has two dimensions; the vertical dimension is long and easy to see, while the circular dimension is small and harder to detect.* **(b)** *A gigantic carpet has three dimensions; the north/south and east/west dimensions are big and easy to see, while the circular part, the carpet's pile, is small and therefore harder to detect.*

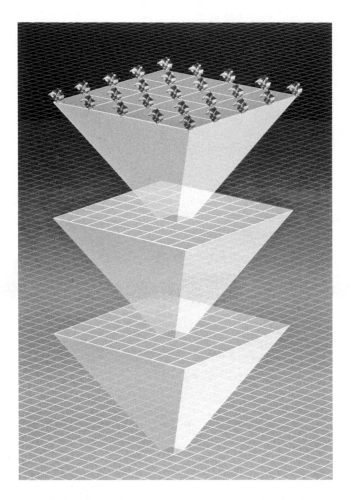

Figure 4.5 *Kaluza-Klein theory posits tiny extra spatial dimensions attached to every point in the familiar three large spatial dimensions. If we could magnify the spatial fabric sufficiently, the hypothesized extra dimensions would become visible. (For the sake of visual clarity, extra dimensions are attached only on grid points in the illustration.)*

to his feet, and erupted into an aria from *The Marriage of Figaro*.[10] Within the modified equations, Kaluza found the ones Einstein had already used successfully to describe gravity in the familiar three dimensions of space and one of time. But because his new formulation included an additional dimension of space, Kaluza found an additional equation. *Lo and behold, when Kaluza derived this equation he recognized it as the very one Maxwell had discovered half a century earlier to describe the electromagnetic field.*

Kaluza revealed that in a universe with an additional dimension of space, gravity and electromagnetism can both be described in terms of spatial ripples. Gravity ripples through the familiar three spatial dimensions, while electromagnetism ripples through the fourth. An outstanding problem with Kaluza's proposal was to explain why we don't see this fourth spatial dimension. It was here that Klein made his mark by suggesting the resolution explained above: dimensions beyond those we directly experience can elude our senses and our equipment if they're sufficiently small.

In 1919, after learning about the extra dimensional proposal for unification, Einstein vacillated. He was impressed by a framework that advanced his dream of unification but was hesitant about such an outlandish approach. After cogitating for a couple of years, in the process holding up publication of Kaluza's paper, Einstein finally warmed to the idea and in time became one of the strongest champions of hidden spatial dimensions. In his own research toward a unified theory, he returned to this theme repeatedly.

Einstein's blessing notwithstanding, subsequent research showed that the Kaluza-Klein program ran up against a number of hurdles, the most difficult being its inability to incorporate the detailed properties of matter particles, such as electrons, into its mathematical structure. Clever ways around this problem, as well as various generalizations and modifications of the original Kaluza-Klein proposal, were pursued on and off for a couple of decades, but as no pitfall-free framework emerged, by the mid-1940s the idea of unification through extra dimensions was largely dropped.

Thirty years later, along came string theory. Rather than allowing for a universe with more than three dimensions, the mathematics of string theory *required* it. And so string theory provided a new, ready-made setting for invoking the Kaluza-Klein program. In response to the question "If string theory is the long-sought unified theory, then why haven't we seen the extra dimensions it needs?" Kaluza-Klein echoed across the decades, answering that the dimensions are all around us but are just too small to be seen. String theory resurrected the Kaluza-Klein program, and by the mid-1980s researchers worldwide were inspired to believe that it was only a matter of time—according to the most enthusiastic proponents, a short time—before string theory would provide a complete theory of all matter and all forces.

Great Expectations

During the early days of string theory, progress came at such a rapid clip that it was nearly impossible to keep up with all the developments. Many compared the atmosphere to that of the 1920s, when scientists stormed into the newly discovered realm of the quantum. With such excitement it's understandable that some theoreticians spoke of a swift resolution to the major problems of fundamental physics: the merger of gravity and quantum mechanics; the unification of all of nature's forces; an explanation of the properties of matter; a determination of the number of spatial dimensions; the elucidation of black hole singularities; and the unraveling of the origin of the universe. As more seasoned researchers anticipated, though, these expectations were premature. String theory is so rich, wide ranging, and mathematically difficult that research to date, nearly three decades after the initial euphoria, has taken us but partway down the road of exploration. And given that the realm of quantum gravity is some hundred billion billion times smaller than anything we can currently access experimentally, levelheaded assessments expect that the road will be long.

Where are we along it? In the rest of the chapter, I'll survey the most advanced understanding in a number of key areas (saving those relevant to the theme of parallel universes for more detailed discussion in subsequent chapters), and I'll appraise the achievements to date and the challenges still looming.

String Theory and the Properties of Particles

One of the deepest questions in all of physics is why nature's particles have the properties they do. Why, for example, does the electron have its particular mass and the up-quark its particular electric charge? The question commands attention not only for its intrinsic interest but also because of a tantalizing fact we alluded to earlier. Had the particles' properties been different—had, say, the electron been moderately heavier or lighter, or had the electric repulsion between electrons been stronger or weaker—the nuclear processes that power stars like our sun would have been disrupted. Without stars, the universe would be a very different place.[11] Most pointedly, without the sun's heat and light, the complex chain of events that led to life on earth would have failed to happen.

This leads to a grand challenge: using pen, paper, possibly a computer, and one's best understanding of the laws of physics, calculate the particle properties and find results in agreement with the measured values. If we could meet this challenge, we'd take one of the most profound steps ever toward understanding why the universe is as it is.

In quantum field theory, the challenge is insurmountable. Permanently. Quantum field theory requires the measured particle properties as *input*—these features are part of the theory's definition—and so can happily accommodate a broad range of values for their masses and charges.[12] In an imaginary world where the electron's mass or charge was larger or smaller than it is in ours, quantum field theory could cope without blinking an eye; it would simply be a matter of adjusting the value of a parameter within the theory's equations.

Can string theory do better?

One of the most beautiful features of string theory (and the facet that most impressed me when I learned the subject) is that particle properties are *determined by the size and shape of the extra dimensions*. Because strings are so tiny, they don't just vibrate within the three big dimensions of common experience; they also vibrate into the tiny, curled-up dimensions. And much as air streams flowing through a wind instrument have vibrational patterns dictated by the instrument's geometrical form, the strings in string theory have vibrational patterns dictated by the geometrical form of the curled-up dimensions. Recalling that string vibrational patterns determine particle properties such as mass and electrical charge, we see that these properties are determined by the geometry of the extra dimensions.

So, if you knew exactly what the extra dimensions of string theory looked like, you'd be well on your way to predicting the detailed properties of vibrating strings, and hence the detailed properties of the elementary particles the strings vibrate into existence. The hurdle is, and has been for some time, that no one has been able to figure out the exact geometrical form of the extra dimensions. The equations of string theory place mathematical restrictions on the geometry of the extra dimensions, requiring them to belong to a particular class called *Calabi-Yau shapes* (or, in mathematical jargon, *Calabi-Yau manifolds*), named after the mathematicians Eugenio Calabi and Shing-Tung Yau, who investigated their properties well before their important role in string theory was discovered (Figure 4.6). The problem is that there's not a single, unique Calabi-Yau shape. Instead, like musical instruments, the shapes come in a wide variety of sizes and contours. And just as different instruments generate different sounds, extra dimensions that differ in size and shape (as well as with respect to more detailed features we'll come upon in the next chapter) generate different string vibrational patterns and hence different sets of particle properties. *The lack of a unique specification of the extra dimensions is the main stumbling block preventing string theorists from making definitive predictions.*

When I started working on string theory, back in the mid-1980s, there were only a handful of known Calabi-Yau shapes, so one could imagine studying each, looking for a match to known physics. My doctoral dissertation was one of the earliest steps in this direction. A few years later, when I was a postdoctoral fellow (working for the Yau of Calabi-Yau), the number of Calabi-Yau shapes had grown to a few thousand, which presented more of a challenge to exhaustive analysis—but that's what graduate students are for. As time passed, however, the pages of the Calabi-Yau catalog continued to multiply; as we will see in Chapter 5, they have now grown more numer-

Figure 4.6 *A close-up of the spatial fabric in string theory, show-ing an example of extra dimensions curled up into a Calabi-Yau shape. Like the pile and backing of a carpet, the Calabi-Yau shape would be attached to every point in the familiar three large spatial dimensions (represented by the two-dimensional grid), but for visual clarity the shapes are shown only on grid points.*

ous than grains of sand on a beach. Every beach. Everywhere. By a long shot. To analyze mathematically each possibility for the extra dimensions is out of the question. String theorists have therefore continued the search for a mathematical directive from the theory that might single out a particular Calabi-Yau shape as "the one." To date, no one has succeeded.

And so, when it comes to explaining the properties of the fundamental particles, string theory has yet to realize its promise. In this regard, it so far offers no improvement over quantum field theory.[13]

Bear in mind, however, that string theory's claim to fame is its ability to resolve *the* central dilemma of twentieth-century theoretical physics: the raging hostility between general relativity and quantum mechanics. Within string theory, general relativity and quantum mechanics finally join together harmoniously. *That's* where string theory provides a vital advance, taking us beyond a critical obstacle that confounded the standard methods of quantum field theory. Should a better understanding of the mathematics of string theory enable us to pick out a unique form for the extra dimensions, one that furthermore allows us to explain all observed particle properties, that would be a phenomenal triumph. But there's no guarantee that string theory can rise to the challenge. There's also no necessity for it to do so. Quantum field theory has been rightly lauded as hugely successful, and yet it can't explain the fundamental particle properties. If string theory also can't explain the particle properties but goes beyond quantum field theory in one key measure, by embracing gravity, that alone would be a monumental achievement.

Indeed, in Chapter 6 we'll see that in a cosmos replete with parallel worlds—as suggested by one modern reading of string theory—it may be plainly wrongheaded to hope that mathematics would pick out a unique form for the extra dimensions. Instead, much as the many different forms for DNA provide for the abundant variety of life on earth, so the many different forms for the extra dimensions may provide for the abundant variety of universes populating a string-based multiverse.

String Theory and Experiment

If a typical string is as small as Figure 4.2 suggests, to probe its extended structure—the very characteristic that distinguishes it from a point—you'd need an accelerator some million billion times more powerful than even the Large Hadron Collider. Using known technology, such an accelerator would need to be about as large as the galaxy, and would consume enough energy each second to power the entire world for a millennium. Barring a spectacular technological breakthrough, this ensures that at the comparatively low energies our accelerators can reach, strings will appear as though they are point particles. This is the experimental version of the theoretical fact I emphasized earlier: at low energy, the mathematics of string theory transforms into the mathematics of quantum field theory. And so, even if string theory is the true fundamental theory, it will impersonate quantum field theory in a wide range of accessible experiments.

That's a good thing. Although quantum field theory is not

equipped to combine general relativity and quantum mechanics, nor to predict the fundamental properties of nature's particles, it can explain a great many other experimental results. It does this by taking the measured properties of particles as input (input that dictates the choice of fields and energy curves in the quantum field theory) and then uses the mathematics of quantum field theory to predict how such particles will behave in other experiments, generally accelerator-based. The results are extremely accurate, which is why generations of particle physicists have made quantum field theory their primary approach.

The choice of fields and energy curves in quantum field theory is tantamount to the choice of the extra dimensional shape in string theory. The particular challenge facing string theory, though, is that the mathematics linking particle properties (such as their masses and charges) to the shape of the extra dimensions is extraordinarily intricate. This makes it difficult to work backwards—to use experimental data to guide the choice of the extra dimensions, much as such data guide the choices of fields and energy curves in quantum field theory. One day we may have the theoretical dexterity to use experimental data to fix the form of string theory's extra dimensions, but not yet.

For the foreseeable future, then, the most promising avenue for linking string theory with data are predictions that, while open to explanations using more traditional methods, are far more naturally and convincingly explained using string theory. Just as you might theorize that I'm typing these words with my toes, a far more natural and convincing hypothesis—and one I can attest to as correct—is that I'm using my fingers. Analogous considerations applied to the experiments summarized in Table 4.1 have the capacity to build a circumstantial case for string theory.

The undertakings range from particle physics experiments at the Large Hadron Collider (searching for supersymmetric particles and for evidence of extra dimensions), to tabletop experiments (measuring the gravitational strength of attraction on scales of a millionth of a meter and smaller), to astronomical observations (looking for particular kinds of gravitational waves and fine temperature variations in the cosmic microwave background radia-

EXPERIMENT/ OBSERVATION	EXPLANATION
Supersymmetry	The "super" in superstring theory refers to *supersymmetry*, a mathematical feature with a straightforward implication: for every known particle species there should be a partner species that has the same electrical and nuclear force properties. Theorists surmise that these particles have so far evaded detection because they are heavier than their known counterparts, and so lie beyond the reach of well-worn accelerators. The Large Hadron Collider may have enough energy to produce them, so there's broad anticipation that we may be on the threshold of revealing nature's supersymmetric quality.
Extra Dimensions and Gravity	Because space is the medium for gravity, more dimensions supply a larger domain within which gravity can spread. And just as a drop of ink grows more diluted when it spreads in a vat of water, the strength of gravity would become diluted as it spreads through the additional dimensions—offering an explanation for why gravity appears weak (when you pick up a coffee cup, your muscles beat out the gravitational pull of the entire earth). If we could measure gravity's strength over distances smaller than the size of the extra dimensions, we'd catch it before it's fully spread and so we should find its strength to be stronger. To date, measurements on scales as short as a micron (10^{-6} meters) have found no deviation from expectations based on a world with three spatial dimensions. Should a deviation be found as physicists push these experiments to ever-shorter distances, that would provide convincing evidence for additional dimensions.
Extra Dimensions and Missing Energy	If the extra dimensions exist but are far smaller than a micron, they will be inaccessible to experiments that directly measure gravity's strength. But the Large Hadron Collider provides another means of revealing their existence. Debris created by head-on collisions between fast-moving protons can be ejected from our familiar large dimensions and squeezed into the others (where, for reasons we'll get to later, the debris would likely be particles of gravity, or *gravitons*). Were this to happen, the debris would carry away energy, and as a result our detectors would register a little less energy after the collision than was present before. Such missing energy signals could provide strong evidence for the existence of extra dimensions.

EXPERIMENT/ OBSERVATION	EXPLANATION
Extra Dimensions and Mini Black Holes	Black holes are usually described as the remains of massive stars that have exhausted their nuclear fuel and collapsed under their own weight, but this is an unduly limited description. *Anything* would become a black hole if compressed sufficiently. Moreover, if there are extra dimensions that result in gravity being stronger when acting over short distances, it would be easier to form black holes, since a stronger gravitational force implies that it takes less compression to generate the same gravitational pull. Even just two protons, if slammed together at the velocities mustered by the Large Hadron Collider, may be able to cram enough energy into a sufficiently small volume to trigger the formation of a black hole. It would be a wisp of a black hole, but it would yield an unmistakable signature. Mathematical analysis, going back to the work of Stephen Hawking, shows that tiny black holes would quickly disintegrate into a spray of lighter particles whose tracks would be picked up by the collider's detectors.
Gravitational Waves	Although strings are tiny, if you could somehow grab hold of one, you could stretch it large. You'd need to apply a force in excess of 10^{20} tons, but stretching a string is merely a matter of exerting enough energy. Theorists have found exotic situations in which the energy for such stretching might be provided by astrophysical processes, generating long strings wafting through space. Even if they were very distant, these strings might be detectable. Calculations show that as a long string vibrates, it creates ripples in spacetime—known as *gravitational waves*—of a highly distinctive shape, and hence they offer a clear observational signature. Within the next few decades, if not sooner, highly sensitive detectors based on earth and, funding permitting, in space, may be able to measure these ripples.
Cosmic Microwave Background Radiation	The cosmic microwave background radiation has already proved itself capable of probing quantum physics: the measured temperature differences in the radiation arise from quantum jitters stretched large by spatial expansion. (Recall the analogy of a tiny message scribbled on a shriveled balloon becoming visible once the balloon is inflated.) In inflation, the stretching of space is so enormous that even tinier imprints, perhaps laid down by strings, might also be stretched sufficiently to be detectable—perhaps by the European Space Agency's Planck satellite. Success or failure turns on details of how strings would have behaved in the earliest moments of the universe—the nature of the message they would have imprinted on the deflated cosmic balloon. Various ideas have been developed and calculations made. Theorists are now waiting for the data to speak for themselves.

Table 4.1. Experiments and Observations with the Capacity to Link String Theory to Data

tion). The table explains the individual approaches, but the overall assessment is readily summarized. A positive signature from any of these experiments could be explained without invoking string theory. For example, although the mathematical framework of supersymmetry (see the first entry in Table 4.1) was initially discovered through theoretical studies of string theory, it has since been incorporated into non-string theoretic approaches. Discovering supersymmetric particles would thus confirm a piece of string theory, but would not constitute a smoking gun. Similarly, although extra spatial dimensions have a natural home within string theory, we've seen that they too can be part of non-string theoretic proposals—Kaluza, as a case in point, was not thinking about string theory when he proposed the idea. The most favorable outcome from the approaches in Table 4.1, therefore, would be a series of positive results that would show pieces of the string theory puzzle falling into place. Like touting touch-typing toes, non-string explanations would become overwrought when faced with such a collection of positive results.

Negative experimental results would provide much less useful information. The failure to find supersymmetric particles might mean they don't exist, but it also might mean they are too heavy for even the Large Hadron Collider to produce; the failure to find evidence for extra dimensions might mean they don't exist, but it also might mean they are too small for our technologies to access; the failure to find microscopic black holes might mean that gravity does not get stronger on short scales, but it also might mean that our accelerators are too weak to burrow deeply enough into the microscopic terrain where the increase in strength is substantial; the failure to find stringy signatures in observations of gravitational waves or the cosmic microwave background radiation might mean string theory is wrong, but it might also mean that the signatures are too meager for current equipment to measure.

As of today, then, the most promising positive experimental results would most likely not be able to definitively prove string theory right, while negative results would most likely not be able to prove string theory wrong.[14] Yet, make no mistake. If we find evidence of extra dimensions, supersymmetry, mini black holes, or any of the other potential signatures, that will be a huge moment in the search for a unified theory. It would bolster confidence, and justifiably so, that the mathematical road we've been paving is headed in the right direction.

String Theory, Singularities, and Black Holes

In the vast majority of situations, quantum mechanics and gravity happily ignore each other, the former applying to small things like molecules and atoms and the latter to big things like stars and galaxies. But the two theories are forced to shed their isolation in the realms known as *singularities*. A singularity is any physical setting, real or hypothetical, that is so extreme (huge mass, small size, enormous spacetime curvature, punctures or rips in the spacetime fabric) that quantum mechanics and general relativity go haywire, generating results akin to the error message displayed on a calculator when you divide any number by zero.

A prize achievement of any purported quantum theory of gravity is to meld quantum mechanics and gravity in a manner that cures singularities. The resulting mathematics should never break down—even at the moment of the big bang or in the center of a black hole,[15] thus providing a sensible description of situations that have long baffled researchers. It is here that string theory has made its most impressive strides, taming a growing list of singularities.

In the mid-1980s, the team of Lance Dixon, Jeff Harvey, Cumrun Vafa, and Edward Witten realized that certain punctures in the spatial fabric (known as *orbifold singularities*), which leave Einstein's mathematics in shambles, pose no problem for string theory. The key to this success is that whereas point particles can fall into punctures, strings can't. Because strings are extended objects, they can bang into a puncture, they can wrap around it, or they can get stuck to it, but these mild interactions leave the equations of string theory perfectly sound. This is important not because such ruptures in space actually happen—they may or may not—but because string theory is delivering just what we want from a quantum theory of gravity: a means of making sense of a situation that lies beyond what general relativity and quantum mechanics can handle on their own.

In the 1990s, work I did with Paul Aspinwall and David Morrison, and independent results of Edward Witten, established that yet more intense singularities (known as *flop singularities*) in which a spherical portion of space is compressed to an infinitesimal size can also be handled by string theory. The intuitive reasoning here is that as a string moves it can sweep across the compressed chunk of space, like a hula hoop across a soap bubble, and thus act as an encircling protective barrier. The calculations showed that such a "string shield" nullifies any potentially disastrous consequences, ensuring that string theory's equations suffer no ill effect—no "1 divided by 0" type errors—even though the equations of conventional general relativity would fall apart.

In the years since, researchers have shown that a variety of other more complicated singularities (with names like *conifolds, orientifolds, enhancons* . . .) are also under full control within string theory. So there's a growing list of situations that would have left Einstein, Bohr, Heisenberg, Wheeler, and Feynman saying, "We just don't know what's going on," and yet for which string theory gives a complete and consistent description.

This is great progress. But a remaining challenge for string theory is to cure the singularities of black holes and the big bang, which are more severe than those so far addressed. Theorists have expended much effort trying to reach this goal, and they've taken significant strides. But the executive summary is that there is still a way to go before these most puzzling and most relevant of singularities are fully understood.

Nevertheless, one major advance has illuminated a related aspect of black holes. As I will discuss in Chapter 9, the work of Jacob Bekenstein and Stephen Hawking in the 1970s established that black holes contain a very particular quantity of disorder, technically known as *entropy*. According to basic physics, much as the disorder within a sock drawer reflects the many possible haphazard rearrangements of its contents, the disorder of a black hole reflects the many possible haphazard rearrangements of the black hole's innards. But try as they might, physicists were unable to understand black holes well enough to identify their innards, let alone analyze the possible ways they could be rearranged. The string theorists Andrew Strominger and Cumrun Vafa broke through the impasse. Using a mélange of string theory's fundamental ingredients (some of which we will encounter in Chapter 5), they created a mathematical model for a black hole's disorder, a model transparent enough to enable them to extract a numerical measure of the entropy. The result they found agreed spot-on with the Bekenstein-Hawking answer. While the work left open many deep issues (such as explicitly identifying a black hole's microscopic constituents), it provided the first firm quantum mechanical accounting of a black hole's disorder.[16]

The remarkable advances in dealing with both singularities and black hole entropy give the community of physicists well-grounded confidence that in time the remaining challenges of black holes and the big bang will be conquered.

String Theory and Mathematics

Making contact with data, experimental or observational, is the only way to determine if string theory correctly describes nature. It's a goal that's proved elusive. String theory, for all its advances, is still a wholly mathematical undertaking. But string theory isn't just a consumer of math. Some of its most important contributions have been *to* mathematics.

When he was developing the general theory of relativity in the early twentieth century, Einstein famously mined the mathematical archives in search of rigorous language for describing curved spacetime. The earlier geometrical insights of mathematicians such as Carl Friedrich Gauss, Bernhard Riemann, and Nikolai Lobachevsky provided an important foundation for his success. In a sense, string theory is now helping to repay Einstein's intellectual debt by driving the development of new mathematics. There are numerous examples, but let me give one that captures the flavor of string theory's mathematical achievements.

General relativity established a tight link between the geometry of spacetime and the physics we observe. Einstein's equations, together with the distribution of matter and energy in a region, tell you the resulting shape of spacetime. Different physical environments (different configurations of mass and energy) yield differently shaped spacetimes; different spacetimes correspond to physically distinct environments. What would it feel like to fall into a black hole? Calculate with the spacetime geometry that Karl Schwarzschild discovered in his study of spherical solutions to Einstein's equations. And if the black hole is rapidly spinning? Calculate with the spacetime geometry found in 1963 by the New Zealand mathematician Roy Kerr. In general relativity, geometry is the yin to physics' yang.

String theory provides a twist to this conclusion by establishing that there can be *different* shapes for spacetime that nevertheless yield physically indistinguishable descriptions of reality.

Here's one way to think about it. From antiquity to the modern mathematical era, we've modeled geometrical spaces as collections of points. A Ping-Pong ball, for example, is the collection of points that constitute its surface. Prior to string theory, the basic constituents making up matter were also modeled as points, point particles, and this commonality of basic ingredients spoke to an alignment between geometry and physics. But in string theory, the basic ingredient is not a point. It's a string. This suggests that a new kind of geometry, based not on points but rather on loops, should be linked to string physics. The new geometry is called *stringy geometry*.

To get a feel for stringy geometry, picture a string moving through a geometrical space. Notice that the string can behave much like a point particle, innocently gliding from here to there, bumping into walls, navigating chutes and valleys, and so on. But in certain situations, a string can also do something novel. Imagine that space (or a piece of space) is shaped like a cylinder. A string can wrap itself around such a piece of space, much like a rubber band stretched around a can of soda, realizing a configuration that's simply unavailable to a point particle. Such "wrapped" strings, and their "unwrapped" cousins, probe a geometrical space in different ways. Should a cylinder grow fatter, a string encircling it will respond by stretching, while an unwrapped string sliding on its surface won't.

In this way, wrapped and unwrapped strings are sensitive to different features of a shape through which they're moving.

This observation is of great interest because it gives rise to a striking and thoroughly unexpected conclusion. String theorists have found special pairs of geometrical shapes for space that have completely different features when each is probed by unwrapped strings. They also have completely different features when each is probed by wrapped strings. But—and this is the punch line—when probed both ways, with wrapped and unwrapped strings, the shapes become indistinguishable. What the unwrapped strings see on one space, the wrapped strings see on the other, and vice versa, rendering identical the collective picture gleaned from the full physics of string theory.

Shapes that form such pairs provide a powerful mathematical tool. In general relativity, if you're interested in one or another physical feature, you must complete a mathematical calculation using the unique geometrical space relevant to the situation being studied. But in string theory, the existence of pairs of physically equivalent geometrical shapes means that you have a newfound choice: you can choose to perform the necessary calculation using either shape. And the extraordinary thing is that while you're guaranteed to get the same answer using either shape, the mathematical details en route to the answer can be vastly different. In a variety of situations, overwhelmingly difficult mathematical calculations on one geometrical shape translate into exceedingly easy calculations on the other. And any framework that makes hard mathematical calculations easy is, clearly, of great value.

Over the years, mathematicians and physicists have leveraged this hard-to-easy dictionary to make headway on a number of outstanding mathematical problems. One that I'm particularly fond of has to do with counting the number of spheres that can be packed (in a particular mathematical way) within a given Calabi-Yau shape. Mathematicians had been interested in this question for a long time but found the calculations in all but the simplest cases impenetrable. Take the Calabi-Yau shape of Figure 4.6. When a sphere is packed into this shape, it can wrap around a portion of the Calabi-Yau multiple times, much like a lasso can wrap multiple times around a beer barrel. So, how many ways can you pack a sphere into this shape if it wraps around, say, five times? When asked a question like this, mathematicians had to clear their throats, glance at their shoes, and quickly depart for pressing appointments. String theory flattened the hurdles. By translating such calculations into far easier ones on a paired Calabi-Yau shape, string theorists produced answers that knocked mathematicians back on their heels. The number of five-times-wrapped spheres packed into the Calabi-Yau in Figure 4.6? 229,305,888,887,625. And if the spheres wrap around themselves ten times? 704,288,164,978,454,686,113,488,249,750. Twenty times? 53,126,882,649,923,577,113,917,814,483,472,714,066,922,267,923, 866,471,451,936,000,000. These numbers proved to be harbingers for a spectrum of results that have opened a whole new chapter in mathematical discovery.[17]

So, whether or not string theory offers a correct approach to describing the physical universe, it has already established itself as a potent tool for investigating the mathematical one.

The State of String Theory: An Evaluation

Building on the last four sections, Table 4.2 provides a status report for string theory, including some additional observations that I didn't explicitly call out in the text above. It paints a picture of a theory in progress, one that has produced stunning achievements but has not yet been tested on the most important scale: experimental confirmation. The theory will remain speculative until a convincing link to experiment or observation is forged. Establishing such a link is the great challenge. But it's not a challenge that's peculiar to string theory. Any attempt to unite gravity and quantum mechanics enters a domain that's far beyond the cutting-edge of experimental research. It's part and parcel of taking on such a supremely ambitious goal. Pushing the fundamental boundaries of knowledge, seeking answers to some of the deepest questions contemplated during the past few thousand years of human thought, is a formidable undertaking, one that won't likely be completed overnight. Nor in a handful of decades.

In evaluating the state of the art, many string theorists argue that a crucial next step is to articulate the theory's equations in their most exact, useful, and comprehensive form. Much of the research during the theory's first couple of decades, through the mid-1990s, was carried out using approximate equations that many were convinced could reveal the theory's gross features but were too coarse to yield refined predictions. Recent advances, to which we will now turn, have catapulted understanding far beyond what could be achieved by the approximate methods. While definitive predictions have remained elusive, a new perspective has emerged. It's come from a series of breakthroughs that has opened grand new vistas on the theory's potential implications, among which are new varieties of parallel worlds.

Table 4.2. *A summary status report for string theory.*

GOAL	IS GOAL REQUIRED?	STATUS
Unite gravity and quantum mechanics	**Yes.** The primary goal is to meld general relativity and quantum mechanics.	**Excellent.** A wealth of calculations and insights attest to string theory's successful merger of general relativity and quantum mechanics.[18]
Unify all forces	**No.** Unification of gravity and quantum mechanics does not require a further unification with the other forces of nature.	**Excellent.** While not required, a fully unified theory has long been a goal of physics research. String theory achieves this goal by describing all forces in the same manner— their quanta are strings executing particular vibrational patterns.
Incorporate key breakthroughs from past research	**No.** In principle, a successful theory need bear little resemblance to successful theories from the past.	**Excellent.** Though progress isn't necessarily incremental, history shows that it usually is; successful new theories typically embrace past successes as limiting cases. String theory incorporates the essential key breakthroughs from previously successful physical frameworks.

GOAL	IS GOAL REQUIRED?	STATUS
Explain particle properties	**No.** This is a noble goal, and if achieved would provide a profound level of explanation—but it is not required of a successful theory of quantum gravity.	**Indeterminate; no predictions.** Going beyond quantum field theory, string theory offers a framework for explaining particle properties. But to date, this potential remains unrealized; the many different possible forms for the extra dimensions imply many different possible collections of particle properties. There is no currently available means to pick one shape from the many.
Experimental confirmation	**Yes.** This is the only way to determine whether a theory is a correct description of nature.	**Indeterminate; no predictions.** This is the most important criterion; to date, string theory has not been tested on it. Optimists hope that experiments at the Large Hadron Collider and observations by satellite-borne telescopes have the capacity to bring string theory much closer to data. But there's no guarantee that current technology is sufficiently refined to reach this goal.
Cure singularities	**Yes.** A quantum theory of gravity should make sense of singularities arising in situations that are, even just in principle, physically realizable.	**Excellent.** Tremendous progress; many kinds of singularities have been resolved by string theory. The theory still needs to address black hole and big bang singularities.
Black hole entropy	**Yes.** A black hole's entropy provides a hallmark context in which general relativity and quantum mechanics interface.	**Excellent.** String theory has succeeded in explicitly calculating, and confirming, the entropy formula proposed in the 1970s.
Mathematical contributions	**No.** There's no requirement that correct theories of nature yield mathematical insights.	**Excellent.** Although mathematical insights aren't necessary to validate string theory, significant ones have emerged from the theory, revealing the profound reach of its mathematical underpinnings.

CHAPTER 5

Hovering Universes in Nearby Dimensions

The Brane and Cyclic Multiverses

Late one night many years ago, I was in my office at Cornell University putting together the freshman physics final exam that would be given the following morning. Since this was the honors class, I wanted to enliven things a little by giving them one somewhat more challenging problem. But it was late and I was hungry, so rather than carefully working through various possibilities, I quickly modified a standard problem that most of them had already encountered, wrote it into the exam, and headed home. (The details hardly matter, but the problem had to do with predicting the motion of a ladder, leaning against a wall, as it loses its footing and falls. I modified the standard problem by having the density of the ladder vary along its length.) During the exam the next morning, I sat down to write the solutions, only to find that my seemingly modest modification to the problem had made it exceedingly difficult. The original problem took perhaps half a page to complete. This one took me six pages. I write big. But you get the point.

This little episode represents the rule rather than the exception. Textbook problems are very special, being carefully designed so that they're completely solvable with reasonable effort. But modify textbook problems just a bit, changing this assumption or dropping that simplification, and they can quickly become intricate or intractable. That is, they can quickly become as difficult as analyzing typical real-world situations.

The fact is, the vast majority of phenomena, from the motion of planets to the interactions of particles, are just too complex to be described mathematically with complete precision. Instead, the task of the theoretical physicist is to figure out which complications in a given context can be discarded, yielding a manageable mathematical formulation that still captures essential details. In predicting the course of the earth you'd better include the effects of the sun's gravity; if you include the moon's too, all the better, but the mathematical complexity rises significantly. (In the nineteenth century, the French mathematician Charles-Eugène Delaunay published two 900-page volumes related to intricacies of the sun-earth-moon gravitational dance.) If you try to go further and account fully for the influence of all the other planets, the analysis becomes overwhelming. Luckily, for many applications, you can safely disregard all but the sun's influence, since the effect of other bodies in the solar system on earth's motion is nominal. Such approximations illustrate my earlier assertion that the art of physics lies in deciding what to ignore.

But as practicing physicists know well, approximation is not just a potent means for progress; on occasion it also brings peril. Complications of minimal importance for answering one question can sometimes have a surprisingly significant impact in answering another. A single drop of rain will hardly affect the weight of a boulder. But if the boulder is teetering high on a cliff's edge, that drop of rain could very well coax it to fall, initiating an avalanche. An approximation that disregards the raindrop would miss a crucial detail.

In the mid-1990s, string theorists discovered something akin to a raindrop. They found that various mathematical approximations, widely used to analyze string theory, were overlooking some vital physics. As more precise mathematical methods were developed and applied, string theorists could finally step beyond the approximations; when they did, numerous unanticipated features of the theory came into focus. And among these were new types of parallel universes; one variety in particular may be the most experimentally accessible of all.

Beyond Approximations

Every major established discipline of theoretical physics—such as classical mechanics, electromagnetism, quantum mechanics, and general relativity—is defined by a central equation, or set of equations. (You don't need to know these equations, but I've listed some of them in the notes.)[1] The challenge is that in all but the simplest situations, the equations are extraordinarily difficult to solve. For this reason, physicists routinely use simplifications—like ignoring Pluto's gravity or treating the sun as perfectly round—that make the mathematics easier and bring approximate solutions within reach.

For a long time, research in string theory has faced even bigger challenges. Just finding the central equations proved so difficult that physicists could develop only approximate versions. And even the approximate equations were so intricate that physicists had to make simplifying assumptions to find solutions, thus basing research on approximations of approximations. During the 1990s, however, the situation vastly improved. In a series of advances, a number of string theorists showed how to go well beyond the approximations, offering unmatched clarity and insight.

To get a feel for these breakthroughs, imagine that Ralph is planning to play the next two rounds of the weekly worldwide lottery, and he's proudly worked out the odds of winning. He tells Alice that since he has a 1 in a billion chance each week, if he plays both rounds his chance of winning is 2 in a billion, .000000002. Alice smirks. "Well, that's *close*, Ralph." "Really, wise guy. What do you mean *close*?" "Well," she says, "you've overestimated. Should you win the first round, playing a second time won't increase your chances of winning; you would already have done so. If you win twice, we'll have more money, sure, but since you're working out the odds of winning at all, winning the second lottery after the first just doesn't matter. So, to get the precise answer you'd need to subtract the odds of winning *both* rounds—1 in a billion times 1 in a billion, or .00000000000000001. That yields a final tally of .000000001999999999. Questions, Ralph?"

Minus the smugness, Alice's method is an example of what physicists call a *perturbative approach*. In doing a calculation, it's often easiest to make a first pass that incorporates only the most obvious contributions—that's Ralph's starting point—and then make a second pass that includes finer details, modifying or "perturbing" the first-pass answer, as in Alice's contribution. The approach easily generalizes. If Ralph were planning to play the next ten weekly lotteries, the first-pass approach suggests that his chance of winning is about 10 in a billion, .00000001. But, as in the previous example, this approximation fails to account correctly for multiple wins. When Alice takes over, her second pass would properly account for instances in which Ralph wins twice—say, on the first and second lotteries, or the first and third, or the second and fourth. These corrections, as Alice pointed out above, are proportional to 1 in a billion times 1 in a billion. But there's also an even tinier chance that Ralph wins three times; Alice's third pass takes that, too, into account, producing modifications proportional to 1 in a billion multiplied by itself three times, .000000000000000000000000001. The fourth pass does the same for the even tinier chance of winning four rounds, and so on. Each new contribution is far smaller than the previous, so at some point Alice deems the answer sufficiently accurate and calls it a day.

Calculations in physics, and in many other branches of science too, often proceed in an analogous fashion. If you are interested in how likely it is that two particles heading in opposite directions around the Large Hadron Collider will bang into each other, the first pass imagines they hit once and ricochet (where "hit" doesn't mean they directly touch, but rather that a single force-carrying "bullet," such as a photon, flies from one and is absorbed by the other). The second pass takes into account the chance that the particles hit each other twice (two photons are fired between them); the third pass modifies the previous two by accounting for the chance of the particles hitting each other three times; and so on (Figure 5.1). As with the lottery, this perturbative approach works well if the chance of an ever-greater number of particle interactions—like the chance of an ever-greater number of lottery wins—drops precipitously.

For the lottery, the drop-off is determined by each successive win coming with a factor of 1 in a billion; in the physics example, it's determined by each successive hit coming with a numerical factor, called a *coupling constant*, whose value captures the likelihood that one particle will fire a force-carrying bullet and that the second particle will receive it. For particles such as electrons, governed by the electromagnetic force, experimental measurements have determined that the coupling constant, associated with photon bullets, is

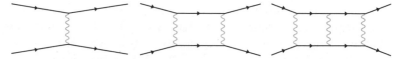

Figure 5.1 *Two particles (represented by the two solid lines on the left in each diagram) interact by firing various "bullets" at each other (the "bullets" are force-carrying particles, represented by the squiggly lines), and then ricochet forward (the two solid lines on the right). Each diagram contributes to the overall likelihood that the particles bounce off each other. The contributions of processes with ever-more bullets are ever smaller.*

about .0073.[2] For neutrinos, governed by the weak nuclear force, the coupling constant is about 10^{-6}. For quarks, components of protons, that are racing around the Large Hadron Collider and whose interactions are governed by the strong nuclear force, the coupling constant is somewhat less than 1. These numbers are not as small as the lottery's .000000001, but if for example we multiply .0073 by itself the result quickly becomes minuscule. After one iteration it's about .0000533, after two it's about .000000389. This explains why theorists only rarely go to the trouble of accounting for electrons hitting each other numerous times. The calculations involving many hits are exceedingly intricate to carry out, and the resulting contributions are so terribly tiny that you can stop at just a few photons fired and still get an extraordinarily accurate answer.

To be sure, physicists would love to have exact results. But for many calculations the mathematics proves too difficult, so the perturbative approach is the best we can do. Fortunately, for small enough coupling constants, the approximate calculations can yield predictions that agree extremely well with experiment.

A similar perturbative approach has long been a mainstay of string theory research. The theory contains a number, called the *string coupling constant* (*string coupling*, for short), that governs the chance that one string bumps off another. If the theory proves correct, the string coupling may one day also be measured, much like the couplings enumerated above. But since such a measurement is at present purely hypothetical, the value of the string coupling is a complete unknown. Over the past few decades, with no guidance from experiment, string theorists have made the key assumption that the string coupling is a small number. To some extent, this has been like the drunkard looking for his keys under a lamppost, because a small string coupling allows physicists to shine the bright lights of perturbative analysis on their calculations. Since many successful approaches prior to string theory *do* have a small coupling, a more favorable version of the analogy notes that the drunkard has been justifiably emboldened by frequently finding his keys in the very location that's illuminated. Either way, the assumption has made possible a vast collection of mathematical calculations that have not only clarified the basic processes of how one string interacts with another, but have also revealed much about the fundamental equations underlying the subject.

If the string coupling *is* small, these approximate calculations are expected to accurately reflect the physics of string theory. But what if it isn't? Unlike what we found with the lottery and with colliding electrons, a large string coupling would mean that successive refinements to first-pass approximations would yield ever-*larger* contributions, so you'd never be justified in stopping a calculation. The thousands of calculations that have used the perturbative scheme would be baseless; years of research would collapse. Adding to the concerns, even with a small yet moderate string coupling, you might also worry that your approximations, at least in some circumstances, were overlooking subtle yet vital physical phenomena, like the raindrop that hits the boulder.

Through the early 1990s, not much could be said about these vexing questions. By the second half of that decade, the silence gave way to a clamor of insight. Researchers found new mathematical methods that could outflank the perturbative approximations by leveraging something called *duality*.

Duality

In the 1980s, theorists realized that there was not one string theory but rather five different versions, to which they gave the catchy names *Type I, Type IIA, Type IIB, Heterotic-O,* and *Heterotic-E.* I've not yet mentioned this complication because although calculations established that the theories differ in detail, all five include the same gross features—vibrating strings and extra spatial dimensions—on which we've so far focused. But we're now at a point where the five variations on the string theory theme come to the fore.

For many years, physicists had relied on perturbative methods to analyze each of the string theories. When working with the Type I string theory, they assumed its coupling was small, and pressed on with multi-pass calculations similar to what Ralph and Alice did in the lottery analysis. When working with the Heterotic-O, or any of the other string theories, they did the same. But outside of this restricted domain of small string couplings, researchers could do nothing more than shrug, throw up their hands, and admit that the math they were using was too feeble to provide any reliable insight.

Until, that is, the spring of 1995, when Edward Witten rocked the string theory community with a series of stunning results. Drawing on the insights of scientists including Joe Polchinski, Michael Duff, Paul Townsend, Chris Hull, John Schwarz, Ashoke Sen, and many others, Witten provided strong evidence that string theorists could safely navigate beyond the shores of small couplings. The central idea was simple and powerful. Witten argued that when the coupling constant in any one formulation of string theory is dialed ever larger, the theory—remarkably—steadily morphs into something thoroughly familiar: one of the other formulations of string theory, but with a coupling constant that's dialed ever smaller. For example, when the Type I string coupling is large, it transforms into the Heterotic-O string theory with a coupling that's small. Which means that the five string theories are not different after all. Each appears different when examined in a limited context—small values of its particular coupling constant—but when this restriction is lifted, each string theory transforms into the others.

I recently encountered a splendid graphic that from close up looks like Albert Einstein, with a bit more distance becomes ambiguous, and from far away resolves into Marilyn Monroe (Figure 5.2). If you saw only the images that come into focus at the two extremes, you'd have every reason to think you were looking at two separate

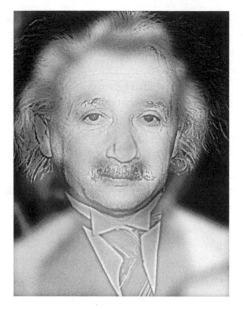

Figure 5.2 *From close up, the image looks like Albert Einstein. From farther away, it looks like Marilyn Monroe. (The image was created by Aude Oliva of the Massachusetts Institute of Technology.)*

pictures. But if you steadily examine the image through the range of intermediate distances, you unexpectedly find that Einstein and Monroe are aspects of a single portrait. Similarly, an examination of two string theories, in the extreme case when each has a small coupling, reveals that they're as different as Albert and Marilyn. If you stopped there, as for years string theorists did, you'd conclude that you were studying two separate theories. But if you examine the theories as their couplings are varied over the range of intermediate values, you find that, like Albert turning into Marilyn, each gradually morphs into the other.

The morphing from Einstein to Monroe is amusing. The morphing of one string theory into another is transformative. It implies that if perturbative calculations in one string theory can't be undertaken because that theory's coupling is too large, the calculations can be faithfully translated into the language of another formulation of string theory, one in which a perturbative approach succeeds because the coupling is small. Physicists call the transition between naïvely distinct theories *duality*. It has become one of the most pervasive themes in modern string theory research. By providing two mathematical descriptions of one and the same physics, duality doubles our calculational arsenal. Calculations that are impossibly difficult from one perspective become perfectly doable from another.*

Witten argued, and others since have filled in important details, that all five string theories are linked through a network of such dualities.[3] Their overarching union, called *M-theory* (we'll see why in a moment), combines insights from all five formulations, stitched together through the various duality relationships, to gain a far more refined understanding of each. One such insight, central to the theme we're pursuing, showed that there's much more to string theory than strings.

Branes

When I started studying string theory, I asked the very question that many in the years since have asked me: Why are strings considered so special? Why focus solely on fundamental ingredients that have only length? After all, the theory itself requires that the arena within which its ingredients exist—the spatial universe—has nine dimensions, so why not consider entities shaped like two-dimensional sheets or three-dimensional blobs or their higher-dimensional cousins? The answer I learned as a graduate student in the 1980s, and explained frequently when I lectured on the subject through the mid-1990s, was that the mathematics describing fundamental constituents with more than one spatial dimension suffered from fatal inconsistencies (such as quantum processes that would have nega-

---

*You can think of this as a grand generalization of the results touched on in Chapter 4, in which different forms for the extra dimensions can give rise to identical physical models.

tive probabilities, a meaningless mathematical result). But when the same mathematics was applied to strings, the inconsistencies canceled themselves out, leaving a cogent description.*[4] Strings were definitely in a class of their own.

Or so it seemed.

Armed with the newfound calculational methods, physicists started analyzing their equations much more precisely and produced a range of unexpected results. One of the most surprising established that the rationale for excluding anything but strings was rickety. Theorists realized that the mathematical problems encountered when studying higher-dimensional ingredients, such as discs and blobs, were artifacts of the approximations being used. Using the more precise methods, a small army of theorists established that ingredients with various numbers of spatial dimensions *do* lurk in string theory's mathematical shadows.[5] The perturbative techniques were too coarse to expose these ingredients but the new methods finally could. By the late 1990s, it was abundantly clear that string theory was not just a theory that contained strings.

The analyses revealed objects, shaped like Frisbees or flying carpets, with two spatial dimensions: *membranes* (one meaning of the "M" of M-theory), also called *two-branes*. But there was more. The analyses revealed objects with three spatial dimensions, so-called *three-branes*; objects with four spatial dimensions, *four-branes*, and so on, all the way up to *nine-branes*. The mathematics made clear that all of these entities could vibrate and wiggle, much like strings; indeed, in this context, strings are best thought of as *one-branes*— a single entry on an unexpectedly long list of the theory's basic building blocks.

An allied revelation, just as flabbergasting to those who'd spent the better part of their professional lives working on the subject, was that the number of spatial dimensions the theory requires is not actually nine. It's ten. And if we fold in the dimension of time, the total number of spacetime dimensions is eleven. How could this be?

---

*This wasn't the result of a mysterious mathematical coincidence. Instead, in a precise mathematical sense, strings are highly symmetric shapes, and it was this symmetry that wiped away the inconsistencies. See note 4 for details.

Remember the "$(D-10)$ times *Trouble*" consideration, recounted in Chapter 4, underlying the conclusion that string theory needs ten spacetime dimensions. The mathematical analysis that produced that equation was, once again, based on a perturbative approximation scheme that assumed the string coupling was small. Surprise, surprise, that approximation missed one of the theory's spatial dimensions. The reason, Witten showed, is that the size of the string coupling directly controls the size of the hitherto unknown tenth spatial dimension. By taking the coupling small, researchers had unwittingly made this spatial dimension small, too—so small as to be invisible to the mathematics itself. The more precise methods rectified this failing, revealing a string/M-theory universe with ten dimensions of space and one of time, for a total of eleven spacetime dimensions.

I remember well the dazed and wide-eyed looks everywhere at the international string theory conference, held at the University of Southern California in 1995, at which Witten first announced some of these results, the first shot in what is now called the Second String Theory Revolution.* For the multiverse story, it is the branes that are central. Using them, researchers have been led by the hand to another variety of parallel universes.

Branes and Parallel Worlds

We typically imagine that strings are ultra-small; that very feature makes testing the theory such a challenge. However, I noted in Chapter 4 that strings are not necessarily minute. Rather, a string's length is controlled by its energy. The energies associated with the masses of electrons, quarks, and other known particles are so tiny that the corresponding strings would indeed be minuscule. But inject enough energy into a string, and you could cause it to stretch large. We don't have anywhere near the capacity to do this here on earth, but that's a limitation of our technological development. If

---

*The first revolution was the 1984 results of John Schwarz and Michael Green, which launched the modern version of the subject.

string theory is right, an advanced civilization would be able to pump strings up to whatever size it liked. Natural cosmological phenomena also have the capacity to produce long strings; for example, strings can wrap around a portion of space and get caught up in the cosmological expansion, stretching long in the process. One of the possible experimental signatures outlined in Table 4.1 looks for gravitational waves that such long strings may emit as they vibrate far away in space.

Like strings, higher-dimensional branes can be big. And this opens up a wholly new way in which string theory can describe the cosmos. To grasp what I mean, picture first a long string, as long as an overhead electric cable that runs as far as the eye can see. Next, picture a large two-brane, like an enormous tablecloth or a gargantuan flag, whose surface extends indefinitely. These are both easy to visualize because we can picture them located within the three dimensions of common experience.

If a three-brane is enormous, perhaps infinitely big, the situation changes. A three-brane of this sort would *fill* the space we occupy, like water filling a huge fish tank. Such ubiquity suggests that rather than think of the three-brane as an object that happens to be situated within our three spatial dimensions, we should envision it as the very substrate of space itself. Just as fish inhabit the water, we would inhabit a space-filling three-brane. Space, at least the space we directly inhabit, would be far more corporeal than generally imagined. Space would be a thing, an object, an entity—a three-brane. As we run and walk, as we live and breathe, we move in and through a three-brane. String theorists call this the *braneworld scenario*.

It is here that parallel universes make their stringy entrance.

I've been focusing on the relationship between three-branes and three spatial dimensions because I wanted to make contact with the familiar domain of everyday reality. But in string theory, there are more than just three spatial dimensions. And a higher-dimensional expanse offers ample room for accommodating more than one three-brane. Starting conservatively, imagine that there are two enormous three-branes. You may find it difficult to picture this. I certainly do. Evolution has prepared us to identify objects, those presenting opportunity as well as danger, that sit squarely *within* three-dimensional space. Consequently, although we can easily picture two ordinary three-dimensional objects inhabiting a region of space, few of us can picture two coexisting but separate three-dimensional entities, each of which could fully fill three-dimensional space. For ease in discussing the braneworld scenario, then, let's suppress one spatial dimension in our visualizations and think about life on a giant two-brane. And for a definite mental image, think of the two-brane as a giant, extraordinarily thin slice of bread.*

To use this metaphor effectively, imagine that the slice of bread includes the entirety of what we've traditionally called the universe—the Orion, Horsehead, and Crab nebulae; the entire Milky Way; the Andromeda, Sombrero, and Whirlpool Galaxies; and so on—everything within our three-dimensional spatial expanse, however distant, as sketched in Figure 5.3a. To visualize a second three-brane we just need to picture a second enormous slice of bread. Where? Place it next to ours, just shifted slightly away in the extra dimensions (Figure 5.3b). To visualize three or four or any other number of three-branes is equally easy. Just add slices to the cosmic loaf. And while the loaf metaphor emphasizes a collection of branes all aligned with one another, it's easy to imagine yet more general possibilities. The branes can be oriented any which way, and branes of any other dimensionality, higher or lower, can be included just the same.

The same fundamental laws of physics would apply all across the collection of branes, since they all emerge from a single theory, string/M-theory. But, much as with the bubble universes in the Inflationary Multiverse, environmental details such as the value of this or that field permeating a brane, or even the number of spatial dimensions defining a brane, can profoundly affect its physical features. Some braneworlds might be much like our own, filled with galaxies, stars, and planets, while others might be very different. On

---

*If you're being careful, you'll note that a slice of bread is really three-dimensional (width and height on the slice's surface, but also depth from the slice's thickness), but don't let that trouble you. The thickness of the bread will remind us that our slices are visual stand-ins for large three-branes.

(a)

(b)

Figure 5.3 (a) *In the braneworld scenario, what we have traditionally thought to be the entire cosmos is imagined to reside within a three-dimensional brane. For visual ease, we suppress one dimension and show the braneworld as having two spatial dimensions; we also show only a finite piece of branes that may extend infinitely far.* **(b)** *The higher-dimensional expanse of string theory can accommodate many parallel braneworlds.*

one or more of those branes there might be self-aware beings who, like us, once thought that their slice—their expanse of space—was the entirety of the cosmos. In string theory's braneworld scenario, we would now recognize this as a parochial perspective. In the braneworld scenario, our universe is just one of many that populate the *Brane Multiverse.*

When the Brane Multiverse was first floated in the string theory community, the immediate response focused on an obvious question. If there are giant branes right next door, entire parallel universes hovering nearby like slices of rye cozying up to their neighbors, why don't we see them?

Sticky Branes and Gravity's Tentacles

Strings come in two shapes, loops and snippets. I haven't addressed this distinction because it's not essential for understanding many of the theory's overarching features. But for braneworlds the distinction between loops and snippets is crucial, and a simple question reveals why. Can strings fly off a brane? The answer: A loop can. A snippet can't.

As first realized by renowned string theorist Joe Polchinski, it all has to do with the endpoints of a string snippet. The equations that convinced physicists that branes were part of string theory also revealed that strings and branes have a particularly intimate relationship. Branes are the only locations where the endpoints of string snippets can reside, as in Figure 5.4. The math showed that if you try to remove a string's endpoint from a brane, you are attempting the impossible, like seeking to make π smaller or the square root of 2 bigger. Physically, it's like trying to remove the north or south pole from the ends of a bar magnet. It just can't be done. String snippets can freely move within and through a brane, effortlessly gliding from here to there, but they can't leave it.

If these ideas are more than just interesting mathematics and we are in fact all living on a brane, you're right now directly experiencing the viselike grip our brane exerts on string endpoints. Try to jump off our three-brane. Try again, harder. I suspect you're still

Figure 5.4 *Branes are the only locations where the endpoints of string snippets can reside.*

here. In a braneworld, the strings that make up you, and the rest of ordinary matter, are snippets. While you can jump up and down, throw a baseball from first to second, and send a sound wave from radio to ear, all with absolutely no resistance from the brane, *you can't depart the brane.* When you try to jump off, the endpoints of your string snippets anchor you to the brane, unalterably. Our reality could be a floating slab in a higher-dimensional expanse, but we'd be permanently imprisoned, unable to venture out and explore the grander cosmos.

The same picture holds for the particles that transmit the three nongravitational forces. The analysis shows that they, too, arise from string snippets. Most notable among these are photons, the purveyors of the electromagnetic force. Visible light, which is a stream of photons, can therefore travel freely through the brane, from this text to your eyes, or from the Andromeda Galaxy to the Wilson Observatory, but it too is unable to escape. Another braneworld could be hovering millimeters away, but because light can't travel across the gap, we would never see the slightest hint of its presence.

The one force that's different in this regard is gravity. The distinguishing feature of gravitons, noted in Chapter 4, is that they have spin-2, twice that of the particles arising from string snippets (such as photons) that convey the nongravitational forces. That gravitons have twice the spin of an individual string snippet means you can think of gravitons as being built of two such snippets, the two ends of one melding with those of the other, yielding a loop. And since loops have no endpoints, branes can't trap them. Gravitons can therefore leave and reenter a braneworld. In a braneworld scenario, then, gravity provides our only means of probing beyond our three-dimensional spatial expanse.

This realization plays a central role in some of the potential tests of string theory mentioned in Chapter 4 (Table 4.1). In the 1980s and 1990s, before branes entered the picture, physicists imagined that string theory's extra dimensions were roughly Planck-sized (a radius of about 10^{-33} centimeters), the natural scale for a theory involving gravity and quantum mechanics. But the braneworld scenario encourages more expansive thinking. With our only probe beyond the three common dimensions being gravity—the feeblest of all forces—the extra dimensions can be a good deal larger and have still avoided detection. So far.

If the extra dimensions exist, and are *much* larger than previously thought—perhaps a billion billion billion times larger (about 10^{-4} centimeters across)—then experiments that measure the strength of gravity, described in the second row of Table 4.1, stand a chance of detecting them. When objects attract each other gravitationally, they exchange streams of gravitons; the gravitons are invisible messengers that communicate gravity's influence. The more gravitons the objects exchange, the stronger the mutual gravitational pull. When some of these streaming gravitons leak off our brane and flow into the extra dimensions, the gravitational attraction between objects will be diluted. The larger the extra dimensions, the more the dilution, and the weaker gravity appears. By carefully measuring the gravitational pull between two objects brought closer together than the size of the extra dimensions, experimenters envision intercepting the gravitons before they leak from our brane; if so, the experimenters should measure a strength for gravity that's proportionately larger. Thus, although I didn't mention it in Chapter 4, this approach for unmasking the extra dimensions relies on the braneworld scenario.

A more modest increase in the size of the extra dimensions, to only about 10^{-18} centimeters across, would still make them potentially accessible to the Large Hadron Collider. As discussed in the third entry of Table 4.1, high-energy collisions between protons can eject debris into the extra dimensions, resulting in an apparent loss of energy in our dimensions that might be detectable. This experiment, too, relies on the braneworld scenario. Data attesting to missing energy would be explained by positing that our universe exists on a brane and arguing that debris with the capacity to fly off our brane—gravitons—had carried the energy away.

The prospect of mini black holes, the fourth entry of Table 4.1, is yet another braneworld by-product. The Large Hadron Collider stands a chance of producing mini black holes in proton-proton collisions only if the intrinsic strength of gravity grows large when probed over short distances. As above, it is the braneworld scenario that makes this possible.

The details cast these three experiments in a new light. Not only are these experiments seeking evidence of exotic structures such as extra dimensions of space and tiny black holes, they are also seeking evidence that we're living on a brane. In turn, a positive result would not only build a case for string theory's braneworld scenario, but would also provide indirect evidence for universes beyond our own. If we can establish that we're living on a brane, the mathematics gives us no reason to expect that ours is the only one.

Time, Cycles, and the Multiverse

The multiverses we've so far encountered, however different in detail, share one basic trait. In the Quilted, Inflationary, and Brane Multiverses, the other universes are all "out there" in space. For the Quilted Multiverse "out there" means far away in the everyday sense; for the Inflationary Multiverse it means beyond our bubble universe and across the rapidly expanding intervening realm; for the Brane Multiverse it means a possibly short distance away but the separation is through another dimension. Evidence supporting the braneworld scenario would lead us to consider seriously another variety of multiverse, one that leverages not the opportunities afforded by space but those of time.[6]

Since Einstein, we've known that space and time can warp, curve, and stretch. But we generally don't envision the whole universe wafting this way or that. What would it mean for the entirety of space to move ten feet to the "right" or "left"? It's a good brainteaser, but it becomes pedestrian when considered in the braneworld scenario. Like particles and strings, branes can surely move through the surrounding environment they inhabit. And so, if the universe we observe and experience is a three-brane, we could very well be gliding through a higher-dimensional spatial expanse.*

If we are on such a gliding brane, and there are other branes nearby, what would happen if we slammed into one of them? Although there are details that have not yet been fully worked out, you can be certain that a collision between two branes—a collision between two universes—would be violent. The simplest possibility would be two parallel three-branes coming closer and closer together till finally they collided straight-on, much like two cymbals crashing. The tremendous energy harbored in their relative motion would yield a fiery rush of particles and radiation that would obliterate any organized structures that either brane universe contained.

To a group of researchers including Paul Steinhardt, Neil Turok, Burt Ovrut, and Justin Khoury, this cataclysm rang not just of an end but of a beginning. An intensely hot, thoroughly dense environment in which particles stream this way and that sounds much like the conditions just after the big bang. Perhaps, then, when two branes collide they wipe out whatever structures may have coalesced during either of their histories, from galaxies to planets to people, while setting the stage for a cosmic rebirth. Indeed, a three-brane filled with a blistering plasma of particles and radiation responds just as an ordinary three-dimensional spatial expanse would: it expands.

---

*You could still ask whether the entire higher-dimensional spatial expanse can move, but however interesting to contemplate, it's not relevant to the discussion here.

And as it does, the environment cools, allowing particles to clump, ultimately yielding the next generation of stars and galaxies. Some have suggested that an apt name for this reprocessing of universes would be the *big splat*.

Evocative though it may be, "splat" misses a central feature of brane collisions. Steinhardt and his collaborators have argued that when branes collide, they don't stick together. They bounce apart. The gravitational force they exert on each other then gradually slows their relative motion; eventually, they reach a maximum separation from which they start approaching once again. As the branes fall back together, each builds up speed, they collide, and through the ensuing firestorm the conditions on each brane are reset once again, initiating a new era of cosmological evolution. The essence of this cosmology thus involves worlds that repeatedly cycle through time, generating a new variety of parallel universes called the *Cyclic Multiverse*.

If we are living on a brane in the Cyclic Multiverse, the other member universes (in addition to the partner brane with which we periodically collide) are in our past and future. Steinhardt and his co-workers estimated the time scale for a full cycle of the colliding cosmic tango—birth, evolution, and death—and came up with about a trillion years. In this scenario, the universe as we know it would merely be the latest in a temporal series, some of which may have contained intelligent life and the culture they created, but are now long ago extinguished. In due course, all of our contributions and those of any other life-forms our universe supports would be similarly erased.

The Past and Future of Cyclic Universes

Although the braneworld approach is its most refined incarnation, cyclical cosmologies have enjoyed a long history. Earth's rotation, yielding the predictable pattern of day and night, as well as its orbit, yielding the repetitive sequence of passing seasons, presages the cyclical approaches developed by many traditions in their attempt to explain the cosmos. One of the oldest prescientific cosmologies, the Hindu tradition, envisions a nested complex of cosmological cycles within cycles, which, according to some interpretations, stretch from millions to trillions of years. Western thinkers, from as far back as the pre-Socratic philosopher Heraclitus and the Roman statesman Cicero, also developed various cyclic cosmological theories. A universe consumed by fire and emerging anew from the smoldering embers was a popular scenario among those who considered lofty issues such as cosmic origins. With the spread of Christianity, the concept of genesis as a unique, onetime event gradually gained the upper hand, but cyclic theories continued to sporadically attract attention.

In the modern scientific era, cyclical models have been pursued since the earliest cosmological investigations invoking general relativity. Alexander Friedmann, in a popular book published in Russia in 1923, noted that some of his cosmological solutions to Einstein's gravitational equations suggested an oscillating universe that would expand, reach a maximal size, contract, shrink to a "point," and then might begin expanding anew.[7] In 1931, Einstein himself, having by then dropped his proposal for a static universe, also investigated the possibility of an oscillatory universe. Most detailed of all was a series of papers published from 1931 to 1934 by Richard Tolman at the California Institute of Technology. Tolman undertook thorough mathematical investigations of cyclical cosmological models, initiating a stream of such studies—often swirling in the backwaters of physics but sometimes bubbling up to broader prominence—that have continued to this day.

Part of the appeal of a cyclical cosmology is its apparent ability to avoid the knotty issue of how the universe began. If the universe goes through cycle after cycle, and if the cycles have always happened (and perhaps always will), then the problem of an ultimate beginning is sidestepped. Each cycle has its own beginning, but the theory provides a concrete physical cause: the termination of the previous cycle. And if you ask about the beginning of the entire cycle of universes, the answer is simply that there was no such beginning, because the cycles have been repeating for eternity.

In a sense, then, cyclical models are an attempt to have your cosmological cake and eat it too. Back in the early days of scientific cosmology, the *steady state* theory provided its own end run around the question of cosmic origin by suggesting that although the universe is expanding, it did not have a beginning: as the universe expands, new matter is continually created to fill the additional space, ensuring that constant conditions are maintained throughout the cosmos for all eternity. But the steady state theory ran afoul of astronomical observations pointing strongly toward earlier epochs whose conditions differed markedly from those we experience today. Most pointed of all were observations zeroing in on an earliest cosmological phase that was far from steady and stately, being instead chaotic and combustible. A big bang undermines dreams of steady state, bringing the question of origin back to center stage. It's here that cyclical cosmologies offer a compelling alternative. Each cycle *can* incorporate a big-bang-like past, in alignment with the astronomical data. But by stringing together an infinite number of cycles the theory still avoids having to supply an ultimate beginning. Cyclical cosmologies, so it would seem, thereby meld the most attractive features of the steady state and big bang models.

Then in the 1950s, the Dutch astrophysicist Herman Zanstra called attention to a problematic feature of cyclical models, one that was implicit in Tolman's analysis a couple of decades earlier. Zanstra showed that there couldn't have been an infinite number of cycles preceding our own. The wrench in the cosmological works was the Second Law of Thermodynamics. This law, which we'll discuss more fully in Chapter 9, establishes that disorder— *entropy*—increases over time. It's something we routinely experience. Kitchens, however ordered in the morning, have a way of becoming disordered by nightfall; the same goes for laundry bins, desktops, and playrooms. In these everyday settings, the increase in entropy is a mere nuisance; in cyclic cosmology, the increase in entropy is pivotal. As Tolman himself had realized, the equations of general relativity link the entropy content of the universe with the duration of a given cycle. More entropy means more disordered particles squeezed together when the universe shrinks; that generates a more powerful rebound, space expands further, and so the cycle lasts longer. Looking back from today, the Second Law then implies that ever-earlier cycles would have had ever-less entropy (because the Second Law says that entropy increases toward the future, it must decrease toward the past),* and would thus have had ever-shorter durations. Working this out mathematically, Zanstra showed that sufficiently far back in time the cycles would have been so short that they would have ceased. They *would* have had a beginning.

Steinhardt and company claim that their new version of cyclical cosmology avoids this pitfall. In their approach, the cycles arise not from a universe expanding, contracting, and expanding again but rather from the *separation* between braneworlds expanding, contracting, and expanding again. The branes themselves continually expand—and they do so throughout each and every cycle. Entropy builds from one cycle to the next, just as the Second Law requires, but because the branes expand the entropy is spread over ever-larger spatial volumes. The total entropy goes up, but the entropy *density* goes down. By the end of each cycle, the entropy is so diluted that its density is driven very nearly to zero—a full reset. And so, unlike what happens in the analysis of Tolman and Zanstra, the cycles can continue indefinitely toward the future as well as the past. The braneworld Cyclic Multiverse has no need for a beginning to time.[8]

Sidestepping an age-old conundrum is a feather in the Cyclic Multiverse's cap. But as its proponents note, the Cyclic Multiverse goes beyond offering resolution to cosmological conundra—it makes a specific prediction that distinguishes it from the widely accepted inflationary paradigm. In inflationary cosmology, the violent burst of expansion in the early universe would have so thoroughly disturbed the spatial fabric that substantial gravitational waves would have been produced. These ripples would have left trace imprints on the cosmic microwave background radiation, and highly sensitive observations are now seeking them out. A brane collision, by contrast, creates a momentary maelstrom—but without the spectacular inflationary stretching of space, any gravitational waves produced would almost certainly be too weak to create a lasting signal.

---

*For readers familiar with the puzzle of time's arrow, note that I am assuming, in keeping with observations, that entropy decreases toward the past. See *The Fabric of the Cosmos*, Chapter 6, for a detailed discussion.

So evidence of gravitational waves produced in the early universe would be strong evidence against the Cyclic Multiverse. On the other hand, failure to observe any evidence of these gravitational waves would severely challenge a great many inflationary models and make the cyclic framework all the more attractive.

The Cyclic Multiverse is widely known within the physics community but is viewed, almost as widely, with much skepticism. Observations have the capacity to change this. If evidence for braneworlds emerges from the Large Hadron Collider, and if signs of gravitational waves from the early universe remain elusive, the Cyclic Multiverse will likely garner increased support.

In Flux

The mathematical realization that string theory is not just a theory of strings but also includes branes has had a major impact on research in the field. The braneworld scenario, and the multiverses to which it gives rise, is one resulting area of investigation with the capacity to profoundly remake our perspective on reality. Without the more exact mathematical methods developed over the last decade and a half, most of these insights would have remained beyond reach. Nevertheless, the main problem physicists hoped the more exact methods would address—the need to pick one form for the extra dimensions out of the many candidates that theoretical analyses have uncovered—has not yet been solved. Far from it. The new methods have actually made the problem all the more challenging. They've resulted in the discovery of vast new troves of possible forms for the extra dimensions, increasing the candidate pool enormously while not providing an iota of insight into how to single out one as ours.

Pivotal to these developments is a property of branes called *flux*. Just as an electron gives rise to an electric field, an electric "mist" that permeates the area around it, and a magnet gives rise to a magnetic field, a magnetic "mist" that permeates its region, so a brane gives rise to a *brane field*, a brane "mist" that permeates its region, as in Figure 5.5. When Faraday was performing the first experi-

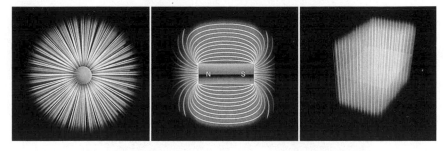

Figure 5.5 *Electric flux produced by an electron; magnetic flux produced by a bar magnet; brane flux produced by a brane.*

ments with electric and magnetic fields, in the early 1800s, he imagined quantifying their strength by delineating the density of field lines at a given distance from the source, a measure he called the field's *flux*. The word has since become ensconced in the physics lexicon. The strength of a brane's field is also delineated by the flux it generates.

String theorists, including Raphael Bousso, Polchinski, Steven Giddings, Shamit Kachru, and many others realized that a full description of string theory's extra dimensions requires not only specifying their shape and size—which researchers in this area, including me, had focused on more or less exclusively in the 1980s and early 1990s—but *also* specifying the brane fluxes that permeate them. Let me take a moment to flesh this out.

Since the earliest mathematical work investigating string theory's extra dimensions, researchers have known that Calabi-Yau shapes typically contain a great many open regions, like the space inside a beach ball, a doughnut's hole, or within a blown glass sculpture. But it wasn't until the early years of the new millennium that theorists realized that these open regions needn't be completely empty. They can be wrapped by one or another brane, and threaded by flux piercing through them, as in Figure 5.6. Previous research (as summarized, for instance, in *The Elegant Universe*) had for the most part considered only "naked" Calabi-Yau shapes, from which all such adornments were absent. When researchers realized that a given Calabi-Yau shape could be "dressed up" with

Figure 5.6 *Parts of the extra dimensions in string theory can be wrapped by branes and threaded by fluxes, yielding "dressed-up" Calabi-Yau shapes. (The figure uses a simplified version of a Calabi-Yau shape—a "three-hole doughnut"—and represents wrapped branes and flux lines schematically with glowing bands encircling portions of the space.)*

these additional features, they uncovered a gargantuan collection of modified forms for the extra dimensions.

A rough count gives a sense of scale. Focus on fluxes. Just as quantum mechanics establishes that photons and electrons come in discrete units—you can have 3 photons and 7 electrons, but not 1.2 photons or 6.4 electrons—so quantum mechanics shows that flux lines also come in discrete bundles. They can penetrate a surrounding surface once, twice, three times, and so on. But apart from this restriction to whole numbers, there's in principle no other limit. In practice, when the amount of flux is large, it tends to distort the surrounding Calabi-Yau shape, rendering previously reliable mathematical methods inaccurate. To avoid venturing into these more turbulent mathematical waters, researchers typically consider only flux numbers that are about 10 or less.[9]

This means that if a given Calabi-Yau shape contains one open region, we can dress it up with flux in ten different ways, yielding ten new forms for the extra dimensions. If a given Calabi-Yau has two such regions, there are $10 \times 10 = 100$ different flux dressings (10 possible fluxes through the first paired with 10 through the second); with three open regions there are 10^3 different flux dressings, and so on. How large can the number of these dressings get? Some Calabi-Yau shapes have on the order of five hundred open regions. *The same reasoning yields on the order of 10^{500} different forms for the extra dimensions.*

In this way, rather then winnowing the candidates to a few specific shapes for the extra dimensions, the more refined mathematical methods have led to a cornucopia of new possibilities. All of a sudden, Calabi-Yau spaces can clothe themselves with far more outfits than there are particles in the observable universe. For some string theorists, this caused great distress. As emphasized in the previous chapter, without a means of choosing the exact form for the extra dimensions—which we now realize means also selecting the flux outfit that shape wears—the mathematics of string theory loses its predictive power. Much hope had been placed on mathematical methods that could go beyond the limitations of perturbation theory. Yet, when some of those methods materialized, the problem of fixing the form for the extra dimensions only got worse. Some string theorists lost heart.

Others, more sanguine, believe it's too early to give up hope. One day—perhaps a day that's just around the corner, perhaps a day that's far off—we will discover the missing principle that determines what the extra dimensions look like, including the fluxes the shape may be sporting.

Others still have taken a more radical tack. Maybe, they suggest, the decades of fruitless attempts to pin down the form for the extra dimensions are telling us something. Maybe, these radicals brazenly continue, we need to take seriously *all* of the possible shapes and fluxes emerging from string theory's mathematics. Maybe, they urge, the reason the mathematics contains all these possibilities is that *they're all real,* each shape being the extra-dimensional part of its own separate universe. And maybe, grounding a seemingly wild flight of fancy in observational data, this is just what's needed to address perhaps the thorniest problem of all: the cosmological constant.

CHAPTER 6

New Thinking About an Old Constant

The Landscape Multiverse

The difference between 0 and .000000000000000000000000000000
0000000000000000000000000000000000000000000000000000000000
00000000000000000000000000000000000000001 might not seem
like much. And by any familiar measure it's not. Yet there's growing
suspicion that this tiny difference may be responsible for a radical
shift in how we envision the landscape of reality.

The tiny number printed above was first measured in 1998 by
two teams of astronomers making meticulous observations of
exploding stars in distant galaxies. Since then, the work of many has
corroborated the teams' result. What is the number, and why such a
fuss? Evidence is mounting that it's what I referred to earlier as the
entry on the third line of the general relativity tax form: Einstein's
cosmological constant, which specifies the amount of invisible
dark energy permeating the fabric of space.

As the result continues to hold up under intense scrutiny, physicists are becoming increasingly confident that decades of previous
observations and theoretical deductions, which had convinced the
vast majority of researchers that the cosmological constant was 0,
have been overthrown. Theorists scurried to figure out where
they'd gone wrong. But not all had. Years earlier, a contentious line
of thought had suggested that a nonzero cosmological constant
might one day be found. The key supposition? We're living in one
of many universes. *Many* universes.

The Return of the Cosmological Constant

Remember that the cosmological constant, if it exists, fills space with a uniform invisible energy—dark energy—whose iconic feature would be its repulsive gravitational force. Einstein latched on to the idea in 1917, invoking the cosmological constant's antigravity to balance the otherwise attractive gravitational pull of the universe's ordinary matter, and thus allow for a cosmos that neither expanded nor contracted.*

Many have reported that upon learning of Hubble's 1929 observations, which established that space is expanding, Einstein called the cosmological constant his "greatest blunder." George Gamow recounted a conversation in which Einstein is purported to have said this, but given Gamow's penchant for playful hyperbole, some have questioned the accuracy of the story.[1] What's certain is that Einstein dropped the cosmological constant from his equations when the observations showed that his belief in a static universe was misguided, noting years later that had "Hubble's expansion been discovered at the time of the creation of the general theory of relativity, the cosmological constant would never have been introduced."[2] But hindsight is not always 20–20; it can sometimes blur earlier clarity. In 1917, in a letter he wrote to the physicist Willem de Sitter, Einstein expressed a more nuanced perspective:

> In any case, one thing stands. The general theory of relativity *allows* the inclusion of the cosmological constant in the field equations. One day, our actual knowledge of the composition of the fixed star sky, the apparent motions of fixed stars, and the position of spectral lines as a function of distance, will probably have come far enough for us to be

---

*One point of language. For the most part, I use the terms "cosmological constant" and "dark energy" interchangeably. When I need a little more precision, I take the value of the cosmological constant to denote the *amount* of dark energy suffusing space. As noted earlier, physicists often use the term "dark energy" a bit more liberally, to mean anything that can look like or masquerade as a cosmological constant over reasonably long time scales, but might slowly change and hence not truly be constant.

able to decide empirically the question of whether or not the cosmological constant vanishes. Conviction is a good motive, but a bad judge.[3]

Some eight decades later, the Supernova Cosmology Project, led by Saul Perlmutter, and the High-Z Supernova Search Team, led by Brian Schmidt, took this very approach. They carefully studied an abundance of *spectral lines*—light emitted by distant stars—and, just as Einstein had anticipated, they were able to address empirically the question of whether the cosmological constant vanishes.

To the shock of many, they found strong evidence that it doesn't.

Cosmic Destiny

When these astronomers began their work, neither group was focused on measuring the cosmological constant. Instead, the teams had set their sights on measuring another cosmological feature, the rate at which the expansion of space is slowing. Ordinary attractive gravity acts to pull every object closer to every other, so it causes the expansion speed to decrease. The precise rate of slowdown is central to predicting what the universe will be like in the far future. A big slowdown would mean that the expansion of space would diminish all the way to zero and then reverse its motion, leading to a period of spatial contraction. Unabated, this might result in a *big crunch*—a reverse of the big bang—or perhaps a bounce, as in the cyclical models introduced in the previous chapter. A small slowdown would yield a very different outcome. Much as a ball with a high speed can escape the earth's gravity and head ever farther outward, if the speed of spatial expansion were high enough, and the rate of its slowdown sufficiently meager, space could expand forever. By measuring the cosmic slowdown, the two groups sought the ultimate fate of the cosmos.

The approach of each team was straightforward: measure how fast space was expanding at various times in the past, and by comparing those speeds determine the rate at which the expansion has been slowing over the course of cosmic history. Okay. But how would you do this? As with many questions in astronomy, the answer comes down to careful measurements of light. Galaxies are luminous beacons whose motion traces the spatial expansion. If we could determine how fast galaxies at a range of distances were receding from us when, long ago, they emitted the light we now see, we could determine how fast space was expanding at a variety of moments in the past. By comparing those speeds, we'd learn the rate of cosmic slowdown. That's the essential idea.

To fill in the details, we need to address two primary questions. From today's observations of faraway galaxies, how can we determine their distances, and how can we determine their speeds? Begin with distance.

Distance and Brightness

One of the oldest and most important problems in astronomy is to determine the distances to celestial objects. And one of the first techniques for doing so, *parallax*, is an approach with which five-year-olds routinely experiment. Children can be fascinated (momentarily) by looking at an object while alternately closing their left and right eyes because the object appears to jump from side to side. If you haven't been five for some time, try the experiment by holding up this book and looking at one of its corners. The jump occurs because your left and right eyes, being spaced apart, have to point at different angles to focus on the same spot. For objects that are farther away, the jumping is less noticeable, because the difference in angle gets smaller. This simple observation can be made quantitative, providing a precise correlation between the difference in angle between the lines of sight of your two eyes ╺the parallax╺ and the distance of the object you're viewing. But don't worry about working out the details; your visual system does it automatically. It's why you see the world in 3D.*

Parallax

---

*It's also how 3D movie technology works: by suitably choosing the spatial offsets on the screen of nearly duplicate images, the filmmaker causes your brain to interpret the resulting parallaxes as different distances, creating the illusion of a 3D environment.

When you look at stars in the night sky, the parallax is too small to be reliably measured; your eyes are just too close together to yield a significant difference in angle. But there's a clever way around this: measure the position of a star on two occasions, some six months apart, thus using the two locations of the earth in place of the two locations of your eyes. The larger separation of the observing locations increases the parallax; it's still small, but in some cases is big enough to be measured. Back in the early 1800s there was an intense competition among a group of scientists to be the first to measure such stellar parallax; in 1838, the German astronomer and mathematician Friedrich Bessel won the bragging rights, successfully measuring the parallax to a star called 61 Cygni, in the constellation Cygnus. The angular difference turned out to be .000084 degrees, placing the star about 10 light-years away.

Since then, the technique has been steadily refined and is now undertaken by satellites that can measure parallax angles far smaller than what Bessel achieved. Such advances have allowed for accurate distance measurements of stars that are up to a few thousand light-years away, but much beyond that the angular differences again become too small, and the method is thwarted.

Another approach, which has the capacity to measure yet greater celestial distances, is based on an even simpler idea: the farther away you move a light-emitting object, be it a car's headlights or a blazing star, the more the emitted light will spread out during its journey toward you, and so the dimmer it will appear. By comparing an object's *apparent* brightness (how bright it appears when observed from earth) with its *intrinsic* brightness (how bright it would appear if observed from close by), you can thus work out its distance.

The hitch, and it's not a small one, lies in establishing the intrinsic brightness of astrophysical objects. Is a star dim because it's especially distant or because it just doesn't give off much light? This makes clear why a long-standing effort has been to find a relatively common astronomical species whose intrinsic brightness can be reliably determined without the need to stand right next to it. If you could find such so-called *standard candles*, you'd have a uniform benchmark for judging distances. The degree to which one standard candle appeared dimmer than another would tell you directly how much farther away it is.

For over a century, a variety of standard candles have been proposed and used, with varying success. In recent times, the most fruitful method has made use of a kind of stellar explosion called a Type Ia supernova. A Type Ia supernova occurs when a white dwarf star pulls material from the surface of a companion, typically a nearby red giant that it's orbiting. Well-developed physics of stellar structure establishes that if the white dwarf pulls away enough material (so that its total mass increases to about 1.4 times that of the sun), it can no longer support its own weight. The bloated dwarf star collapses, setting off an explosion so violent that the light generated rivals the combined output of the other 100 billion or so stars residing in the galaxy it inhabits.

These supernovae are ideal standard candles. Because the explosions are so powerful, we can see them out to fantastically large distances. And, crucially, because the explosions are all the result of the same physical process—a white dwarf's mass increasing to about 1.4 times that of the sun's, resulting in stellar collapse—the ensuing supernovae flare to a very similar peak intrinsic brightness. The challenge in using Type Ia supernovae, however, is that in a typical galaxy they take place only once every few hundred years: How do you catch them in the act? Both the Supernova Cosmology Project and the High-Z Supernova Search Team tackled this obstacle in a manner reminiscent of epidemiological studies: accurate information about even relatively rare conditions can be gained if you study large populations. Similarly, by using telescopes equipped with wide-field-of-view detectors capable of simultaneously examining thousands of galaxies, the researchers were able to locate dozens of Type Ia supernovae, which could then be closely observed with more conventional telescopes. On the basis of how bright each appeared, the teams were able to calculate the distance to dozens of galaxies situated billions of light-years away—thus accomplishing the first step in the task they'd set for themselves.

Whose Distance Is It, Anyway?

Before moving on to the next step, the determination of how fast the universe was expanding when each of these distant supernovae happened, let me briefly untangle a potential knot of confusion. When we're talking about distances on such fantastically large scales, and in the context of a universe that's continually expanding, the question inevitably arises of which distance the astronomers are actually measuring. Is it the distance between the locations we and a given galaxy each occupied eons ago, when the galaxy emitted the light we're just now seeing? Is it the distance between our current location and the location the galaxy occupied eons ago, when it emitted the light we're just now seeing? Or is it the distance between our current location and the galaxy's current location?

Here's what I consider the most insightful way of thinking about these and a whole slew of similarly confusing cosmological questions.

Imagine you want to know the distances, as the crow flies, among three cities, New York, Los Angeles, and Austin, so you measure their separation on a map of the United States. You find that New York is 39 centimeters from Los Angeles; Los Angeles is 19 centimeters from Austin; and Austin is 24 centimeters from New York. You then convert these measurements into real-world distances by looking at the map's legend, which provides a conversion factor—1 centimeter = 100 kilometers—which allows you to conclude that the three cities are about 3,900 kilometers, 1,900 kilometers, and 2,400 kilometers apart, respectively.

Now imagine that the earth's surface swells uniformly, doubling all separations. This would certainly be a radical transformation, but even so your map of the United States would continue to be perfectly valid as long as you made one important change. You'd need to modify the legend so that the conversion factor read "1 centimeter = 200 kilometers." Thirty-nine centimeters, 19 centimeters, and 24 centimeters on the map would now correspond to 7,800 kilometers, 3,800 kilometers, and 4,800 kilometers across the expanded United States. Were the expansion of the earth to continue, your static, unchanging map would remain accurate, as long as you continually updated its legend with the conversion factor relevant at each moment—1 centimeter = 200 kilometers at noon; 1 centimeter = 300 kilometers at two p.m.; 1 centimeter = 400 kilometers at four p.m.—to reflect how locations were being dragged apart by the expanding surface.

The expanding earth proves a useful conceit because similar considerations apply to the expanding cosmos. Galaxies don't move under their own power. Rather, like the cities on our expanding earth, they race apart because the substrate in which they're embedded—space itself—is swelling. This means that had some cosmic cartographer mapped galaxy locations billions of years ago, the map would be as valid today as it was then.[4] But, like the legend for the map of an expanding earth, the cosmic map's legend must be updated to ensure that the conversion factor, from map distances to real distances, remains accurate. The cosmological conversion factor is called the universe's *scale factor*; in an expanding universe, the scale factor increases with time.

Whenever you think about the expanding universe, I urge you to picture an unchanging cosmic map. Think of it as if it were any ordinary map lying flat on a table, and account for the cosmic expansion by updating the map's legend over time. With a little practice, you'll see that this approach vastly simplifies conceptual hurdles.

As a case in point, consider light from a supernova explosion in the distant Noa Galaxy. When we compare the supernova's apparent brightness with its intrinsic brightness, we are measuring the dilution of the light's intensity between emission (Figure 6.1a) and reception (Figure 6.1c), arising from its having spread out on a large sphere (drawn as a circle in Figure 6.1d) during the journey. By measuring the dilution, we determine the size of the sphere—its surface area—and then, with a little high school geometry, we can determine the sphere's radius. This radius traces the light's entire trajectory, and so its length equals the distance the light has traveled. Now the question that initiated this section pops up: To which of the three candidate distances, if any, does the measurement correspond?

During the light's journey, space has continually expanded. But the only change this requires to the static cosmic map is a regular updating of the scale factor recorded in the legend. And since we have just *now* received the supernova's light, since it has just *now* completed its journey, we must use the scale factor that's just *now* written in the map's legend to translate the separation on the map— the trajectory from the supernova to us, traced in Figure 6.1d—into the physical distance traveled. The procedure makes clear that the

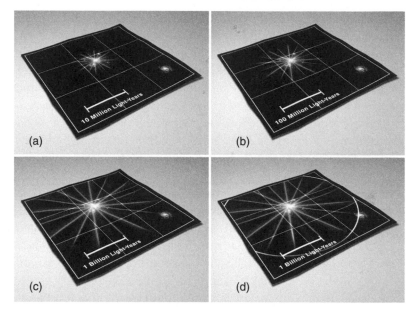

Figure 6.1 (a) *Light from a distant supernova spreads as it travels toward us (we are situated in the galaxy on the map's right-hand side).* (b) *During the light's journey, the universe expands, which is reflected in the map's legend.* (c) *When we receive the light, its intensity has been diluted through the spreading.* (d) *When we compare the supernova's apparent brightness to its intrinsic brightness, we are measuring the area of the sphere on which it has spread (drawn as a circle), and hence also its radius. The radius of the sphere traces the light's trajectory. Its length is the distance now between us and the galaxy that contained the supernova, so that's what the observations determine.*

result is the distance *now* between us and the current location of the Noa Galaxy: the third of our multiple-choice options.

Notice, too, that because the universe is continually expanding, earlier segments of a photon's journey continue to stretch long after the photon has sped past. If a photon painted a line on space that traced its path, the length of that line would increase as space expanded. By applying the map's scale factor at the time of reception to the light's entire journey, the third answer directly incorporates all such expansion. This is the right approach, because the amount by which the light's intensity is diluted depends on the size of the sphere over which the light *now* spreads—and this sphere's radius is the length of the light's trajectory *now*, including all post facto stretching.[5]

When we compare the intrinsic brightness of a supernova with its apparent brightness, we are therefore determining the distance now between us and the galaxy it occupied. Those are the distances the two groups of astronomers measured.[6]

The Colors of Cosmology

So much for measuring distances to faraway galaxies containing brilliant Type Ia supernovae. How do we learn about the rate of the universe's expansion ages ago, when each of those cosmic beacons momentarily ignited? The physics involved isn't much more complex than that at work in neon signs.

A neon sign glows red because when a current runs through the sign's gaseous interior, orbiting electrons in the neon atoms are momentarily knocked into higher-energy states. Then, as the neon atoms calm, the excited electrons jump down to their normal state of motion, relinquishing the extra energy by emitting photons. The color of the photons—their wavelength—is determined by the energy they carry. A key discovery, fully established by quantum mechanics in the early decades of the twentieth century, is that atoms of a given element have a unique collection of possible electron energy jumps; this translates into a unique collection of colors for released photons. For neon atoms, a dominant color is red (or, really, reddish orange), which accounts for the appearance of neon signs. Other elements—helium, oxygen, chlorine, and so on— exhibit similar behavior, the main difference being the wavelengths of the photons emitted. A "neon" sign of a color other than red is more than likely filled with mercury (if it's blue) or helium (if it's gold), or is made from glass tubes coated with substances, typically phosphors, whose atoms can emit light of yet other wavelengths.

Much of observational astronomy relies on the very same considerations. Astronomers use telescopes to gather light from distant objects, and from the colors they find—the particular wavelengths of light they measure—they can identify the chemical composition of the sources. An early demonstration occurred during the solar eclipse of 1868, when the French astronomer Pierre Janssen and, independently, the English astronomer Joseph Norman Lockyer examined light from the outermost shell of the sun, peeking just beyond the moon's rim, and found a mysterious bright emission with a wavelength that no one could reproduce in the laboratory using known substances. This led to the bold—and correct—suggestion that the light was emitted by a new, hitherto unknown element. The unknown substance was helium, which thus claims the singular distinction of being the only element discovered in the sun before it was found on earth. Such work established convincingly that, much as you can be uniquely identified by the pattern of lines making up your fingerprint, so an atomic species is uniquely identified by the pattern of wavelengths of the light it emits (and also absorbs).

In the decades that followed, astronomers who examined the wavelengths of light gathered from more and more distant astrophysical sources became aware of a peculiar feature. Although the collection of wavelengths resembled those familiar from laboratory experiments with well-known atoms such as hydrogen and helium, they were all somewhat longer. From one distant source, the wavelengths might be 3 percent longer; from another source, 12 percent longer; from a third 21 percent longer. Astronomers named this effect *redshift*, in recognition that ever longer wavelengths of light, at least in the visible part of the spectrum, become ever redder.

Naming is a good start, but what causes the wavelengths to stretch? The well-known answer, which emerged most clearly from the observations of Vesto Slipher and Edwin Hubble, is that the universe is expanding. The static map framework introduced earlier is tailor-made for providing an intuitive explanation.

Picture a light wave undulating its way from the Noa Galaxy toward earth. As we plot the light's progress across our unchanging map, we see a uniform succession of wave crests, one following another, as the undisturbed wave train heads toward our telescope. The uniformity of the waves might lead you to think that the wavelength of the light when emitted (the distance between successive wave crests) will be the same as when it's received. But the delightfully interesting part of the story comes into focus when we use the map's legend to convert map distances into real distances. Because the universe is expanding, the map's conversion factor is larger when the light concludes its journey than it was at inception. The implication is that although the light's wavelength as measured on the map is unchanging, when converted to real distances, the wavelength *grows*. When we finally receive the light, its wavelength is longer than when it was emitted. It's as if light waves are threads stitched through a piece of spandex. Just as stretching the spandex stretches the stitching, so expanding the spatial fabric stretches the light waves.

We can be quantitative. If the wavelength appears stretched by 3 percent, then the universe is 3 percent larger now than it was when the light was emitted; if the light appears 21 percent longer, then the universe has stretched 21 percent since the light began its journey. Redshift measurements thus tell us about the *size* of the universe when the light we're now examining was emitted, as compared with the size of the universe today.*

It's a straightforward final step to parlay a *series* of such redshift measurements into a determination of the universe's expansion profile over time.

A pencil mark drawn long ago on your child's wall records how

---

*If space is infinitely big, you might wonder what it means to say that the universe is larger now than it was in the past. The answer is that "larger" refers to the distances between galaxies today compared with the distances between those same galaxies in the past. The expansion of the universe means the galaxies are now farther apart, which is reflected mathematically in the universe's scale factor being larger. In the case of an infinite universe, "larger" does not refer to the overall size of space, since once infinite always infinite. But for ease of language, I will continue to refer to the changing size of the universe, even in the case of infinite space, with the understanding that I'm referring to the changing distances between galaxies.

tall she was at the date specified. A series of pencil marks gives her height at a series of dates. Given enough marks, you can determine how quickly she was growing at various times in the past. A growth spurt at nine, a slower period until eleven, another rapid spurt at thirteen, and so on. When astronomers measure a Type Ia supernova's redshift, they're determining an analogous "pencil mark" for space. Much like your child's height marks, a series of such redshift measurements of various Type Ia supernovae would enable them to calculate how quickly the universe was growing over various intervals in the past. With those data, in turn, the astronomers could determine the rate at which the expansion of space has been slowing. That was the plan of attack laid out by the research teams.

To execute it, they would have to complete one remaining step: dating the universe's pencil marks. The teams needed to determine when the light from a given supernova was emitted. This is a straightforward task. Since the difference between a supernova's apparent and intrinsic brightness reveals its distance, and since we know light's speed, we should be able to immediately calculate how long ago the supernova's light was emitted. The reasoning is right, but there is one essential subtlety, to do with the "post-facto" stretching of light's trajectory mentioned above, that's worth emphasizing.

When light travels in an expanding universe, it covers a given distance partly because of its intrinsic speed through space, but partly also because of the stretching of space itself. You can compare this with what happens on an airport's moving walkway. Without increasing your intrinsic speed, you travel farther than you otherwise would because the moving walkway augments your motion. Similarly, without increasing its intrinsic speed, light from a distant supernova travels farther than it otherwise would because during its journey the stretching space augments its motion. To judge correctly when the light we now see was emitted, we must take account of both contributions to the distance it covers. The math gets a little involved (see the notes if you are curious), but it is by now thoroughly understood.[7]

Being careful about this point, as well as numerous other theoretical and observational details, both groups were able to work out the size of the universe's scale factor at various identifiable times in the past. They were able, that is, to find a series of dated pencil marks delineating the universe's size, and therefore to determine how the expansion rate has been changing over the history of the cosmos.

Cosmic Acceleration

After checking, and rechecking, and checking again, both teams released their conclusions. For the last 7 billion years, contrary to long-held expectations, the expansion of space has not been slowing down. *It's been speeding up.*

A summary of this pioneering work, together with subsequent observations that cinched the case even more tightly, is given in Figure 6.2. The observations revealed that until about 7 billion years ago, the scale factor did indeed behave as expected: its growth gradually slowed down. Had this continued, the graph would have leveled off or even turned downward. But the data show that at about the 7-billion-year mark, something dramatic happened. The

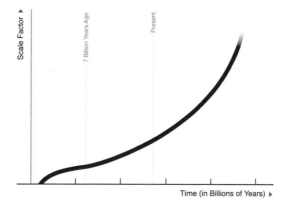

Figure 6.2 *The scale factor of the universe over time, showing that cosmic expansion slowed down until about 7 billion years ago, when it began to speed up.*

graph turned upward, which means that the growth rate of the scale factor began to *increase*. The universe kicked into high gear as the expansion of space started to accelerate.

Our cosmic destiny turns on the shape of this graph. With accelerated expansion, space will continue to spread indefinitely, dragging away distant galaxies ever farther and ever faster. A hundred billion years from now, any galaxies not now resident in our neighborhood (a gravitationally bound cluster of about a dozen galaxies called our "local group") will exit our cosmic horizon and enter a realm permanently beyond our capacity to see. Unless future astronomers have records handed down to them from an earlier era, their cosmological theories will seek explanations for an island universe, with galaxies numbering no more than students in a backwoods school, floating in a static sea of darkness. We live in a privileged age. Insights the universe giveth, accelerated expansion will taketh away.

As we will see in the pages that follow, the limited view on offer for future astronomers is all the more striking when compared with the enormity of the cosmic expanse to which our generation has been led in attempting to explain the accelerated expansion.

The Cosmological Constant

If you saw a ball's speed *increase* after someone threw it upward, you'd conclude that something was pushing it away from the earth's surface. The supernova researchers similarly concluded that the unexpected speeding up of the cosmic exodus required something to push outward, something to overwhelm the inward pull of attractive gravity. As we're now amply familiar, this is the very job description which makes the cosmological constant, and the repulsive gravity to which it gives rise, the ideal candidate. The supernova observations thus ushered the cosmological constant back into the limelight, not through the "bad judgment of conviction" to which Einstein had alluded in his letter decades earlier, but through the raw power of data.

The data also allowed the researchers to fix the numerical value of the cosmological constant—the amount of dark energy suffusing space. Expressing the result in terms of an equivalent amount of mass, as is conventional among physicists (using $E = mc^2$ in the less familiar form, $m = E/c^2$), the researchers showed that the supernova data required a cosmological constant of just under 10^{-29} grams in every cubic centimeter.[8] The outward push of such a small cosmological constant would have been trumped for the first 7 billion years by the inward pull of ordinary matter and energy, in keeping with the observational data. But the expansion of space would have diluted ordinary matter and energy, ultimately allowing the cosmological constant to gain the upper hand. Remember, the cosmological constant does not dilute; the repulsive gravity supplied by a cosmological constant is an intrinsic feature of space—every cubic meter of space contributes the same outward push, dictated by the cosmological constant's value. And so the more space there is between any two objects, arising from cosmic expansion, the stronger the force driving them apart. By about the 7-billion-year mark, the cosmological contant's repulsive gravity would have carried the day; the universe's expansion has been speeding up ever since, just as the data in Figure 6.2 attest.

To conform more fully to convention, I should re-express the cosmological constant's value in the units physicists more typically use. Much as it would be strange to ask a grocer for 10^{15} picograms of potatoes (instead, you'd ask for 1 kilogram, an equivalent measure in more sensible units), or tell a waiting friend that you'll be with her in 10^9 nanoseconds (instead, you'd say 1 second, an equivalent measure in more sensible units); it is similarly odd for a physicist to quote the energy of the cosmological constant in grams per cubic centimeter. Instead, for reasons that will shortly become apparent, the natural choice is to express the cosmological constant's value as a multiple of the so-called Planck mass (about 10^{-5} grams) per cubic Planck length (a cube that measures about 10^{-33} centimeters on each side and so has a volume of 10^{-99} cubic centimeters). In these units, the cosmological constant's measured value is about 10^{-123}, the tiny number that opened this chapter.[9]

How sure are we of this result? The data establishing accelerated expansion have only become more conclusive in the years since the first measurements were made. Moreover, complementary measurements (focusing on, for example, detailed features of the microwave background radiation; see *Fabric of the Cosmos*, Chapter 14) dovetail spectacularly well with the supernova results. If there's room for maneuvering, it lies in what we accept as an explanation for the accelerated expansion. Taking general relativity as the mathematical description of gravity, the only option is indeed the antigravity of a cosmological constant. Other explanations emerge if we modify this picture by including additional exotic quantum fields (which, much as we found in inflationary cosmology, can for periods of time masquerade as a cosmological constant),[10] or alter the equations of general relativity (so that attractive gravity drops off in strength with separation more precipitously than it does according to Newton's or Einstein's mathematics, thus allowing distant regions to rush away more quickly, without requiring a cosmological constant). But to date, the simplest and most convincing explanation for the observations of accelerated expansion is that the cosmological constant doesn't vanish, and so space is suffused with dark energy.

To many researchers, the discovery of a nonzero cosmological constant is the single most surprising observational result to have emerged in their lifetimes.

Explaining Zero

When I first caught wind of the supernova results suggesting a nonzero cosmological constant, my reaction was typical of many physicists. "It just can't be." Most (but not all) theoreticians had concluded decades before that the value of the cosmological constant was zero. This view initially arose from the "Einstein's greatest blunder" lore, but, over time, a variety of compelling arguments emerged to support it. The most potent came from considerations of quantum uncertainty.

Because of quantum uncertainty and the attendant jitters experienced by all quantum fields, even empty space is home to frenetic microscopic activity. And much like atoms bouncing around a box or kids jumping around a playground, quantum jitters harbor energy. But unlike atoms or kids, quantum jitters are ubiquitous and inevitable. You can't declare a region of space closed and send the quantum jitters home; the energy supplied by quantum jitters permeates space and can't be removed. Since the cosmological constant is nothing but energy that permeates space, quantum field jitters provide a microscopic mechanism that *generates* a cosmological constant. That's a pivotal insight. You'll recall that when Einstein introduced the notion of a cosmological constant, he did so abstractly—he didn't specify what it might be, where it might come from, or how it might arise. The link to quantum jitters makes it inevitable that had Einstein not dreamed up the cosmological constant, someone engaged with quantum physics subsequently would have. Once quantum mechanics is taken into account, you are forced to confront an energy contribution provided by fields that's uniformly spread through space, and so you are led directly to the notion of a cosmological constant.

The question this raises is one of numerical detail. *How much energy is contained in these omnipresent quantum jitters?* When theorists calculated the answer, they got a well-nigh ridiculous result: there should be an *infinite* amount of energy in every volume of space. To see why, think of a field jittering inside an empty box of any size. Figure 6.3 shows some sample shapes the jitters can assume. Every such jitter contributes to the field's energy content (in fact, the shorter the wavelength, the more rapid the jitter and hence the greater the energy). And since there are infinitely many possible wave shapes, each with a shorter wavelength than the previous, the total energy contained in the jitters is infinite.[11]

Although clearly unacceptable, the result did not engender fits of apoplexy because researchers recognized it as a symptom of the larger, well-recognized problem that we discussed earlier: the hostility between gravity and quantum mechanics. Everyone knew that you can't trust quantum field theory on super-small distance scales. Jitters with wavelengths as small as the Planck scale, 10^{-33} centimeters, and smaller, have energy (and from $m = E/c^2$, mass equivalent) so large that the gravitational force matters. To describe them properly requires a framework that melds quantum mechanics and

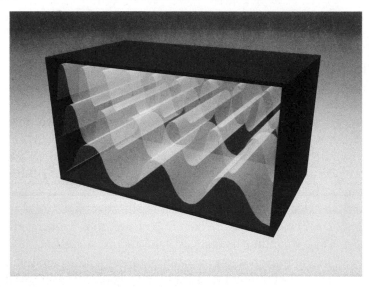

Figure 6.3 *There are infinitely many wave shapes in any volume and hence infinitely many distinct quantum jitters. This yields the problematic result of an infinite energy contribution.*

general relativity. Conceptually, this shifts the discussion to string theory, or to any other proposed quantum theory that includes gravity. But the immediate and more pragmatic response among researchers was simply to declare that the calculations should disregard jitters on scales smaller than the Planck length. Failure to implement this exclusion would extend a quantum field theory calculation into a realm clearly beyond its range of validity. The expectation was that we will one day understand string theory or quantum gravity well enough to deal with the super-small jitters quantitatively, but the interim stopgap was to mathematically quarantine the most pernicious fluctuations. The import of the directive is clear: if you ignore jitters shorter than the Planck length, you're left with only a finite number, so the total energy they contribute to a region of empty space is also finite.

That's progress. Or, at the very least, it shifts the burden to future insights that would, fingers crossed, tame the super-small-wavelength quantum fluctuations. But even so, researchers found that the resulting answer for the energy jitters, while finite, was still gargantuan, about 10^{94} grams per cubic centimeter. This is far larger than what you'd get from compressing all the stars in all the known galaxies into a thimble. Focusing on an infinitesimal cube, one that measures a Planck length on each side, this stupendous density amounts to 10^{-5} grams per cubic Planck length, or 1 Planck mass per Planck volume (which is why these units, like kilos for potatoes and seconds for waiting, are the natural and sensible choice). A cosmological constant of this magnitude would drive such an enormously fast outward burst that everything from galaxies to atoms would be ripped apart. More quantitatively, astronomical observations had established a tight limit on how large a cosmological constant could be, if there were one at all, and the theoretical results exceeded the limit by a staggering factor of more than a hundred orders of magnitude. While a large finite number for the energy that suffuses space is better than an infinite one, physicists realized the dire need for dramatically reducing the result from their calculations.

Here's where theoretical prejudice came to the fore. Assume for the moment that the cosmological constant is not just small. Assume it's zero. Zero is a favorite number of theoreticians because there's a tried and true way for it to emerge from calculations: symmetry. For example, imagine that Archie has enrolled in a continuing education course and for homework has to add together the sixty-third power of each of the first ten positive numbers, $1^{63} + 2^{63} + 3^{63} + 4^{63} + 5^{63} + 6^{63} + 7^{63} + 8^{63} + 9^{63} + 10^{63}$, and then add the result to the sum of the sixty-third power of each of the first ten negative numbers, $(-1)^{63} + (-2)^{63} + (-3)^{63} + (-4)^{63} + (-5)^{63} + (-6)^{63} + (-7)^{63} + (-8)^{63} + (-9)^{63} + (-10)^{63}$. What's the final tally? As he laboriously calculates, getting ever-more frustrated, multiplying and then adding together numbers with more than five dozen digits, Edith chimes in: "Use symmetry, Archie." "Huh?" What she means is that each term in the first collection has a symmetric balancing term in the second: 1^{63} and $(-1)^{63}$ sum to 0 (a negative raised to an odd power remains negative); 2^{63} and $(-2)^{63}$ sum to 0, and so on. The symmetry between the expressions results in a total cancellation, as if they were children of equal weight balancing on opposite sides of a seesaw. Needing no calculations at all, Edith shows that the answer is 0.

Many physicists believed—or, I should really say, hoped—that a similar total cancellation due to an as yet unidentified symmetry in the laws of physics would rescue the calculation of the energy contained in quantum jitters. Physicists surmised that the huge energies from quantum jitters would cancel against some as yet unidentified huge balancing contributions, once the physics was sufficiently well understood. This was about the only strategy physicists could come up with for tamping down the unruly results of the rough calculations. And that's why many theorists concluded that the cosmological constant had to be zero.

Supersymmetry provides a concrete example of how this could play out. Recall from Chapter 4 (Table 4.1) that supersymmetry entails a pairing of species of particles, and hence species of fields: electrons are paired with species of particles called supersymmetric electrons, or selectrons for short; quarks with squarks; neutrinos with sneutrinos, and so on. All of these "sparticle" species are currently hypothetical, but experiments in the next few years at the Large Hadron Collider may change that. In any event, an intriguing fact came to light when theoreticians examined mathematically the quantum jitters associated with each of the paired fields. For every jitter of the first field, there's a corresponding jitter of its partner that has the same size but opposite sign, much as in Archie's math homework. And just as in that example, when we add together all such contributions pair by pair, they cancel out, yielding a final result of zero.[12]

The catch, and it's a big one, is that the total cancellation occurs only if both members of a pair have not only the same electric and nuclear charges (which they do), but also the same mass. Experimental data have ruled this out. Even if nature makes use of supersymmetry, the data show that it can't be realized in its most potent form. The as yet unknown particles (selectrons, squarks, sneutrinos, and so on) must be much heavier than their known counterparts—only this can explain why they haven't been seen in accelerator experiments. When the different particle masses are accounted for, the symmetry is disturbed, the balancing is unbalanced, and the cancellations are imperfect; the result is once again huge.

Over the years, many analogous proposals were put forward, invoking a range of additional symmetry principles and cancellation mechanisms, but none achieved the goal of establishing theoretically that the cosmological constant should vanish. Even so, most researchers took this merely as a sign of our incomplete understanding of physics, not as a clue that belief in a vanishing cosmological constant was misguided.

One physicist who challenged the orthodoxy was the Nobel laureate Steven Weinberg.* In a paper published in 1987, more than a decade before the revolutionary supernova measurements, Weinberg suggested an alternative theoretical scheme that yielded a decidedly different outcome: a cosmological constant that is small *but not zero.* Weinberg's calculations were based on one of the most polarizing concepts to have gripped the physics community in decades—a principle some revere and others vilify, a principle some call profound and others call silly. Its official, if misleading, name is the *anthropic principle.*

Cosmological Anthropics

Nicolaus Copernicus' heliocentric model of the solar system is acknowledged as the first convincing scientific demonstration that we humans are not the focal point of the cosmos. Modern discoveries have reinforced the lesson with a vengeance. We now realize that Copernicus' result is but one of a series of nested demotions overthrowing long-held assumptions regarding humanity's special status: we're not located at the center of the solar system, we're not located at the center of the galaxy, we're not located at the center of the universe, we're not even made of the dark ingredients constitut-

*The Cambridge astrophysicist George Efstathiou was also one of the early pioneers who argued strongly and convincingly for a nonzero cosmological constant.

ing the vast majority of the universe's mass. Such cosmic downgrading, from headliner to extra, exemplifies what scientists now call the *Copernican principle*: in the grand scheme of things, everything we know points toward human beings not occupying a privileged position.

Nearly five hundred years after Copernicus' work, at a commemorative conference in Kraków, one presentation in particular—given by the Australian physicist Brandon Carter—provided a tantalizing twist to the Copernican principle. Carter expounded his belief that an overadherence to the Copernican perspective might, in certain circumstances, divert researchers from significant opportunities for making progress. Yes, Carter agreed, we humans are not central to the cosmic order. Yet, he continued, aligning with similar insights articulated by scientists such as Alfred Russel Wallace, Abraham Zelmanov, and Robert Dicke, there is one arena in which we *do* play an absolutely indispensable role: our own observations. However far we have been demoted by Copernicus and his legacy, we top the bill when credits are conferred for the gathering and analyzing of the data that mold our beliefs. Because of this unavoidable position, we must take account of what statisticians call *selection bias*.

It's a simple and widely applicable idea. If you are investigating trout populations but only canvass the Sahara Desert, your data will be biased by your focusing on an environment particularly inhospitable to your subject. If you are studying the general public's interest in opera, but send your survey solely to the database collected by the journal *Can't Live Without Opera*, your results won't be accurate because the respondents are not representative of the population as a whole. If you are interviewing a group of refugees who have endured astoundingly harsh conditions during their trek to safety, you might conclude that they are among the hardiest ethnicities on the planet. Yet, when you learn the devastating fact that you are speaking with less than 1 percent of those who started out, you realize that such a deduction is biased because only the phenomenally strong survived the journey.

Addressing these biases is vital for getting meaningful results and for avoiding the futile search to explain conclusions based on unrepresentative data. Why are trout extinct? What's the cause of the public's surging interest in opera? Why is it that a particular ethnicity is so astoundingly resilient? Biased observations can launch you on meaningless quests to explain things that a broader, more representative view renders moot.

In most cases, these types of biases are easily identified and corrected. But there's a related variety of bias that's more subtle, one so basic it can easily be overlooked. It's the kind in which limitations on when and where we are *able* to live can have a profound impact on what we are able to see. If we fail to take proper account of the impact such intrinsic limitations have on our observations, then, as in the examples above, we can draw wildly erroneous conclusions, including some that may impel us on fruitless journeys to explain meaningless MacGuffins.

For instance, imagine that you're intent on understanding (as was the great scientist Johannes Kepler) why the earth is 93 million miles from the sun. You want to find, deep within the laws of physics, something that will explain this observational fact. For years you struggle mightily but are unable to synthesize a convincing explanation. Should you keep trying? Well, if you reflect on your efforts, taking account of selection bias, you will soon realize that you're on a wild goose chase.

The laws of gravity, Newton's as well as Einstein's, allow a planet to orbit a star at any distance. If you were to grab hold of the earth, move it to some arbitrary distance from the sun, and then set it in motion again at the right velocity (a velocity easy to work out with basic physics), it would happily go into orbit. The only thing special about being 93 million miles from the sun is that it yields a temperature range on earth conducive to our being here. If earth were much closer or much farther away from the sun, the temperature would be much hotter or colder, eliminating an essential ingredient for our form of life: liquid water. This reveals the in-built bias. The very fact that *we* measure the distance from our planet to the sun mandates that the result we find must be within the limited range compatible with our own existence. Otherwise, we wouldn't be here to contemplate the earth's distance from the sun.

If earth were the only planet in the solar system, or the only planet in the universe, you still might feel compelled to carry your investigations further. Yes, you might say, I understand that my own existence is tied to the earth's distance from the sun, yet this only heightens my urge to explain why the earth happens to be situated at such a cozy, life-compatible position. Is it just a lucky coincidence? Is there a deeper explanation?

But the earth is not the only planet in the universe, let alone in the solar system. There are many others. And this fact casts such questions in a very different light. To see what I mean, imagine that you mistakenly think a particular shop carries only a single shoe size, and so are gleefully surprised when the salesman brings you a pair that fits perfectly. "Of all possible shoe sizes," you reflect, "it's amazing that the single one they carry is mine. Is that just a lucky coincidence? Is there a deeper explanation?" But when you learn that the shop actually carries a wide range of sizes, the questions evaporate. A universe with many planets, situated at a range of distances from their host stars, provides a similar situation. Just as it's no big surprise that among all the shoes in the shop there's at least one pair that fits, so it's no big surprise that among all the planets in all the solar systems in all the galaxies there's at least one at the right distance from its host star to yield a climate conducive to our form of life. And it's on one of those planets, of course, that we live. We simply couldn't evolve or survive on the others.

So there is no fundamental reason why the earth is 93 million miles from the sun. A planet's orbital distance from its host star is due to the vagaries of historical happenstance, the innumerable detailed features of the swirling gas cloud from which a particular solar system coalesced; it's a contingent fact that's unavailable for fundamental explanation. Indeed, these astrophysical processes have produced planets throughout the cosmos, orbiting their respective suns at a vast assortment of distances. We find ourselves on one such planet situated 93 million miles from our sun because that's a planet on which our form of life *could* evolve. Failure to take account of this selection bias would lead one to search for a deeper answer. But that's a fool's errand.

Carter's paper emphasized the importance of paying heed to such bias, an accounting he called the anthropic principle (an unfortunate name, because the idea would apply equally well to any form of intelligent life that makes and analyzes observations, not just to humans). No one took exception to this element of Carter's argument. The controversial part was his suggestion that the anthropic principle might cast its net not just over things in the universe, like planetary distances, but over the universe itself.

What would that mean?

Imagine you're puzzling over some fundamental feature of the universe, say the mass of an electron, .00054 (expressed as a fraction of the proton's mass), or the strength of the electromagnetic force, .0073 (expressed by its coupling constant), or, of primary interest to us here, the value of the cosmological constant, 1.38×10^{-123} (expressed in Planck units). Your intention is to explain why these constants have the particular values they do. You try and try but come up emptyhanded. Take a step back, Carter says. Maybe you're failing for the same reason you'd fail to explain the earth-sun distance: there is no fundamental explanation. Just as there are many planets at many distances and we necessarily inhabit one whose orbit yields hospitable conditions, maybe there are many universes with many different values for the "constants" and we necessarily inhabit the one in which the values are conducive to our existence.

In this way of thinking, to ask why the constants have their particular values is to ask the wrong kind of question. There is no law dictating their values; their values can and do vary across the multiverse. Our intrinsic selection bias ensures that we find ourselves in that part of the multiverse in which the constants have the values with which we're familiar simply because we're unable to exist in the parts of the multiverse where the values are different.

Note that the reasoning would fall flat if our universe were unique because you could still ask the "lucky coincidence" or "deeper explanation" questions. Much as a potent explanation for why the shop has your shoe size requires that the shelves be stocked with many different sizes, and much as a potent explanation for why there's a planet situated at a bio-friendly distance from its host star requires planets orbiting their stars at many different distances, so a potent explanation of nature's constants requires a vast assortment of universes endowed with many different values for those constants. Only in this setting—a multiverse, and a robust one at that—does anthropic reasoning have the capacity to make the mysterious mundane.*

Clearly, then, the degree to which you are swayed by the anthropic approach depends on the degree to which you are convinced of its three essential assumptions: (1) our universe is part of a multiverse; (2) from universe to universe in the multiverse, the constants take on a broad range of possible values; and (3) for most variations of the constants away from the values we measure, life as we know it would fail to take hold.

In the 1970s, when Carter put forward these ideas, the notion of parallel universes was anathema to many physicists. Certainly, there's still ample reason to be skeptical. But we've seen in the previous chapters that although the case for any particular version of the multiverse is surely tentative, there's reason for giving this new view of reality serious consideration, Assumption 1. Many scientists now are. Regarding Assumption 2, we've also seen that, for example, in the Inflationary and Brane Multiverses, we would indeed expect physical features, such as the constants of nature, to vary from universe to universe. Later in this chapter we'll look at this point more closely.

But what about Assumption 3, concerning life and the constants?

Life, Galaxies, and Nature's Numbers

For many of nature's constants, even modest variations would render life as we know it impossible. Make the gravitational constant stronger, and stars burn up too quickly for life on nearby planets to evolve. Make it weaker and galaxies don't hold together. Make the electromagnetic force stronger, and hydrogen atoms repel each

---

*In Chapter 7, we will examine more thoroughly and more generally the challenges of testing theories that involve a multiverse; we will also more closely analyze the role of anthropic reasoning in yielding potentially testable outcomes.

other too strongly to fuse and supply power to stars.[13] But what about the cosmological constant? Does life's existence depend on its value? This is the issue Steven Weinberg took up in his 1987 paper.

Because the formation of life is a complex process about which our understanding is in its earliest stages, Weinberg recognized that it was hopeless to determine how one or another value of the cosmological constant directly impacts the myriad steps that breathe life into matter. But rather than give up, Weinberg introduced a clever proxy for the formation of life: the formation of galaxies. Without galaxies, he reasoned, the formation of stars and planets would be thoroughly compromised, with a devastating impact on the chance that life might emerge. This approach was not only eminently reasonable but also useful: it shifted the focus to determining the impact that cosmological constants of various sizes would have on galaxy formation, and that was a problem Weinberg could solve.

The essential physics is elementary. While precise details of galaxy formation are an active area of research, the broad-brush process involves a kind of astrophysical snowball effect. A clump of matter forms here or there, and by virtue of being more dense than its surroundings, it exerts a greater gravitational pull on nearby matter and thus grows larger still. The cycle continues feeding on itself to ultimately produce a swirling mass of gas and dust, from which stars and planets coalesce. Weinberg's realization was that a cosmological constant with a value large enough would disrupt the clumping process. The repulsive gravity it would generate, if sufficiently strong, would thwart galactic formation by making the initial clumps—which were small and fragile—stream apart before they had time to become robust by attracting surrounding matter.

Weinberg worked out the idea mathematically and found that a cosmological constant any larger than a few hundred times the current cosmological density of matter, a few protons per cubic meter, would disrupt the formation of galaxies. (Weinberg also considered the impact of a negative cosmological constant. The constraints in that case are even tighter, because a negative value increases the attractive pull of gravity and makes the whole universe collapse before stars even have time to ignite.). If you imagine, then, that we're part of a multiverse and that the cosmological constant's value varies over a wide range from universe to universe, much as planet-star distances vary over a wide range from solar system to solar system—the only universes that could have galaxies, and hence the only universes we could inhabit, are ones in which the cosmological constant is no larger than Weinberg's limit, which in Planck units is about 10^{-121}.

After years of failed efforts by the community of physicists, this was the first theoretical calculation to result in a value for the cosmological constant that was not absurdly larger than limits inferred from observational astronomy. Nor did it contradict a belief widely held at the time of Weinberg's work, that the cosmological constant vanished. Weinberg took this apparent progress one step further by encouraging an even more aggressive interpretation of his result. He suggested that we should expect to find ourselves in a universe with a cosmological constant whose value is as small as it needs to be for us to exist, but not a whole lot smaller. A much smaller constant, he reasoned, would call for an explanation that goes beyond mere compatibility with our existence. That is, it would require precisely the kind of explanation that physics had valiantly sought but so far failed to find. This led Weinberg to suggest that more refined measurements might one day reveal that the cosmological constant doesn't vanish but, instead, has a value near or at the upper limit that he'd calculated. As we've seen, within a decade of Weinberg's paper, the observations of the Supernova Cosmology Project and the High-Z Supernova Search Team proved this suggestion prophetic.

But to assess fully this unconventional explanatory framework, we need to examine Weinberg's reasoning more closely. Weinberg is imagining a sprawling multiverse so diverse in population that it just *has* to contain at least one universe with the cosmological constant we've observed. But what kind of multiverse will guarantee, or at least make it highly likely, that this is the case?

To think this through, consider first an analogous problem with simpler numbers. Imagine you work for the notorious film producer Harvey W. Einstein, who has asked you to put out a casting call for the lead in his new indie, *Pulp Friction.* "How tall do you want him?" you ask. "I dunno. Taller than a meter, less than two. But you better make sure whatever height I decide, there's someone who fits the bill." You're tempted to correct your boss, noting that because of quantum uncertainty he really doesn't need to have *every* height represented but, thinking back on what happened to the surly little talking fly who tried that, you refrain.

Now you face a decision. How many actors should you have at the audition? You reason: If W. measures heights to a centimeter's accuracy, there are a hundred different possibilities between one and two meters. So you need at least a hundred actors. But since some actors who show up may have the same height, leaving other heights unrepresented, you'd better gather more than a hundred. To be safe, maybe you should put out a call for a few hundred actors. That's a lot, but fewer than what you'd need if W. measured heights to a millimeter's accuracy. In that case, there'd be a thousand different heights between one and two meters, so to be safe you'd need to gather a few thousand actors.

The same reasoning is relevant for the case of universes with different cosmological constants. Assume that all the universes in a multiverse have cosmological constant values between zero and one (in the usual Planck units); smaller values lead to universes that collapse, larger values would strain the applicability of our mathematical formulations, compromising all understanding. So just as the actors' heights had a range of one (in meters), the universes' cosmological constants have a range of one (in Planck units). As for accuracy, the analog of W. using centimeter ticks, or millimeter ticks, is the precision with which we can measure the cosmological constant. Today's accuracy is about 10^{-124} (in Planck units). In the future, our accuracy will no doubt improve, but as we'll see, that will hardly affect our conclusions. Then just as there are 10^2 different possible heights spaced at least 10^{-2} meters apart (1 centimeter) in a one-meter range, and 10^3 different possible heights spaced at least 10^{-3} meters apart (1 millimeter), so there are 10^{124} different values of the cosmological constant spaced at least 10^{-124} apart between the values 0 and 1.

To ensure that every possible cosmological constant is realized, we'd therefore need a multiverse with at least 10^{124} different universes. But as with the actors, we need to account for possible duplicates, universes that may have the same cosmological constant value. And so to play it safe and make it highly likely that every possible cosmological constant value is realized, we should have a multiverse with far more than 10^{124} universes, say a million times more, bringing it to a nice even 10^{130} universes. I'm being cavalier because when we're talking about numbers this large, the exact values hardly matter. No familiar example of anything—not the number of cells in your body (10^{13}); not the number of seconds since the big bang (10^{18}); not the number of photons in the observable part of the universe (10^{88})—comes even remotely close to the number of universes we're contemplating. The bottom line is that Weinberg's approach for explaining the cosmological constant works only if we're part of a multiverse in which there are a huge number of different universes; their cosmological constants must fill out some 10^{124} distinct values. Only with that many different universes is there a high likelihood that there's one with a cosmological constant that matches ours.

Are there theoretical frameworks that naturally yield such a spectacular profusion of universes with different cosmological constants?[14]

From Vice to Virtue

There are. We encountered such a framework in the previous chapter. A count of the different possible forms for the extra dimensions in string theory, when including fluxes that can thread through them, came to about 10^{500}. This dwarfs 10^{124}. Multiply 10^{124} by a few hundred orders of magnitude and 10^{500} still dwarfs it. Subtract 10^{124} from 10^{500}, and then subtract it again, and again, and do so a billion times over, and you'd barely make a dent. The result would still be nearly 10^{500}.

Critically, the cosmological constant does indeed vary from one such universe to another. Just as magnetic flux carries energy (it can move things), so the fluxes threading holes in Calabi-Yau shapes also have energy, whose quantity is quite sensitive to the shape's geometrical details. If you have two different Calabi-Yau shapes with different fluxes penetrating different holes, their energies will generally be different too. And since a given Calabi-Yau shape is attached to every point in the familiar three large dimensions of space, much as circular loops of pile attach to every point on the large extended base of a carpet, the energy the shape contains would uniformly fill the three large dimensions, much as soaking the individual fibers in a carpet's pile would make the entire carpet backing uniformly heavy. Thus, should one or another of the 10^{500} different dressed-up Calabi-Yau shapes constitute the requisite extra dimensions, *the energy it contains would contribute to the cosmological constant.* Results obtained by Raphael Bousso and Joe Polchinski made this observation quantitative. They argued that the various cosmological constants supplied by the 10^{500} or so different possible forms for the extra dimensions are distributed uniformly across a broad range of values.

This is just what the doctor ordered. Having 10^{500} tick marks distributed across a range from 0 to 1 ensures that many of them lie extremely close to the value of the cosmological constant astronomers have measured during the past decade. It may be hard to find the explicit examples among the 10^{500} possibilities, because even if today's fastest computers took a single second to analyze each form for the extra dimensions, after a billion years only a paltry 10^{32} examples would have been examined. But this reasoning suggests strongly that they exist.

Certainly, a collection of 10^{500} different possible forms for the extra dimensions is about as far from a unique universe as anyone imagined string theory research would ever take us. And for those who've held strongly to Einstein's dream of finding a unified theory describing one single universe—ours—these developments came with significant discomfort. But analysis of the cosmological constant casts the situation in a different light. Rather than despair because a unique universe seems not to emerge, we are encouraged to celebrate: string theory makes the least plausible part of Weinberg's explanation of the cosmological constant—the requirement that there be many more than 10^{124} different universes—suddenly seem plausible.

The Final Step, in Brief

The elements of a tantalizing story seem to be coming together. But a gap remains in the reasoning. It's one thing for string theory to allow for a huge number of possible distinct universes. It's another to claim that string theory ensures that all of the possible universes to which it can give rise are actually out there, parallel worlds populating a vast multiverse. As emphasized most emphatically by Leonard Susskind—who was inspired by the pioneering work of Shamit Kachru, Renata Kallosh, Andrei Linde, and Sandip Trivedi—if we weave eternal inflation into the tapestry, the gap can be filled.[15]

I'll now explain this final step, but in case you're reaching saturation and just want the punch line, here's a three-sentence summary. The Inflationary Multiverse—the ever-expanding Swiss cheese cosmos—contains a vast, ever-increasing number of bubble universes. The idea is that when inflationary cosmology and string theory are melded, the process of eternal inflation sprinkles string theory's 10^{500} possible forms for the extra dimensions across the bubbles—one form for the extra dimensions per bubble universe—providing a cosmological framework that realizes all possibilities. By this reasoning, we live in that bubble whose extra dimensions yield a universe, cosmological constant and all, that's hospitable to our form of life and whose properties agree with observations.

In the remainder of the chapter, I will flesh out the details, but if you're ready to move on, feel free to jump ahead to the chapter's last section.

The String Landscape

In explaining inflationary cosmology back in Chapter 3, I used a variation on a common metaphor. A mountain's peak represents the highest value of energy contained in an inflaton field suffusing space. The act of rolling down the mountain and coming to rest at a low point in the terrain represents the inflaton shedding this energy, which in the process is converted to particles of matter and radiation.

Let's revisit three aspects of the metaphor, updating them with insights we've since acquired. First, we've learned that the inflaton is only one source of the energy that may fill space; other contributions come from the quantum jitters of any and all fields—electromagnetic, nuclear, and so on. To revise the metaphor accordingly, altitude will now reflect the combined energy uniformly suffusing space contributed by all sources.

Second, the original metaphor envisioned the base of the mountain, where the inflaton finally comes to rest, as being at "sea level," altitude zero, meaning the inflaton has shed all its energy (and pressure). But with our revised metaphor, the height of the mountain's base should represent the combined energy suffusing space from all sources after inflation has drawn to a close. This is another name for that bubble universe's cosmological constant. The mystery in explaining our cosmological constant thus translates into the mystery of explaining the altitude of our mountain's base—why is it so close to, but not exactly at, sea level?

Finally, we initially considered the simplest of mountainous terrains, a peak leading smoothly to a base, where the inflaton would ultimately settle (see Figure 3.1, page 61). We then went a step further, taking account of other ingredients (Higgs fields) whose evolution and final resting places would influence the physical features manifest in the bubble universes (see Figure 3.6, page 74). In string theory, the range of possible universes is richer still. The shape of the extra dimensions determines the physical features within a given bubble universe, and so the possible "resting places," the various valleys in Figure 3.6b, now represent the possible shapes the extra dimensions can take. To accommodate the 10^{500} possible forms for these dimensions, the mountain terrain therefore needs a lush assortment of valleys, ledges, and outcroppings, as represented in Figure 6.4. Any such feature in the terrain where a ball could come to rest represents a possible shape into which the extra dimensions could relax; the altitude at that location represents the cosmological constant of the corresponding bubble universe. Figure 6.4 illustrates what's called the *string landscape*.

Figure 6.4 *The string landscape can be visualized schematically as a mountainous terrain in which different valleys represent different forms for the extra dimensions, and altitude represents the cosmological constant's value.*

With this more refined understanding of the mountain—or landscape—metaphor, we now consider how quantum processes affect the form of the extra dimensions in this setting. As we will see, quantum mechanics lights up the landscape.

Quantum Tunneling in the Landscape

While Figure 6.4 is necessarily schematic (each of the different Higgs fields in Figure 3.6 has its own axis; similarly each of the roughly 500 different field fluxes that can thread a Calabi-Yau shape should also have its own axis—but sketching mountains in a 500-dimensional space is a challenge), it correctly suggests that universes with different forms for the extra dimensions are part of a connected terrain.[16] And when quantum physics is taken into account, using results discovered independently of string theory by the legendary physicist Sidney Coleman in collaboration with Frank De Luccia, the connections between the universes allow for dramatic transmutations.

The core physics relies on a process known as *quantum tunneling*. Imagine a particle, an electron for instance, encountering a solid barrier, say a slab of steel ten feet thick, that classical physics predicts it can't penetrate. A hallmark of quantum mechanics is that the rigid classical notion of "can't penetrate" often translates into the softer quantum declaration of "has a small but nonzero probability of penetrating." The reason is that the quantum jitters of a particle allow it, every so often, to suddenly materialize on the other side of an otherwise impervious barrier. The moment at which such quantum tunneling happens is random; the best we can do is predict the likelihood that it will take place during one interval or another. But the math says that if you wait long enough, penetration through just about any barrier will happen. And it does happen. If it didn't, the sun wouldn't shine: for hydrogen nuclei to get close enough to fuse, they must tunnel through the barrier created by the electromagnetic repulsion of their protons.

Coleman and De Luccia, and many who have since followed their lead, scaled quantum tunneling up from single particles to an entire universe that's faced with a similar "impenetrable" barrier separating its current configuration from another that's possible. To get a feel for their result, imagine two possible universes that are otherwise identical save for a field, uniformly suffusing each, whose energy is higher in one, lower in the other. In the absence of a barrier, the higher energy-field value rolls to the lower, like a ball rolling down a hill as we've seen in the discussion of inflationary cosmology. But what happens if the field's energy curve has a "mountainous bump" separating its current value from the one it seeks, as in Figure 6.5? Coleman and De Luccia found that much as is the case for a single particle, a universe can do what classical physics forbids: it can jitter its way—it can quantum tunnel—through the barrier and reach the lower energy configuration.

But because we are talking about a universe and not just a sin-

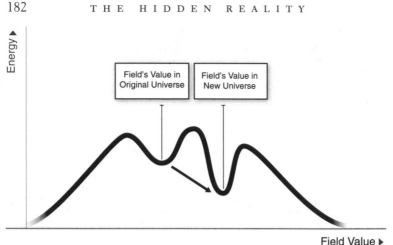

Figure 6.5 *An example of a field's energy curve that has two values—two troughs or valleys—where the field naturally comes to rest. A universe suffused with the higher-energy field value can quantum tunnel to the lower value. The process involves a small randomly located region of space in the original universe acquiring the lower field value; the region then expands, converting an ever-wider domain from the higher to the lower energy.*

gle particle, the tunneling process is more involved. It's not that the field's value throughout all of space tunnels simultaneously through the barrier, Coleman and De Luccia argued; rather, a "seed" tunneling event would create a small, randomly located bubble suffused with the smaller field energy. The bubble would then grow, much like Vonnegut's ice-nine, ever enlarging the domain in which the field had tunneled to the lower energy.

These ideas can be applied directly to the string landscape. Imagine that the universe has a particular form for the extra dimensions, which corresponds to the left valley in Figure 6.6a. Because of this valley's high altitude, the three familiar spatial dimensions are permeated by a large cosmological constant—yielding strong repulsive gravity—and so are rapidly inflating. This expanding universe, together with its extra dimensions, is illustrated on the left side of Figure 6.6b. Then, at some random location and moment, a tiny region of space tunnels through the intervening mountain to

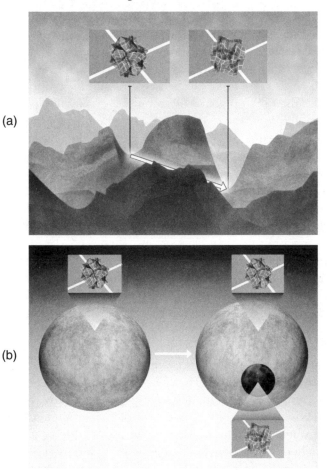

Figure 6.6 (a) *A quantum tunneling event, within the string landscape.* **(b)** *The tunneling creates a small region of space—represented by the smaller and darker bubble—within which the form of the extra dimensions has changed.*

the valley on the right side of Figure 6.6a. Not that the tiny region of space moves (whatever that would mean); rather, the form of the extra dimensions (its shape, size, fluxes it carries) in this little region changes. The extra dimensions in the tiny region transmute, acquiring the form associated to the right valley in Figure 6.6a.

This new bubble universe lies within the original, as illustrated in Figure 6.6b.

The new universe will rapidly expand and continue to transform the extra dimensions as it spreads. But since the new universe's cosmological constant has decreased—its altitude in the landscape is lower than the original—the repulsive gravity it experiences is weaker, and so it won't expand as fast as the original universe. We thus have an expanding bubble universe, with the new form for the extra dimensions, contained in an even faster expanding bubble universe, with the original form for the extra dimensions.[17]

The process can repeat. At other locations inside the original universe as well as inside the new one, further tunneling events cause additional bubbles to open up, creating regions with yet different forms for the extra dimensions (Figure 6.7). In due course, the expanse of space will be riddled with bubbles inside of bubbles inside of bubbles—each undergoing inflationary expansion, each with a different form for the extra dimensions, and each with a smaller cosmological constant than the larger bubble universe within which it formed.

The result is a more intricate version of the Swiss cheese multiverse we found in our earlier encounter with eternal inflation. In that version, we had two types of regions: the "cheesy" ones that were undergoing inflationary expansion and the "holes" that weren't. This was a direct reflection of the simplified landscape with a single mountain whose base we assumed to be at sea level. The richer string theory landscape, with its sundry peaks and valleys corresponding to different values of the cosmological constant, gives rise to the many different regions in Figure 6.7—bubbles inside of bubbles inside of bubbles, like a sequence of Matryosta dolls, each painted by a different artist. Ultimately, the relentless series of quantum tunnelings through the mountainous string landscape realizes every possible form for the extra dimensions in one or another bubble universe. This is the *Landscape Multiverse*.

The Landscape Multiverse is just what we need for Weinberg's explanation of the cosmological constant. We've argued that the string landscape ensures that there are, in principle, *possible* forms for the extra dimensions that would have a cosmological constant in the ballpark of the observed value: there are valleys in the string

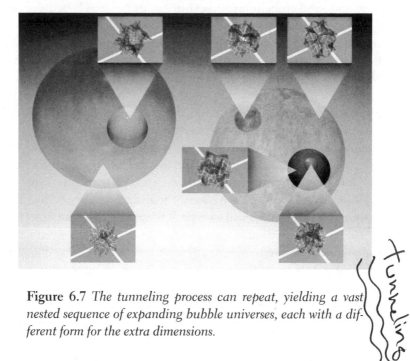

Figure 6.7 *The tunneling process can repeat, yielding a vast nested sequence of expanding bubble universes, each with a different form for the extra dimensions.*

landscape whose tiny altitude is on par with the tiny but nonzero cosmological constant that the supernovae observations revealed. When the string landscape combines with eternal inflation, all possible forms for the extra dimensions, including those with such a small cosmological constant, are brought to life. Somewhere within the vast nested sequence of bubbles constituting the Landscape Multiverse, there are universes whose cosmological constant is about 10^{-123}, the minuscule number that launched this chapter. And according to this line of thought, it is in one of those bubbles that we live.

The Rest of Physics?

The cosmological constant is but one feature of the universe we inhabit. It is arguably among the most puzzling, since its small measured value is so famously at odds with the numbers that emerge from the most straightforward estimates using established theory. This chasm draws singular focus to the cosmological constant and underlies the urgency of finding a framework, however exotic, with the capacity to explain it. Proponents of the interlocking set of ideas laid out above argue that the string multiverse does just that.

But what about all the other features of our universe—the existence of three kinds of neutrinos, the particular mass of the electron, the strength of the weak nuclear force, and so on? While we can at least imagine calculating these numbers, no one has as yet managed to do so. You might wonder whether their values, too, are ripe for a multiverse-based explanation. Indeed, researchers surveying the string landscape have found that these numbers, like the cosmological constant, also vary from place to place, and hence—at least in our current understanding of string theory—are not uniquely determined. This leads to a perspective very different from what dominated in the early days of research on the subject. It suggests that trying to calculate the properties of the fundamental particles, like trying to explain the distance between the earth and the sun, may be misguided. Like planetary distances, some or all of the properties would vary from one universe to the next.

For this line of thinking to be credible, though, we need at a bare minimum to know not only that there are bubble universes in which the cosmological constant has the right value, but also that in at least one such bubble the forces and the particles agree with what scientists in our universe have measured. We need to be sure that our universe, in all its detail, is somewhere in the landscape. This is the goal of a vibrant field called *string model building*. The research program amounts to hunting around the string landscape and examining possible forms for the extra dimensions mathematically, in search of universes that most resemble ours. It's a formidable challenge, because the landscape is too large and intricate to be fully studied in any systematic way. Progress requires sharp calculational skills as well as intuition regarding which pieces to assemble—the extra-dimensional shape, its size, the field fluxes cycling its holes, the presence of various branes, and so on. Those who lead this charge combine the best of rigorous science with an artistic sensibility. To date, no one has found an example that reproduces the features of our universe exactly. But with some 10^{500} possibilities awaiting exploration, the consensus is that our universe has a home somewhere in the landscape.

Is This Science?

In this chapter we've turned a logical corner. Until now, we've been exploring the implications for reality, writ large, of various developments in fundamental physics and cosmology research. I delight in the possibility that copies of the earth exist in the far reaches of space, or that our universe is one of many bubbles in an inflating cosmos, or that we live on one of many braneworlds constituting a giant cosmic loaf. These are undeniably provocative and alluring ideas.

But with the Landscape Multiverse, we've invoked parallel universes in a different way. In the approach we've just followed, the Landscape Multiverse is not merely broadening our view of what might be out there. Instead, an array of parallel universes, worlds that may be beyond our ability to visit or see or test or influence, now and perhaps always, are directly invoked to provide insight into observations we make here, in this universe.

Which raises an essential question: Is this science?

CHAPTER 7

Science and the Multiverse

On Inference, Explanation, and Prediction

When David Gross, co-recipient of the 2004 Nobel Prize in physics, inveighs against string theory's Landscape Multiverse, there's a fair chance he'll quote Winston Churchill's speech of October 29, 1941: "Never give in. . . . Never, never, never, never— in nothing, great or small, large or petty—never give in." When Paul Steinhardt, the Albert Einstein Professor in Science at Princeton University and co-discoverer of the modern form of inflationary cosmology, speaks of his distaste for the Landscape Multiverse, the rhetorical flourishes are more subdued, but you can be pretty sure a comparison to religion, an unfavorable one at that, will at some point appear. Martin Rees, the United Kingdom's Astronomer Royal, sees the multiverse as the natural next step in our deepening grasp of all there is. Leonard Susskind says those who ignore the possibility that we're part of a multiverse are merely averting their eyes from a vision they find overwhelming. And these are just a few examples. There are many others on both sides, vehement naysayers and enthusiastic devotees, and they don't always express their opinions in terms so lofty.

In the quarter century I've been working on string theory, I've never seen passions run quite so high, or language turn quite so sharp, as in discussions of string theory's landscape and the multiverse to which it may give rise. And it's clear why. Many see these developments as a battleground for the very soul of science.

The Soul of Science

While the Landscape Multiverse has been the catalyst, the arguments turn on issues central to any theory in which a multiverse plays a role. Is it scientifically justifiable to speak of a multiverse, an approach that invokes realms inaccessible not just in practice but, in many cases, even in principle? Is the notion of a multiverse testable or falsifiable? Can invoking a multiverse provide explanatory power of which we'd otherwise be deprived?

If the answer to these questions is no, as detractors insist is the case, then multiverse proponents are assuming an unusual stance. Nontestable, nonfalsifiable proposals, invoking hidden realms beyond our capacity to access—these seem a far cry from what most of us would want to call science. And therein lies the spark that makes passions flare. Proponents counter that although the manner in which a given multiverse connects with observation may be different from what we're used to—it may be more indirect; it may be less explicit; it may require fortune to shine favorably on future experiments—in respectable proposals, such connections are not fundamentally absent. Unapologetically, this line of argument takes an expansive view of what our theories and observations can reveal, and how the insights can be verified.

Where you come down on the multiverse also depends on your view of science's core mandate. General summaries often emphasize that science is about finding regularities in the workings of the universe, explaining how the regularities both illuminate and reflect underlying laws of nature, and testing the purported laws by making predictions that can be verified or refuted through further experiment and observation. Reasonable though the description may be, it glosses over the fact that the actual process of science is a much messier business, one in which asking the right questions is often as important as finding and testing the proposed answers. And the questions aren't floating in some preexisting realm in which the role of science is to pick them off, one by one. Instead, today's questions are very often shaped by yesterday's insights. Breakthroughs generally answer some questions but then give rise to a host of others that previously could not even be imagined. In judging any development, including multiverse theories, we must take account not only of its capacity for revealing hidden truths but also of its impact on the questions we are led to address. The impact, that is, on the very practice of science. As will become clear, multiverse theories have the capacity to reshape some of the deepest questions scientists have wrestled with for decades. That prospect invigorates some and infuriates others.

Having set the scene, let's now systematically think through the legitimacy, testability, and utility of frameworks that imagine ours to be one of many universes.

Accessible Multiverses

It's hard to achieve consensus on these issues partly because the multiverse concept isn't monolithic. We've already come upon five versions—Quilted, Inflationary, Brane, Cyclic, and Landscape—and in the chapters that follow we will encounter four more. Understandably, the *generic* notion of a multiverse has a reputation for lying beyond testability. After all, the typical assessment goes, we're considering universes other than our own, but since we have access only to this one, we might as well be talking about ghosts or the tooth fairy. Indeed, this is the central problem, with which we'll shortly grapple, but note first that some multiverses *do* allow for interactions between member universes. We've seen that in the Brane Multiverse untethered string loops can travel from one brane to another. And in the Inflationary Multiverse, bubble universes can find themselves in even more direct contact.

Recall that the space between two bubble universes in the Inflationary Multiverse is permeated by an inflaton field whose energy and negative pressure remain high and which therefore undergoes inflationary expansion. This expansion drives the bubble universes apart. Even so, if the rate at which the bubbles themselves expand exceeds the rate at which the swelling space propels them to separate, the bubbles will collide. Bearing in mind that inflationary expansion is cumulative—the more swelling space there is between two bubbles, the faster they're driven apart—we come to an interesting realization. If two bubbles form *really* close together, there will be so little intervening space that their rate of separation will be slower than their rate of expansion. That puts the bubbles on a collision course.

This reasoning is borne out by the mathematics. In the Inflationary Multiverse, universes can collide. Moreover, a number of research groups (including Jaume Garriga, Alan Guth, and Alexander Vilenkin; Ben Freivogel, Matthew Kleban, Alberto Nicolis, and Kris Sigurdson; as well as Anthony Aguirre and Matthew Johnson) have established that whereas some collisions may violently disrupt each bubble universe's internal structure—not good for possible bubble dwellers like us—gentler brushups may also occur, avoiding disastrous consequences yet still yielding observable signatures. The calculations show that if we had such a fender-bender with another universe, the impact would send shock waves rippling through space, generating modifications to the pattern of hot and cold regions in the microwave background radiation.[1] Researchers are now working out the detailed fingerprint such a disruption would leave, laying the groundwork for observations that could one day provide evidence that our universe has collided with others— evidence that other universes are out there.

But, however exciting the prospect may be, what if no test seeking evidence of an interaction or an encounter with another universe proves successful? Taking a hardheaded perspective, where does the concept of a multiverse stand if we never find any experimental or observational signatures of other universes?

Science and the Inaccessible I:
Can it be scientifically justifiable to invoke unobservable universes?

Every theoretical framework comes with an assumed architecture— the theory's fundamental ingredients, and the mathematical laws that govern them. Besides defining the theory, this architecture also establishes the kinds of questions we can ask within the theory. Isaac Newton's architecture was tangible. His mathematics dealt with the positions and velocities of objects we directly encounter or can easily see, from rocks and balls to the moon and sun. A great many observations confirmed Newton's predictions, giving us confidence that his mathematics did indeed describe how familiar objects move. James Clerk Maxwell's architecture introduced a significant step of abstraction. Vibrating electric and magnetic fields are not the kinds of things for which our senses have evolved a direct affinity. Although we see "light"—electromagnetic undulations whose wavelengths lie in the range our eyes can detect—our visual experiences don't directly trace the undulating fields the theory posits. Even so, we can build sophisticated equipment that measures these vibrations and that, together with the theory's abundance of confirmed predictions, builds an overwhelming case that we're immersed in a pulsating ocean of electromagnetic fields.

In the twentieth century, fundamental science came to increasingly rely on inaccessible features. Space and time, through their melded union, provide the scaffolding for special relativity. When subsequently endowed with Einsteinian malleability, they become the flexible backdrop of the general theory of relativity. Now, I've seen watches tick and I've used rulers to measure, yet I've never grasped spacetime in the same way I grasp the arms of my chair. I feel the effects of gravity, but if you pressed me on whether I can directly affirm that I'm immersed in curved spacetime, I find myself back in the Maxwellian situation. I'm convinced that the theories of special and general relativity are correct not because I have tangible access to their core ingredients but rather because when I accept their assumed frameworks, the mathematics makes predictions about things I can measure. And the predictions turn out to be extraordinarily accurate.

Quantum mechanics takes such inaccessibility still further. The central ingredient of quantum mechanics is the probability wave, governed by an equation discovered in the mid-1920s by Erwin Schrödinger. Even though such waves are its hallmark feature, we will see in Chapter 8 that the architecture of quantum physics ensures that they're permanently and completely unobservable. Probability waves give rise to predictions for where this or that particle is likely to be found, but the waves themselves slither outside the arena of everyday reality.[2] Nevertheless, because the predictions succeed so well, generations of scientists have accepted such an odd situation: a theory introduces a radically new and vital construct that, according to the theory itself, is unobservable.

The common theme running through these examples is that a theory's success can be used as an after-the-fact justification for its basic architecture, even when that architecture remains beyond our ability to access directly. This is so thoroughly part of the daily experience of theoretical physicists that the language used and the questions formulated regularly refer, without the slightest hesitation, to things that are at the very least far less accessible than tables and chairs and some of which lie permanently outside the bounds of direct experience.*

When we go further and use a theory's architecture to learn about the phenomena it entails, yet other kinds of inaccessibility present themselves. Black holes emerge from the mathematics of general relativity, and astronomical observations have provided substantial evidence that they're not only real but commonplace. Even so, the interior of a black hole is an exotic environment. According to Einstein's equations, the black hole's edge, its event horizon, is a surface of no return. You can cross in, but you can't cross out. We committed exterior dwellers will never observe a black hole's interior, not just because of practical considerations but as a consequence of the very laws of general relativity. Yet, there's full consensus that the region on the other side of a black hole's event horizon is real.

---

*Because there are differing perspectives regarding the role of scientific theory in the quest to understand nature, the points I'm making are subject to a range of interpretations. Two prominent positions are *realists*, who hold that mathematical theories can provide direct insight into the nature of reality, and *instrumentalists*, who believe that theory provides a means for predicting what our measuring devices should register but tells us nothing about an underlying reality. Over decades of exacting argument, philosophers of science have developed numerous refinements of these and related positions. As no doubt is clear, my perspective, and the approach I take in this book, is decidedly in the realist camp. This chapter in particular, examining the scientific validity of certain types of theories, and assessing what those theories might imply for the nature of reality, is one in which various philosophical orientations would approach the topic with considerable differences.

The application of general relativity to cosmology provides even more extreme instances of inaccessibility. If you don't mind a one-way journey, the interior of a black hole is at least a possible destination. But realms lying beyond our cosmic horizon are unreachable, even if we were able to travel at nearly light speed. In an accelerating universe such as ours, this point becomes forcefully evident. Given the measured value of the cosmological speedup (and assuming it will never change), any object more distant from us than about 20 billion light-years lies permanently outside what we can see, visit, measure, or influence. Farther than that distance, space will always be receding from us so quickly that any attempt to breach the separation would be as fruitless as a kayaker navigating against a current flowing faster than she can paddle.

Objects that have always been beyond our cosmic horizon are objects that we have never observed and never will observe; conversely, they have never observed us, and never will. Objects that at some time in the past were within our cosmic horizon but have been dragged beyond it by spatial expansion are objects that we once could see but never will again. Yet I think we can agree that such objects are as real as anything tangible, and so are the realms they inhabit. It would surely be peculiar to argue that a galaxy that we could once see but that has since slipped over our cosmic horizon has entered a realm that's nonexistent, a realm that because of its permanent inaccessibility needs to be wiped off reality's map. Even though we can't observe or influence such realms, nor they us, they are properly included in our picture of what exists.[3]

These examples make clear that science is no stranger to theories that include elements, from basic ingredients to derived consequences, that are inaccessible. Our confidence in such intangibles relies on our confidence in the theory. When quantum mechanics invokes probability waves, its impressive ability to describe things we can measure, such as the behavior of atoms and subatomic particles, compels us to embrace the ethereal reality it posits. When general relativity predicts the existence of places we can't observe, its phenomenal successes in describing those things we can observe, such as the motion of planets and the trajectory of light, compels us to take the predictions seriously.

So for confidence in a theory to grow we don't require that all of its features be verifiable; a robust and varied assortment of confirmed predictions is enough. Scientific work going back well over a century has accepted that a theory may invoke hidden, inaccessible elements—provided it also makes interesting, novel, and testable predictions about an abundance of observable phenomena.

This suggests that it's possible to mount a convincing argument for a theory involving a multiverse even if we can't obtain any direct evidence for universes beyond our own. If the experimental and observational evidence supporting a theory compels you to embrace it, and if the theory is founded on such a tight mathematical structure that there's no room for cherry-picking among its features, then you have to embrace all of it. And if the theory implies the existence of other universes, then that's the reality the theory requires you to take on board.

In principle, then—and make no mistake, my point here is one of principle—the mere invocation of inaccessible universes does not consign a proposal to stand outside science. To amplify this, imagine that one day we assemble a convincing experimental and observational case for string theory. Perhaps a future accelerator is able to detect sequences of string vibrational patterns and evidence for extra dimensions, while astronomical observations detect stringy features in the microwave background radiation, as well as the signatures of long stretched strings undulating through space. Suppose further that our understanding of string theory has progressed substantially, and we've learned that the theory absolutely, positively, incontrovertibly generates the Landscape Multiverse. Notwithstanding calls to the contrary, a theory with strong experimental and observational support, whose internal structure requires a multiverse, would lead us to conclude inexorably that the time for "giving in" had arrived.*

---

*In a multiverse containing an enormous number of different universes, a reasonable concern is that regardless of what experiments and observations reveal, there is some universe in the theory's gargantuan collection that's compatible with the results. If so, there'd be no experimental evidence that could prove the theory wrong; in turn, no data could be properly interpreted as evidence supporting the theory. I will consider this issue shortly.

So to address the question heading this section, in the right scientific context it would not merely be respectable to invoke a multiverse; *failing* to do so would evidence nonscientific prejudice.

Science and the Inaccessible II:
So much for principle; where do we stand in practice?

The skeptic will rightly respond that it's one thing to make a point of principle about how the case for a given multiverse theory might be fashioned. It's another to assess whether any of the multiverse proposals we've described qualify as experimentally confirmed theories that come equipped with an absolute prediction of other universes. Do they?

The Quilted Multiverse arises from an infinite spatial expanse, a possibility that fits squarely within general relativity. The snag is that general relativity allows for an infinite spatial expanse but doesn't *require* it, which in turn explains why, even though general relativity is an accepted framework, the Quilted Multiverse remains tentative. An infinite spatial expanse does emerge directly from eternal inflation—recall that each bubble universe when viewed from the inside appears infinitely large—but in this setting the Quilted Multiverse is rendered uncertain because the underlying proposal, eternal inflation, remains hypothetical.

The same consideration affects the Inflationary Multiverse, which also emerges from eternal inflation. Astronomical observations over the past decade have bolstered the physics community's confidence in inflationary cosmology but have nothing to say about whether the inflationary expansion is eternal. Theoretical studies show that although many versions are eternal, yielding bubble universe upon bubble universe, some entail but a single ballooning spatial expanse.

The Brane, Cyclic, and Landscape Multiverses are based on string theory, so they suffer multiple uncertainties. Remarkable as string theory may be, rich as its mathematical structure may have become, the dearth of testable predictions, and the concomitant absence of contact with observations or experiments, relegates it to the realm of scientific speculation. Moreover, with the theory still very much a work in progress, it's unclear which features will continue to play a primary role in future refinements. Will branes, the basis of the Brane and Cyclic Multiverses, remain central? Will the copious choices for the extra dimensions, the basis for the Landscape Multiverse, persist, or will we eventually find a mathematical principle that picks out one particular shape? We just don't know.

So, although it's conceivable that we could fashion a convincing argument for a multiverse theory that made little or no reference to its prediction of other universes, for the multiverse scenarios we've encountered that approach won't fly. At least not yet. To assess any of them, we will need to tackle their prediction of a multiverse head-on.

Can we? Can a theory's invocation of other universes yield testable predictions even if those universes lie beyond the reach of experiments and observations? Let's address this key question through a number of steps. We'll follow the pattern above, progressing from an "in principle" to an "in practice" perspective.

Predictions in a Multiverse I:
If the universes constituting a multiverse are inaccessible, can they nevertheless meaningfully contribute to making predictions?

Some scientists who resist multiverse theories see the enterprise as an admission of failure, a full-fledged retreat from the long-sought goal of understanding why the universe we see has the properties it does. I empathize, being one of many who have worked for decades to realize string theory's tantalizing promise of calculating every fundamental observable feature of the universe, including the values of nature's constants. If we accept that we're part of a multiverse in which some or perhaps even all of the constants vary from one universe to another, then we accept that this goal is misguided. If the fundamental laws allow, say, the strength of the electromagnetic force to have many different values across the multiverse, then the very notion of calculating *the* strength is meaningless, like asking a pianist to pick out *the* note.

But here's the question: Does variation in features mean that we lose all power to predict (or postdict) those intrinsic to our own universe? Not necessarily. Even though a multiverse precludes uniqueness, it's possible that a degree of predictive capability can be retained. It comes down to statistics.

Consider dogs. They don't have a unique weight. There are very light dogs, such as Chihuahuas, that can weigh under two pounds; there are very heavy dogs, such as Old English mastiffs, that can tip the scales at over two hundred pounds. Were I to challenge you to predict the weight of the next dog you pass in the street, it might seem that the best you could do would be to pick a random number within the range I've given. Yet, with a little information, you can make a more refined guess. If you get ahold of the dog population data in your neighborhood, such as the number of people who have this or that breed, the distribution of weights within each breed, and perhaps even information on the number of times per day different breeds typically need to be taken for a walk, you can figure out the weight of the dog you are most *likely* to encounter.

This wouldn't be a sharp prediction; statistical insights often aren't. But depending on the distribution of dogs, you may be able to do much better than just pulling a number out of a hat. If your neighborhood has a highly skewed distribution, with 80 percent of the dogs being Labrador retrievers whose average weight is sixty pounds, and the other 20 percent composed of a range of breeds from Scottish terriers to poodles whose average weight is thirty pounds, then something in the fifty-five- to sixty-five-pound range would be a good bet. The dog you next encounter may be a fluffy shih tzu, but odds are it won't be. For distributions that are even more skewed, your predictions can be more precise. If 95 percent of the dogs in your area were sixty-two-pound Labrador retrievers, then you'd be on firmer ground in predicting that the next dog you pass will be one of these.

A similar statistical approach can be applied to a multiverse. Imagine we are investigating a multiverse theory that allows for a wide range of different universes—different values of force strengths, particle properties, cosmological constant values, and so on. Imagine further that the cosmological process by which these universes form (such as the creation of bubble universes in the Landscape Multiverse) is sufficiently well understood that we can determine the distribution of universes, with various properties, across the multiverse. This information has the capacity to yield significant insights.

To illustrate the possibilities, suppose our calculations yield a particularly simple distribution: some physical features vary widely from universe to universe, but others are unchanging. For example, imagine the math reveals that there's a collection of particles, common to all the universes in the multiverse, whose masses and charges have the same values in each universe. A distribution like this generates absolutely firm predictions. If experiments undertaken in our single lone universe don't find the predicted collection of particles, we'd rule out the theory, multiverse and all. Knowledge of the distribution thus makes this multiverse proposal falsifiable. Conversely, if our experiments were to find the predicted particles, that would increase our confidence that the theory is right.[4]

For another example, imagine a multiverse in which the cosmological constant varies across a huge range of values, but it does so in a highly nonuniform manner, as illustrated schematically in Figure 7.1. The graph denotes the fraction of universes within the multiverse (vertical axis) that have a given value of the cosmological constant (horizontal axis). If we were part of such a multiverse, the mystery of the cosmological constant would take on a decidedly different character. Most universes in this scenario have a cosmological constant close to what we've measured in our universe, so while the range of *possible* values would be huge, the skewed distribution implies that the value we've observed is nothing special. For such a multiverse, you should be no more mystified by our universe's having a cosmological constant value 10^{-123} than you should be surprised by encountering a sixty-two-pound Labrador retriever during your next stroll around the neighborhood. Given the relevant distributions, each is the most likely thing that could happen.

Here's a variation on the theme. Imagine that, in a given multiverse proposal, the cosmological constant's value varies widely, but unlike in the previous example, it varies uniformly; the number of

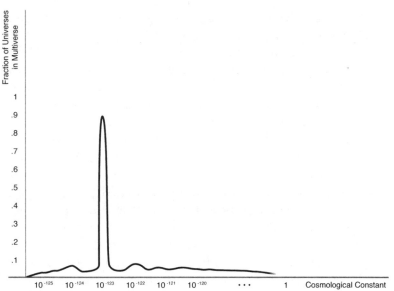

Figure 7.1 *A possible distribution of cosmological constant values across a hypothetical multiverse, illustrating that highly skewed distributions can make otherwise puzzling observations understandable.*

universes that have a given value of the cosmological constant is on a par with the number of universes that have any other value of the cosmological constant. But imagine further that a close mathematical study of the proposed multiverse theory reveals an unexpected feature in the distribution. For those universes in which the cosmological constant is in the range we've observed, the math shows there's always a species of particle whose mass is, say, five thousand times that of the proton—too heavy to have been observed in accelerators built in the twentieth century, but right within the range of those built in the twenty-first. Because of the tight correlation between these two physical features, this multiverse theory is also falsifiable. If we fail to find the predicted heavy species of particle we would disprove this proposed multiverse; discovery of the particle would strengthen our confidence that the proposal is correct.

Let me underscore that these scenarios are hypothetical. I invoke them because they illuminate a possible profile for scientific insight and verification in the context of a multiverse. I suggested earlier that if a multiverse theory gives rise to testable features beyond the prediction of other universes, it's possible—in principle—to assemble a supporting case even if the other universes are inaccessible. The examples just given make this suggestion explicit. For these kinds of multiverse proposals, the answer to the question heading this section would unequivocally be yes.

The essential feature of such "predictive multiverses" is that they're not composed from a grab-bag of constituent universes. Instead, the capacity to make predictions emerges from the multiverse evincing an underlying mathematical pattern: physical properties are distributed across the constituent universes in a sharply skewed or highly correlated manner.

How might this happen? And, leaving the realm of "in principle," *does* it happen in the multiverse theories we've encountered?

Predictions in a Multiverse II:
So much for principle; where do we stand in practice?

The distribution of dogs in a given area depends on a range of influences, among them cultural and financial factors and plain old happenstance. Because of this complexity, if you were intent on making statistical predictions your best bet would be to bypass considerations of how a given dog distribution came to be and simply use the relevant data from the local dog licensing authority. Unfortunately, multiverse scenarios don't have comparable census bureaus, so the analogous option isn't available. We're forced to rely on our theoretical understanding of how a given multiverse might arise to determine the distribution of the universes it would contain.

The Landscape Multiverse, relying on eternal inflation and string theory, provides a good case study. In this scenario, the twin engines driving the production of new universes are inflationary expansion and quantum tunneling. Remember how this goes: An inflating universe, corresponding to one or another valley in the string landscape, quantum-tunnels through one of the surrounding mountains and settles down in another valley. The first universe—with definite features such as force strengths, particle properties, value of the cosmological constant, and so forth—acquires an expanding bubble of the new universe (see Figure 6.7), with a new set of physical features, and the process continues.

Now, being a quantum process, such tunneling events have a probabilistic character. You can't predict when or where they will happen. But you can predict the *probability* that a tunneling event will happen in any given interval of time and burrow in any given direction—probabilities that depend on detailed features of the string landscape, such as the altitude of the various mountain peaks and valleys (the value, that is, of their respective cosmological constants). The more probable tunneling events will happen more often, and the resulting distribution of universes will reflect this. The strategy, then, is to use the mathematics of inflationary cosmology and string theory to calculate the distribution of universes, with various physical features, across the Landscape Multiverse.

The rub is that so far no one has been able to do so. Our current understanding suggests a lush string landscape with a gargantuan number of mountains and valleys, which makes it a ferociously difficult mathematical challenge to work out the details of the resulting multiverse. Pioneering work by cosmologists and string theorists have contributed significantly to our understanding, but the investigations are still rudimentary.[5]

To go further, multiverse proponents advocate introducing one more important element into the mix. Consideration of the selection effects introduced in the previous chapter: anthropic reasoning.

Predictions in a Multiverse III:
Anthropic reasoning

Many of the universes in a given multiverse are bound to be lifeless. The reason, as we've seen, is that changes to nature's fundamental parameters from their known values tend to disrupt the conditions favorable for life to emerge.[6] Our very existence implies that we could never find ourselves in any of the lifeless domains, and so there's nothing further to explain about why we don't see their particular combination of properties. If a given multiverse proposal implied a unique life-supporting universe, we'd be golden. We would work out that special universe's properties mathematically; if they differed from what we've measured in our own universe, we could rule out that multiverse proposal. If the properties agreed with ours, we'd have an impressive vindication of anthropic multiverse theorizing—and reason to vastly expand our picture of reality.

In the more plausible case that there is not a unique life-supporting universe, a number of theorists (they include Steven Weinberg, Andrei Linde, Alex Vilenkin, George Efstathiou, and many others) have advocated an enhanced statistical approach. Rather than calculate the relative preponderance, within the multiverse, of various kinds of universes, they propose that we calculate the number of inhabitants—physicists usually call them observers—who would find themselves in various kinds of universes. In some universes, conditions might barely be compatible with life, so observers would be rare, like the occasional cactus in a harsh desert; other universes, with more hospitable conditions, would teem with observers. The idea is that, just as canine census data let us predict what kinds of dogs we can expect to encounter, so observer census data let us predict the properties that a typical inhabitant living somewhere in the multiverse—you and I, according to the reasoning of this approach—should expect to see.

A concrete example was worked out in 1997 by Weinberg and his collaborators Hugo Martel and Paul Shapiro. For a multiverse in which the cosmological constant varies from universe to universe, they calculated how abundant life would be in each. This difficult task was made feasible by invoking the Weinberg proxy (Chapter 6): instead of life proper, they considered the formation of galaxies. More galaxies means more planetary systems and hence, the underlying assumption goes, a greater likelihood of life, intelligent life in particular. Now, as Weinberg had found in 1987, even a modest cosmological constant generates enough repulsive gravity to disrupt galaxy formation so only domains of the multiverse that have sufficiently small cosmological constants need be considered.

A cosmological constant that's negative results in a universe that collapses well before galaxies form, so these realms of the multiverse can be omitted from the analysis, too. Anthropic reasoning thus focuses our attention on the portion of the multiverse in which the cosmological constant lies in a narrow window; as discussed in Chapter 6, the calculations show that for a given universe to contain galaxies, its cosmological constant needs to be less than about 200 times the critical density (a mass equivalent of about 10^{-27} grams in each cubic centimeter of space, or about 10^{-121} in Planck units).[7]

For universes whose cosmological constant is in this range, Weinberg, Martel, and Shapiro then undertook a more refined calculation. They determined the fraction of matter in each such universe that would clump together over the course of cosmological evolution, a pivotal step on the road to galaxy formation. They found that if the cosmological constant is very near the window's upper limit, relatively few clumps would form, because the outward push of the cosmological constant acts like a strong wind, blowing most dust accumulations apart. If the cosmological constant's value is near the window's lower limit, zero, they found that many clumps form, because the disrupting influence of the cosmological constant is minimized. Which means there's a large chance you'll be in a universe whose cosmological constant is near zero, since such universes have an abundance of galaxies and, by the reasoning of this approach, life. There's a small chance you'll be in a universe whose cosmological constant is near the window's upper limit, about 10^{-121}, because such universes are endowed with far fewer galaxies. And there's a modest chance you'll be in a universe whose cosmological constant lies at a value between these extremes.

Using the quantitative version of these results, Weinberg and his collaborators calculated the cosmic analog of encountering a sixty-two-pound Labrador on an average walk around the neighborhood—the cosmological constant value, that is, witnessed by an average observer in the multiverse. The answer? Somewhat larger than what the subsequent supernova measurements revealed, but definitely in the same ballpark. They found that roughly 1 in 10 to 1 in 20 inhabitants of the multiverse would have an experience comparable to ours, measuring the cosmological constant's value in their universe to be about 10^{-123}. While a higher percentage would be more satisfying, the result is impressive, nonetheless. Prior to this calculation, physics faced a mismatch between theory and observation of more than 120 orders of magnitude, suggesting strongly that something was profoundly amiss with our understanding. The multiverse approach of Weinberg and his collaborators, however, showed that finding yourself in a universe whose cosmological constant is on a par with the value we've measured is roughly as surprising as running into that shih tzu in a neighborhood dominated by Labs. Which is to say, not that surprising at all. Certainly, when viewed from this multiverse perspective, the observed value of the cosmological constant doesn't suggest a profound lack of understanding, and that's an encouraging step forward.

Subsequent analyses, though, emphasized an interesting facet that some interpret as weakening the result. For simplicity's sake, Weinberg and his collaborators imagined that across their multiverse only the cosmological constant's value varied from universe to universe; other physical parameters were assumed fixed. Max Tegmark and Martin Rees noted that if both the cosmological constant's value and, say, the size of the early universe quantum jitters were imagined to vary from universe to universe, the conclusion would change. Recall that the jitters are the primordial seeds of galaxy formation: tiny quantum fluctuations, stretched by inflationary expansion, yield a random assortment of regions where the density of matter is a little higher or a little lower than average. The higher-density regions exert a greater gravitational pull on nearby matter and so grow yet larger, ultimately coalescing into galaxies. Tegmark and Rees pointed out that much as bigger piles of leaves can better withstand a brisk breeze, so larger primordial seeds can better withstand the disruptive outward push of a cosmological constant. A multiverse in which both the seed size and the value of the cosmological constant vary would therefore contain universes where larger cosmological constants were offset by larger seeds; that combination would be compatible with galaxy formation—

and hence with life. A multiverse of this sort increases the cosmological constant value that a typical observer would see and so results in a decrease—potentially a sharp one—of the fraction of observers who would find their cosmological constant to have as small a value as we've measured.

Staunch multiverse proponents are fond of pointing to the analysis of Weinberg and his collaborators as a success of anthropic reasoning. Detractors are fond of pointing to the issues raised by Tegmark and Rees as making the anthropic result less convincing. In reality, the debate is premature. These are all highly exploratory, first-pass calculations, best viewed as providing insight into the general domain of anthropic reasoning. Under certain restrictive assumptions, they show that the anthropic framework can take us within the ballpark of the measured cosmological constant; relax those assumptions somewhat, and the calculations show that the size of the ballpark grows substantially. Such sensitivity implies that a refined multiverse calculation will require a precise understanding of the detailed properties that characterize the constituent universes, and how they vary, thus replacing arbitrary assumptions with theoretical directives. This is essential if a multiverse is to stand a chance of yielding definitive conclusions.

Researchers are working hard to achieve this goal, but as of today, they have yet to reach it.[8]

Prediction in a Multiverse IV:
What will it take?

What hurdles, then, will we need to clear before we can extract predictions from a given multiverse? There are three that figure most prominently.

First, as pointedly illustrated by the example just discussed, a multiverse proposal must allow us to determine which physical features vary from universe to universe, and for those features that do vary, we must be able to calculate their statistical distribution across the multiverse. Essential for doing so is an understanding of the cosmological mechanism by which the proposed multiverse is populated by universes (such as the creation of bubble universes in the Landscape Multiverse). It is this mechanism that determines how prevalent one kind of universe is relative to another, and so it is this mechanism that determines the statistical distribution of physical features. If we're fortunate, the resulting distributions, either across the entire multiverse or across those universes supporting life, will be sufficiently skewed to yield definitive predictions.

A second challenge, if we do need to invoke anthropic reasoning, comes from the central assumption that we humans are garden-variety average. Life might be rare in the multiverse; intelligent life might be rarer still. But among all intelligent beings, the anthropic assumption goes, we are so thoroughly typical that our observations should be the average of what intelligent beings inhabiting the multiverse would see. (Alexander Vilenkin has called this the *principle of mediocrity*.) If we know the distribution of physical features across life-supporting universes, we can calculate such averages. But typicality is a thorny assumption. If future work shows that our observations fall into the range of calculated averages in a particular multiverse, confidence in our typicality—and in the multiverse proposal—would grow. That would be exciting. But if our observations fall outside the averages that could be evidence that the multiverse proposal is wrong, or it could mean that we are just not typical. Even in a neighborhood that has 99 percent Labs, you can still run into Dobermans, an atypical dog. Distinguishing between a failed multiverse proposal and a successful one in which our universe is atypical may prove difficult.[9]

Progress on this issue will likely require a better understanding of how intelligent life arises in a given multiverse; with that knowledge, we could at least clarify how typical our own evolutionary history has so far been. This, of course, is a major challenge. To date, most anthropic reasoning has completely skirted the issue by invoking Weinberg's assumption—that the number of intelligent life-forms in a given universe is proportional to the number of galaxies it contains. As far as we know, intelligent life needs a warm planet, which requires a star, which is generally part of a galaxy, and so there's reason to believe Weinberg's approach holds water. But since we have only the most rudimentary understanding of even our own genesis, the assumption remains tentative. To refine our calculations, the development of intelligent life needs to be far better understood.

The third hurdle is simple to explain but in the long run may well be the one that's last standing. It has to do with dividing up infinity.

Dividing Up Infinity

To understand the problem, return to dogs. If you live in a neighborhood populated with three Labs and one dachshund, then, ignoring complications such as how often the dogs are walked, you're three times more likely to run into a Lab. The same would apply if there were 300 Labs and 100 dachshunds; 3,000 Labs and 1,000 dachshunds; 3 million Labs and 1 million dachshunds, and so on. But what if these numbers were *infinitely* large? How do you compare an infinity of dachshunds to three times infinity of Labradors? Although this sounds like the tortured math of one-upping seven-year-olds, there's a real question here. Is three times infinity larger than plain old infinity? If so, is it three times as large?

Comparisons involving infinitely large numbers are notoriously tricky. For dogs on earth, of course, the difficulty doesn't arise, because the populations are finite. But for universes constituting particular multiverses, the problem can be very real. Take the Inflationary Multiverse. Looking at the entire block of Swiss cheese from an imaginary outsider's perspective, we would see it continue to grow and produce new universes endlessly. That's what the "eternal" in "eternal inflation" means. Moreover, taking an insider's perspective, we've seen that each bubble universe itself harbors an infinite number of separate domains, filling out a Quilted Multiverse. In making predictions we necessarily confront an infinity of universes.

To grasp the mathematical challenge, imagine that you're a contestant on *Let's Make a Deal* and you've won an unusual prize: an infinite collection of envelopes, the first containing $1, the second $2, the third $3, and so on. As the crowd cheers, Monty chimes in to make you an offer. Either keep your prize as is, or elect to have him double the contents of each envelope. At first it seems obvious that you should take the deal. "Each envelope will contain more money than it previously did," you think, "so this has to be the right move." And if you had only a finite number of envelopes, it *would* be the right move. To exchange five envelopes containing $1, $2, $3, $4, and $5 for envelopes with $2, $4, $6, $8, and $10 makes unassailable sense. But after another moment's thought, you start to waver, because you realize that the infinite case is less clear-cut. "If I take the deal," you think, "I'll wind up with envelopes containing $2, $4, $6, and so on, running through all the even numbers. But as things currently stand, my envelopes run through *all* whole numbers, the evens as well as the odds. So it seems that by taking the deal I'll be *removing* the odd dollar amounts from my total tally. That doesn't sound like a smart thing to do." Your head starts to spin. Compared envelope by envelope, the deal looks good. Compared collection to collection, the deal looks bad.

Your dilemma illustrates the kind of mathematical pitfall that makes it so hard to compare infinite collections. The crowd is growing antsy, you have to make a decision, but your assessment of the deal depends on the way you compare the two outcomes.

A similar ambiguity afflicts comparisons of a yet more basic characteristic of such collections: the number of members each contains. The *Let's Make a Deal* example illustrates this, too. Which are more plentiful, whole numbers or even numbers? Most people would say whole numbers, since only half of the whole numbers are even. But your experience with Monty gives you sharper insight. Imagine that you take Monty's deal and wind up with all even dollar amounts. In doing so, you wouldn't return any envelopes nor would you require any new ones, since Monty would simply double the amount of money in each. You conclude, therefore, that the number of envelopes required to accommodate all whole numbers is the same as the number of envelopes required to accommodate all even numbers—which suggests that the populations of each category are equal (Table 7.1). And that's weird. By one method of comparison—considering the even numbers as a subset of the whole numbers—you conclude that there are more whole numbers. By a different method of comparison—considering how many envelopes are needed to contain the members of each group—you conclude that the set of whole numbers and the set of even numbers have equal populations.

Whole Numbers	1	2	3	4	5	6	7	8	9	10	. . .
	↕	↕	↕	↕	↕	↕	↕	↕	↕	↕	
Even Numbers	2	4	6	8	10	12	14	16	18	20	. . .

Table 7.1 *Every whole number is paired with an even number, and vice versa, suggesting that the quantity of each is the same.*

You can even convince yourself that there are *more* even numbers than there are whole numbers. Imagine that Monty offered to quadruple the money in each of the envelopes you initially had, so there would be $4 in the first, $8 in the second, $12 in the third, and so on. Since, again, the number of envelopes involved in the deal stays the same, this suggests that the quantity of whole numbers, where the deal began, is equal to that of numbers divisible by four (Table 7.2), where the deal wound up. But such a pairing, marrying off each whole number to a number that's divisible by 4, leaves an infinite set of even bachelors—the numbers 2, 6, 10, and so on—and thus seems to imply that the evens are more plentiful than the wholes.

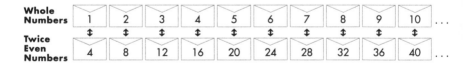

Whole Numbers	1	2	3	4	5	6	7	8	9	10	. . .
	↕	↕	↕	↕	↕	↕	↕	↕	↕	↕	
Twice Even Numbers	4	8	12	16	20	24	28	32	36	40	. . .

Table 7.2 *Every whole number is paired with every other even number, leaving an infinite set of even bachelors, suggesting that there are more evens than wholes.*

From one perspective, the population of even numbers is less than that of whole numbers. From another, the populations are equal. From another still, the population of even numbers is greater than that of the whole numbers. And it's not that one conclusion is right and the others wrong. There simply is no absolute answer to the question of which of these kinds of infinite collections are larger. The result you find depends on the manner in which you do the comparison.[10]

That raises a puzzle for multiverse theories. How do we determine whether galaxies and life are more abundant in one or another type of universe when the number of universes involved is infinite? The very same ambiguity we've just encountered will afflict us just as severely, *unless physics picks out a precise basis on which to make the comparisons.* Theorists have put forward proposals, various analogs of the pairings given in the tables, that emerge from one or another physical consideration—but a definitive procedure has yet to be derived and agreed upon. And, just as in the case of infinite collections of numbers, different approaches yield different results. According to one way of comparing, universes with one array of properties preponderate; according to an alternative way, universes with different properties do.

The ambiguity has a dramatic impact on what we conclude are typical or average properties in a given multiverse. Physicists call this the *measure problem,* a mathematical term whose meaning is well suggested by its name. We need a means for measuring the sizes of different infinite collections of universes. It is this information that we need in order to make predictions. It is this information that we need in order to work out how likely it is that we reside in one type of universe rather than another. Until we find a fundamental dictum for how we should compare infinite collections of universes, we won't be able to foretell mathematically what typical multiverse dwellers—us—should see in experiments and observations. Solving the measure problem is imperative.

A Further Contrarian Concern

I've called out the measure problem in its own section not only because it is a formidable impediment to prediction, but also because it may entail another, disquieting consequence. In Chapter 3, I explained why the inflationary theory has become the de facto cosmological paradigm. A brief burst of rapid expansion during our universe's first moments would have allowed today's distant regions to have communicated early on, which explains the common temperature that measurements have found; rapid expansion also irons out any spatial curvature, rendering the shape of space flat, in line with observations; and finally, such expansion turns quantum jitters into tiny temperature variations across space that are both measurable in the microwave background radiation and essential to galaxy formation. These successes yield a strong case.[11] But the eternal version of inflation has the capacity to undermine the conclusion.

Whenever quantum processes are relevant, the best you can do is predict the likelihood of one outcome relative to another. Experimental physicists, taking this to heart, perform experiments over and over again, acquiring reams of data on which statistical analyses can be run. If quantum mechanics predicts that one outcome is ten times as likely as another, then the data should very nearly reflect this ratio. The cosmic microwave background calculations, whose match to observations is the most convincing evidence for the inflationary theory, rely on quantum field jitters, so they are also probabilistic. But, unlike laboratory experiments, they can't be checked by running the big bang over and over again. So how are they interpreted?

Well, if theoretical considerations conclude, say, that there's a 99 percent probability that the microwave data should take one form and not another, and if the more probable outcome is what we observers see, the data are taken as strongly supporting the theory. The rationale is that if a collection of universes were all produced by this same underlying physics, the theory predicts that about 99 percent of them should look much like what we observe and about 1 percent to deviate significantly.

Now, if the Inflationary Multiverse had a finite population of universes, we could straightforwardly conclude that the number of oddball universes where quantum processes result in data not matching expectations remains, comparatively speaking, very small.

But if, as in the Inflationary Multiverse, the population of universes is not finite, it is far more challenging to interpret the numbers. What's 99 percent of infinity? Infinity. What's 1 percent of infinity? Infinity. Which is bigger? The answer requires us to compare two infinite collections. And as we've seen, even when it seems plain that one infinite collection is larger than another, the conclusion you reach depends on your method of comparison.

The contrarian concludes that when inflation is eternal, *the very predictions that we use to build our confidence in the theory are compromised.* Every possible outcome allowed by the quantum calculations, however unlikely—a .1 percent quantum probability, a .0001 percent quantum probability, a .0000000001 percent quantum probability—would be realized in infinitely many universes simply because any of these numbers times infinity equals infinity. Without a fundamental prescription for comparing infinite collections, we can't possibly say that one collection of universes is larger than the rest and is thus the most likely kind of universe for us to witness, we lose the capacity to make definite predictions.

The optimist concludes that the spectacular agreement between quantum calculations in inflationary cosmology and data, as in Figure 3.5, must reflect a deep truth. With a finite number of universes and observers, the deep truth is that universes in which the data deviate from quantum predictions—those with a .1 percent quantum probability, or a .0001 percent quantum probability, or a .0000000001 percent quantum probability—are indeed rare, and that's why garden-variety multiverse inhabitants like us don't find ourselves living inside one of them. With an infinite number of universes, the optimist concludes, the deep truth must be that the rarity of anomalous universes, in some yet to be established manner, still holds. The expectation is that we will one day derive a measure, a definite means for comparing the various infinite collections of universes, and that those universes emerging from rare quantum aberrations will have a tiny measure compared with those emerging from the likely quantum outcomes. To accomplish this remains an immense challenge, but the majority of researchers in the field are convinced that the agreement in Figure 3.5 means that we will one day succeed.[12]

Mysteries and Multiverses:
Can a multiverse provide explanatory power of which we'd otherwise be deprived?

No doubt you've noticed that even the most sanguine projections suggest that predictions emerging from a multiverse framework will have a different character from those we traditionally expect from physics. The precession of the perihelion of Mercury, the magnetic dipole moment of the electron, the energy released when a nucleus of uranium splits into barium and krypton: *these* are predictions. They result from detailed mathematical calculations based on solid physical theory and produce precise, testable numbers. And the numbers have been verified experimentally. For example, calculations establish that the electron's magnetic moment is 2.0023193043628; measurements reveal it to be 2.0023193043622. Within the tiny margins of error inherent to each, experiment thus confirms theory to better than 1 part in 10 billion.

From where we now stand, it seems that multiverse predictions will never reach this standard of precision. In the most refined scenarios, we might be able to predict that it's "highly likely" that the cosmological constant, or the strength of the electromagnetic force, or the mass of the up-quark lies within some range of values. But to do better, we'll need extraordinarily good fortune. In addition to solving the measure problem, we'll need to discover a convincing multiverse theory with profoundly skewed probabilities (such as a 99.9999 percent probability that an observer will find himself in a universe with a cosmological constant equal to the value we measure) or astonishingly tight correlations (such as that electrons exist only in universes with a cosmological constant equal to 10^{-123}). If a multiverse proposal doesn't have such favorable features, it will lack the precision that for so long has distinguished physics from other disciplines. To some researchers, that's an unacceptable price to pay.

For quite a while, I took that position too, but my view has gradually shifted. Like every other physicist, I prefer sharp, precise, and unequivocal predictions. But I and many others have come to realize that although some fundamental features of the universe are

suited for such precise mathematical predictions, others are not—
or, at the very least, it's logically possible that there *may* be features
that stand beyond precise prediction. From the mid-1980s, when I
was a young graduate student working on string theory, there was
broad expectation that the theory would one day explain the values of particle masses, force strengths, the number of spatial
dimensions, and just about every other fundamental physical
feature. I remain hopeful that this is a goal we will one day reach.
But I also recognize that it is a tall order for a theory's equations
to churn away and produce a number like the electron's mass
(.00000000000000000000000091095 in units of the Planck mass)
or the top quark's mass (.0000000000000000632, in units of the
Planck mass). And when it comes to the cosmological constant,
the challenge appears herculean. A calculation that after pages of
manipulations and megawatts of computer-crunching results in the
very number that highlights the first paragraph of Chapter 6—well,
it's not impossible but it does strain even the optimist's optimism.
Certainly, string theory seems no closer to calculating any of these
numbers today than it did when I first started working on it. This
doesn't mean that it, or some future theory, won't one day succeed.
Maybe the optimist needs to be yet more imaginative. But given the
physics of today, it makes sense to consider new approaches. That's
what the multiverse does.

In a well-developed multiverse proposal, there's a clear delin-
eation of the physical features that need to be approached differently from standard practice: those that vary from universe to
universe. And that's the power of the approach. What you can
absolutely count on from a multiverse theory is a sharp vetting of
which single-universe mysteries persist in the many-universe setting, and which do not.

The cosmological constant is a prime example. If the cosmological constant's value varies across a given multiverse, and does so
in sufficiently fine increments, what was once mysterious—its
value—would now be prosaic. Just as a well-stocked shoe store
surely has your shoe size, an expansive multiverse surely has universes with the value of the cosmological constant we've measured.
What generations of scientists might have struggled valiantly to

explain, the multiverse would have explained away. The multiverse would have shown that a seemingly deep and perplexing issue emerged from the misguided assumption that the cosmological constant has a unique value. It is in this sense that a multiverse theory has the capacity to offer significant explanatory power, and it has the potential to profoundly influence the course of scientific inquiry.

Such reasoning must be wielded with care. What if Newton, after the apple fell, reasoned that we're part of a multiverse in which apples fall down in some universes, up in others, and so the falling apple simply tells us which kind of universe we inhabit, with no need for further investigation? Or, what if he'd concluded that in each universe some apples fall down while others fall up, and the reason we see the falling-down variety is simply the environmental fact that, in our universe, apples that fall up have already done so and have thus long since departed for deep space? This is a fatuous example, of course—there's never been any reason, theoretical or otherwise, for such thinking—but the point is serious. By invoking a multiverse, science could weaken the impetus to clarify particular mysteries, even though some of those mysteries might be ripe for standard, nonmultiverse explanations. When all that was really called for was harder work and deeper thinking, we might instead fail to resist the lure of multiverse temptation and prematurely abandon conventional approaches.

This potential danger explains why some scientists shudder at multiverse reasoning. It's why a multiverse proposal that's taken seriously needs to be strongly motivated from theoretical results, and it must articulate with precision the universes of which it's composed. We must tread carefully and systematically. But to turn away from a multiverse because it *could* lead us down a blind alley is equally dangerous. In doing so, we might well be turning a blind eye to reality.

CHAPTER 8

The Many Worlds of
Quantum Measurement

The Quantum Multiverse

The most reasonable assessment of the parallel universe theories we've so far encountered is that the jury is out. An infinite spatial expanse, eternal inflation, braneworlds, cyclical cosmology, string theory's landscape—these intriguing ideas have emerged from a range of scientific developments. But each remains tentative, as do the multiverse proposals each has spawned. While many physicists are willing to offer their opinions, pro and con, regarding these multiverse schemes, most recognize that future insights— theoretical, experimental, and observational—will determine whether any become part of the scientific canon.

The multiverse we'll now take up, emerging from quantum mechanics, is viewed very differently. Many physicists have already reached a final verdict on this particular multiverse. The thing is, they haven't all reached the same verdict. The differences come down to the deep and as yet unresolved problem of navigating from the probabilistic framework of quantum mechanics to the definite reality of common experience.

Quantum Reality

In 1954, nearly thirty years after the foundations of quantum theory had been set down by luminaries like Niels Bohr, Werner Heisenberg, and Erwin Schrödinger, an unknown graduate student from

Princeton University named Hugh Everett III came to a startling realization. His analysis, which focused on a gaping hole that Bohr, the grand master of quantum mechanics, had danced around but failed to fill, revealed that a proper understanding of the theory might require a vast network of parallel universes. Everett's was one of the earliest mathematically motivated insights suggesting that we might be part of a multiverse.

Everett's approach, which in time would be called the Many Worlds interpretation of quantum mechanics, has had a checkered history. In January 1956, having worked out the mathematical consequences of his new proposal, Everett submitted a draft of his thesis to John Wheeler, his doctoral adviser. Wheeler, one of twentieth-century physics' most celebrated thinkers, was thoroughly impressed. But that May, when Wheeler visited Bohr in Copenhagen and discussed Everett's ideas, the reception was icy. Bohr and his followers had spent decades refining their view of quantum mechanics. To them, the questions Everett raised, and the outlandish ways in which he thought they should be addressed, were of little merit.

Wheeler held Bohr in the highest regard, and so placed particular value on appeasing his elder colleague. In response to the criticisms, Wheeler delayed granting Everett his Ph.D. and compelled him to modify the thesis substantially. Everett was to cut out those parts blatantly critical of Bohr's methodology and emphasize that his results were meant to clarify and extend the conventional formulation of quantum theory. Everett resisted, but he had already accepted a job in the Defense Department (where he would soon play an important behind-the-scenes role in the Eisenhower and Kennedy administrations' nuclear-weapons policy) that required a doctorate, so he reluctantly acquiesced. In March of 1957, Everett submitted a substantially trimmed-down version of his original thesis; by April it was accepted by Princeton as fulfilling his remaining requirements, and in July it was published in the *Reviews of Modern Physics*.[1] But with Everett's approach to quantum theory having already been dismissed by Bohr and his entourage, and with the muting of the grander vision articulated in the original thesis, the paper was ignored.[2]

Ten years later, the renowned physicist Bryce DeWitt plucked Everett's work from obscurity. DeWitt, who was inspired by the results of his graduate student Neill Graham that further developed Everett's mathematics, became a vocal proponent of the Everettian rethinking of quantum theory. Besides publishing a number of technical papers that brought Everett's insights to a small but influential community of specialists, in 1970 DeWitt wrote a general level summary for *Physics Today* that reached a much broader scientific audience. And unlike Everett's 1957 paper, which shied away from talk of other worlds, DeWitt underscored this feature, highlighting it with an unusually candid reflection regarding his "shock" on learning Everett's conclusion that we are part of an enormous "multiworld." The article generated a significant response in a physics community that had become more receptive to tampering with orthodox quantum ideology and ignited a debate, still going on, that concerns the nature of reality when, as we believe they do, quantum laws hold sway.

Let me set the stage.

The upheaval in understanding that took place between roughly 1900 and 1930 resulted in a ferocious assault on intuition, common sense, and the well-accepted laws that the new vanguard soon began calling "classical physics"—a term that carries the weight and respect given to a picture of reality that is at once venerable, immediate, satisfying, and predictive. Tell me how things are now, and I'll use the laws of classical physics to predict how things will be at any moment in the future, or how they were at any moment in the past. Subtleties such as chaos (in the technical sense: slight changes in how things are now can result in huge errors in the predictions) and the complexity of the equations challenge the practicality of this program in all but the simplest situations, but the laws themselves are unwavering in their viselike grip on a definitive past and future.

The quantum revolution required that we give up the classical perspective because new results established that it was demonstrably wrong. For the motion of big objects like the earth and the moon, or of everyday objects like rocks and balls, the classical laws do a fine job of prediction and description. But pass into the microworld of molecules, atoms, and subatomic particles and the classical laws fail. In contradiction of the very heart of classical reasoning, if you run identical experiments on identical particles that have been set up identically, you will generally *not* get identical results.

Imagine, for example, that you have 100 identical boxes, each containing one electron, set up according to an identical laboratory procedure. After exactly 10 minutes, you and 99 cohorts measure the positions of each of the 100 electrons. Despite what Newton, Maxwell, or even a young Einstein would have anticipated—would likely have been willing to bet their lives on—the 100 measurements won't yield the same result. In fact, at first blush the results will look random, with some electrons found near their box's front lower left corner, some near the back upper right, some around the middle, and so on.

The regularities and patterns that make physics a rigorous and predictive discipline become apparent only if you run this same experiment, with 100 boxed electrons, over and over again. Were you to do so, here's what you'd find. If your first batch of 100 measurements found 27 percent of the electrons near the lower left corner, 48 percent near the upper right corner, and 25 percent near the middle, then the second batch will yield a very similar distribution. So will the third batch, the fourth, and those that follow. The regularity, therefore, isn't evident in any single measurement; you can't predict where any given electron will be. Instead, the regularity is found in the *statistical distribution* of many measurements. The regularity, that is, speaks to the likelihood, or *probability*, of finding an electron at any particular location.

The breathtaking achievement of quantum mechanics' founders was to develop a mathematical formalism that dispensed with the absolute predictions intrinsic to classical physics and instead predicted such probabilities. Working from an equation Schrödinger published in 1926 (and an equivalent though somewhat more awkward equation Heisenberg wrote down in 1925), physicists can input the details of how things are now, and then calculate the probability that they will be one way, or another, or another still, at any moment in the future.

But don't be misled by the simplicity of my little electron example. Quantum mechanics applies not just to electrons but to all types of particles, and it tells us not only about their positions but about also their velocities, their angular momenta, their energies, and how they behave in a wide range of situations, from the barrage of neutrinos now wafting through your body, to the frenzied atomic fusions taking place in the cores of distant stars. Across such a broad sweep, the probabilistic predictions of quantum mechanics match experimental data. Always. In the more than eighty years since these ideas were developed, there has not been a single verifiable experiment or astrophysical observation whose results conflict with quantum mechanical predictions.

For a generation of physicists to have confronted such a radical departure from the intuitions formed out of thousands of years of collective experience, and in response to have recast reality within a wholly new framework based on probabilities, is a virtually unmatched intellectual achievement. Yet one uncomfortable detail has been hovering over quantum mechanics since its inception—a detail that eventually opened a pathway to parallel universes. To understand it, we need to look a little more closely at the quantum formalism.

The Puzzle of Alternatives

In April 1925, during an experiment at Bell Labs undertaken by two American physicists, Clinton Davisson and Lester Germer, a glass tube containing a hot chunk of nickel suddenly exploded. Davisson and Germer had been spending their days firing beams of electrons at specimens of nickel to investigate various aspects of the metal's atomic properties; the equipment failure was a nuisance, albeit one all too familiar in experimental work. On cleaning up the glass shards, Davisson and Germer noticed that the nickel had been tarnished during the explosion. Not a big deal, of course. All they had to do was heat the sample, vaporize the contaminant, and start again. And so they did. But that choice, to clean the sample instead of opting for a new one, proved fortuitous. When they directed the electron beam at the newly cleaned nickel, the results were completely different from any they or anyone else had ever encountered. By 1927, it was clear that Davisson and Germer had established a vital feature of the rapidly developing quantum theory. And within a decade, their serendipitous discovery would be honored with the Nobel Prize.

Although Davisson and Germer's demonstration predates talking movies and the Great Depression, it's still the most widely used method for introducing quantum theory's essential ideas. Here's how to think about it. When Davisson and Germer heated the tarnished sample, they caused numerous small nickel crystals to meld into fewer larger ones. In turn, their electron beam no longer reflected off a highly uniform surface of nickel but instead bounced back from a few concentrated locations where the larger nickel crystals were centered. A simplified version of their experiment, the setup of Figure 8.1, in which electrons are fired at a barrier con-

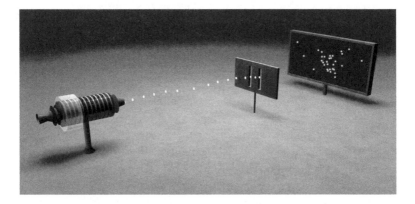

Figure 8.1 *The essence of the Davisson and Germer experiment is captured by the "double-slit" setup in which electrons are fired at a barrier that has two narrow slits. In the Davisson and Germer experiment, two streams of electrons are produced when incident electrons bounce off neighboring nickel crystals; in the double-slit experiment, two similar streams are produced by electrons that pass through the neighboring slits.*

taining two narrow slits, highlights the essential physics. Electrons emanating from one slit or the other are like electrons bouncing back from one nickel crystal or its neighbor. Modeled in this way, Davisson and Germer were carrying out the first version of what's now called the *double-slit experiment.*

To grasp Davisson and Germer's startling result, imagine closing off either the left or the right slit and capturing the electrons that pass through, one by one, on a detector screen. After many such electrons are fired, the detector screens will look like those in Figure 8.2a and Figure 8.2b. A rational, nonquantum trained mind would therefore expect that when both slits are open, the data would be an amalgam of these two results. But the astounding fact is that this is *not* what happens. Instead, Davisson and Germer found data, much like those illustrated in Figure 8.2c, consisting of light and dark bands indicating a series of positions where electrons do and do not land.

These results deviate from expectations in a way that's particularly peculiar. The dark bands are locations where electrons are copiously detected if only the left slit or only the right slit is open (the corresponding regions in Figures 8.2a and 8.2b are *bright*), but which are apparently unreachable when both slits are available. *The presence of the left slit thus changes the possible landing locations of electrons passing through the right slit, and vice versa.* Which is thoroughly perplexing. On the scale of a tiny particle like an electron, the distance between the slits is huge. So when the electron passes through one slit, how could the presence or absence of the other have any effect, let alone the dramatic influence evident in the data? It's as if for many years you happily enter an office building using one door, but when the management finally adds a second door on the building's other side, you can no longer reach your office.

What are we to make of this? The double-slit experiment leads us inescapably to a conclusion hard to fathom. Regardless of which slit it passes through, each individual electron somehow "knows" about both. There's something associated with, or connected to, or part of each individual electron that is affected by both slits.

But what could that something be?

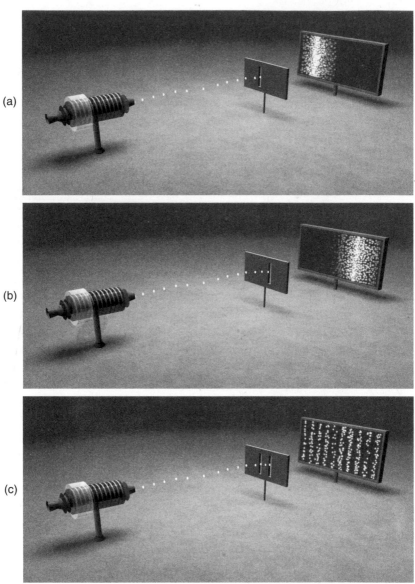

Figure 8.2 (a) *The data obtained when electrons are fired and only the left slit is open.* **(b)** *The data obtained when electrons are fired and only the right slit is open.* **(c)** *The data obtained when electrons are fired and both slits are open.*

Quantum Waves

For a clue as to how an electron traveling through one slit "knows" about the other, look more closely at the data in Figure 8.2c. The light-dark-light-dark pattern is as recognizable to a physicist as a mother's face is to her baby. The pattern says—no, it screams—*waves.* If you've ever dropped two pebbles into a pond and watched as the resulting ripples spread and overlap, you know what I mean. Where the peak of one wave crosses the peak of another, the combined wave height is big; where the trough of one crosses the trough of another, the combined wave depression is deep; and most important of all, where the peak of one crosses the trough of the other, the waves cancel and the water remains level. This is illustrated in Figure 8.3. If you were to insert a detector screen across the top of the figure that recorded the water's agitation at each location—the larger the agitation, the brighter the reading—the result would be a series of alternating bright and dark regions on the screen. Bright regions would be where the waves reinforce each other, yielding much agitation; dark regions would be where the waves cancel, yielding no agitation. Physicists say the overlapping waves *interfere* with one another, and call the bright-dark-bright data they produce an *interference pattern.*

The similarity to Figure 8.2c is unmistakable, so in trying to explain the electron data we're led to think about waves. Good.

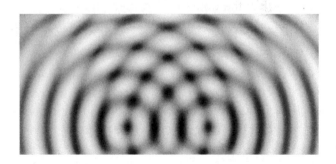

Figure 8.3 *When two water waves overlap, they "interfere," cre-ating alternating regions of more or less agitation called an inter-ference pattern.*

That's a start. But the details are still murky. What kind of waves? Where are they? And what have they to do with particles such as electrons?

The next clue comes from the experimental fact I emphasized at the outset. Reams of data on the motion of particles show that regularities emerge only statistically. The same measurements performed on identically prepared particles will generally reveal them to be in different places; yet many such measurements establish that, on average, the particles have the same probability of being found at any given location. In 1926, the German physicist Max Born joined these two clues together and with them made a leap that nearly three decades later would earn him a Nobel Prize. You've got experimental evidence that waves play a role. You've got experimental evidence that probabilities play a role. Perhaps, Born suggested, the wave associated with a particle is a *probability wave*.

It was an unprecedented and spectacularly original contribution. The idea is that in analyzing the motion of a particle we shouldn't think of it as a rock hurtling from here to there. Instead, we should think of it as a wave *undulating* from here to there. Locations where the wave's values are large, near its peaks and troughs, are locations where the particle is likely to be found. Locations where the probability wave's values are small are locations where the particle is unlikely to be found. Locations where the wave's values vanish are places where the particle won't be found. As the wave rolls onward, the values evolve, going up in some locations, down in others. And since we're interpreting the undulating values as undulating probabilities, the wave is justly called a probability wave.

To flesh out the picture, consider how it explains the double-slit data. As an electron travels toward the barrier in Figure 8.2c, quantum mechanics tells us to think of it as an undulating wave, as in Figure 8.4. When the wave encounters the barrier, two wave fragments make it through the slits and undulate onward toward the detector screen. What happens next is key. Much like overlapping water waves, the probability waves emerging from the two slits overlap and interfere, yielding a combined form that looks much like

Figure 8.4 *When we describe the motion of an electron in terms of an undulating probability wave, the puzzling interference data are explained.*

that in Figure 8.3: a pattern of high and low values that, according to quantum mechanics, corresponds with a pattern of high and low probabilities for where the electron will land. When electron after electron is fired, the cumulative landing positions conform to this probability profile. Many electrons land where the probability is high, few where it's low, and none where the probability vanishes. The result is the bright and dark bands of Figure 8.2c.[3]

And that's how quantum theory explains the data. The description makes manifest that each electron *does* "know" about both slits, since each electron's probability wave passes through both. It's the union of these two partial waves that dictates the probabilities for where the electrons land. That's why the mere presence of a second slit affects the results.

Not So Fast

Although I've focused on electrons, similar experiments have established the same probability-wave picture for all of nature's basic constituents. Photons, neutrinos, muons, quarks—every fundamental particle—all are described by waves of probability. But before we declare victory, three questions immediately present themselves. Two are straightforward. One is a bear. It's the latter that Everett sought to answer back in the 1950s, and it led him to a quantum version of parallel worlds.

First, if quantum theory is right and the world unfolds probabilistically, why is Newton's nonprobabilistic framework so good at predicting the motion of things from baseballs to planets to stars? The answer is that probability waves for big things usually (but not always, as we will shortly see) have a very particular shape. They're extraordinarily narrow, as in Figure 8.5a, meaning there's a huge probability, just shy of 100 percent, that the object is located where the wave is peaked and a minuscule probability, just a shade above 0 percent, that it's located anywhere else.[4] Moreover, the quantum laws show that the peaks of such narrow waves move along the very same trajectories that emerge from Newton's equations. And so, while Newton's laws predict precisely the trajectory of a baseball, quantum theory offers only the most minimal refinement, saying there's a nearly 100 percent probability that the ball will land where Newton says it should, and a nearly 0 percent probability that it won't.

In fact, the words "just shy" and "nearly" don't do the physics justice. The chance of a macroscopic body deviating from Newton's predictions is so fantastically tiny that if you'd been keeping tabs on the cosmos for the last few billion years, the odds are overwhelming that you'd have never seen it happen. But according to quantum theory, the smaller an object, the more spread-out its probability wave typically is. For example, a typical electron's wave might look like that in Figure 8.5b, with substantial probabilities of being at a variety of locations, a totally foreign concept in a Newtonian world. And that's why it's the microrealm where the probabilistic nature of reality comes to the fore.

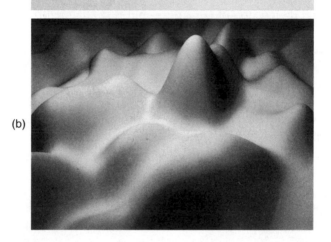

Figure 8.5 (a) *The probability wave for a macroscopic object is generally narrowly peaked.* (b) *The probability wave for a microscopic object, say, a single particle, is typically widely spread.*

Second, can we see the probability waves on which quantum mechanics relies? Is there any way to directly access the unfamiliar probabilistic haze, such as that illustrated schematically in Figure 8.5b, in which a single particle has a chance of being found in a variety of locations? No. The standard approach to quantum mechanics, developed by Bohr and his group, and called the *Copenhagen interpretation* in their honor, envisions that whenever you try to see a probability wave, the very act of observation thwarts your attempt. When you look at an electron's probability wave, where "look" means "measure its position," the electron responds by snapping to attention and coalescing at one definite location. Correspondingly, the probability wave surges to 100 percent at that spot, while collapsing to 0 percent everywhere else, as in Figure 8.6. Look away, and the needle-thin probability wave rapidly spreads, indicating that once again there's a reasonable chance of finding the electron at a variety of locations. Look back, and the electron's wave collapses anew, eliminating the range of possible places the electron might be found in favor of its occupying a single definite spot. In short, every time you attempt to see the probabilistic haze it disappears—it collapses—and is supplanted by familiar reality. The detector screen in Figure 8.2c provides a case in point: it measures the impinging probability wave of an electron and thus immediately causes it to collapse. The detector forces the electron to relinquish the many available options for where it could hit and settle upon a definite landing location, which is then evidenced by a tiny dot on the screen.

I understand full well if this explanation leaves you shaking your head. There's no denying that quantum dogma sounds a lot

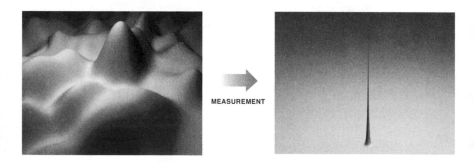

MEASUREMENT

Figure 8.6 *The Copenhagen approach to quantum mechanics envisions that when measured or observed, a particle's probability wave instantaneously collapses at all but one location. The range of possible positions for the particle transforms into one definite outcome.*

like snake oil. I mean, along comes a theory that proposes a startling new picture of reality founded on waves of probability and then, in the very next breath, announces that the waves can't be seen. Imagine Lucille claiming she's a blonde—until someone looks, when she immediately transforms into a redhead. Why would physicists accept an approach that's not only strange but that seems so downright slippery?

Fortunately, for all its mysterious and hidden features, quantum mechanics is testable. According to the Copenhagenists, the larger a probability wave is at a particular location, the greater the chance that when the wave collapses, its sole remaining spike—and hence the electron itself—will be situated there. That statement yields predictions. Run a given experiment over and over again, count how often you find the particle at various locations, and assess whether the frequencies you observe agree with the probabilities dictated by the probability wave. If the wave is 2.874 times as big *here* as it is *there*, do you find the particle *here* 2.874 times as often as you find it *there*? Predictions like these have been enormously successful. Wily as the quantum perspective may seem, it's hard to argue with such phenomenal results.

But not impossible.

Which takes us to the third and most difficult question. The collapsing of probability waves upon measurement, Figure 8.6, is a centerpiece of the Copenhagen approach to quantum theory. The confluence of its successful predictions and Bohr's forceful proselytizing led most physicists to accept it, but even polite prodding quickly reveals an uncomfortable feature. Schrödinger's equation, the mathematical engine of quantum mechanics, dictates how the shape of a probability wave evolves in time. Give me an initial wave shape, say, that of Figure 8.5b, and I can use Schrödinger's equation to draw a picture of what the wave will look like in a minute, or an hour, or at any other moment. But straightforward analysis of the equation reveals that the evolution depicted in Figure 8.6—the instantaneous collapse of a wave at all but one point, like a lone parishioner in a megachurch accidentally standing while everyone else kneels—can't possibly emerge from Schrödinger's math. Waves surely can have a needle-thin spiked shape; we'll make ample use of some spiked waves shortly. But they can't *become* spiked in the manner envisioned by the Copenhagen approach. The math simply doesn't allow it. (We'll see why in just a moment.) Bohr advanced a heavyhanded remedy: evolve probability waves according to Schrödinger's equation whenever you're not looking or performing any kind of measurement. But when you do look, Bohr continued, you should throw Schrödinger's equation aside and *declare* that your observation has caused the wave to collapse.

Now, not only is this prescription ungainly, not only is it arbitrary, not only does it lack a mathematical underpinning, it's not even *clear*. For instance, it doesn't precisely define "looking" or "measuring." Must a human be involved? Or, as Einstein once asked, will a sidelong glance from a mouse suffice? How about a computer's probe, or even a nudge from a bacterium or virus? Do these "measurements" cause probability waves to collapse? Bohr announced that he was drawing a line in the sand separating small things, such as atoms and their constituents, to which Schrödinger's equation would apply, and big things, such as experimenters and their equipment, to which it wouldn't. But he never said where exactly that line would be. The reality is, he couldn't. With each passing year, experimenters confirm that Schrödinger's equation works, without modification, for increasingly large collections of particles, and there's every reason to believe that it works for collections as hefty as those making up you and me and everything else. Like floodwaters slowly rising from your basement, rushing into your living room, and threatening to engulf your attic, the mathematics of quantum mechanics has steadily spilled beyond the atomic domain and has succeeded on ever-larger scales.

So the way to think about the problem is this. You and I and computers and bacteria and viruses and everything else material are made of molecules and atoms, which are themselves composed of particles like electrons and quarks. Schrödinger's equation works for electrons and quarks, and all evidence points to its working for things made of these constituents, regardless of the number of particles involved. This means that Schrödinger's equation should continue to apply during a measurement. After all, a measurement is just one collection of particles (the person, the equipment, the computer . . .) coming into contact with another (the particle or particles being measured). But if that's the case, if Schrödinger's math refuses to bow down, then Bohr is in trouble. Schrödinger's equation doesn't allow waves to collapse. An essential element of the Copenhagen approach would therefore be undermined.

So the third question is this: If the reasoning just recounted is right and probability waves don't collapse, how do we pass from the range of possible outcomes that exist before a measurement to the single outcome the measurement reveals? Or to put it in more general terms, what happens to a probability wave during a measurement that allows a familiar, definite, unique reality to take hold?

Everett pursued this question in his Princeton doctoral dissertation and came to an unforeseen conclusion.

Linearity and Its Discontents

To understand Everett's path of discovery, you need to know a little more about Schrödinger's equation. I've emphasized that it doesn't allow probability waves to suddenly collapse. But why not? And what *does* it allow? Let's get a feel for how Schrödinger's math guides a probability wave as it evolves through time.

This is fairly straightforward, because Schrödinger's is one of the simplest kinds of mathematical equations, characterized by a property known as *linearity*—a mathematical embodiment of the whole being the sum of its parts. To see what this means, imagine that the shape in Figure 8.7a is the probability wave at noon for a given electron (for visual clarity, I will use a probability wave that depends on location in the one dimension represented by the horizontal axis, but the ideas are general). We can use Schrödinger's equation to follow the evolution of this wave forward in time, yielding its shape at, say, one p.m., schematically illustrated in Figure 8.7b. Now notice the following. You can decompose the initial wave shape in Figure 8.7a into two simpler pieces, as in Figure 8.8a; if you combine the two waves in the figure, adding their values point by point, you recover the original wave shape. The lin-

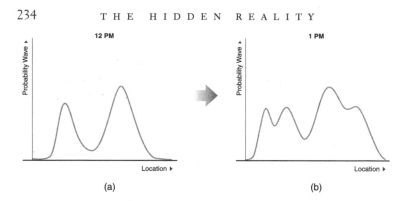

Figure 8.7 (a) *An initial probability wave shape at one moment evolves via Schrödinger's equation to a different shape* (b) *at a later time.*

earity of Schrödinger's equation means that you can use it on each piece in Figure 8.8a separately, determining what each wave fragment will look like at one p.m., and then combine the results as in Figure 8.8b to recover the full result shown in Figure 8.7b. And there's nothing sacred about decomposition into *two* pieces; you can break the initial shape up into any number of pieces, evolve each separately, and combine the results to get the final wave shape.

This may sound like a mere technical nicety, but linearity is an extraordinarily powerful mathematical feature. It allows for an all-important divide-and-conquer strategy. If an initial wave shape is complicated, you are free to divide it up into simpler pieces and analyze each separately. At the end, you just put the individual results back together. We've actually already seen an important application of linearity through our analysis of the double-slit experiment in Figure 8.4. To determine how the electron's probability wave evolves, we divided the task: we noted how the piece passing through the left slit evolves, we noted how the piece passing through the right slit evolves, and we then added the two waves together. That's how we found the famous interference pattern. Look at a quantum theorist's blackboard, and it is this very approach you'll see underlying a great many of the mathematical manipulations.

But linearity not only makes quantum calculations manage-

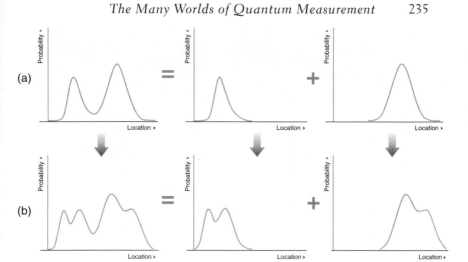

Figure 8.8 (a) *An initial probability wave shape can be decomposed as the union of two simpler shapes.* **(b)** *The evolution of the initial probability wave can be reproduced by evolving the simpler pieces and combining the results.*

able; it's also at the heart of the theory's difficulty in explaining what happens during a measurement. This is best understood by applying linearity to the act of measurement itself.

Imagine you're an experimentalist with great nostalgia for your childhood in New York, so you're measuring the positions of electrons that you inject into a miniature tabletop model of the city. You start your experiments with one electron whose probability wave has a particularly simple shape—it's nice and spiked, as in Figure 8.9, indicating that with essentially 100 percent probability the electron is momentarily sitting at the corner of Thirty-fourth Street and Broadway. (Don't worry about how the electron got this wave shape; just take it as a given.)* If at that very moment you

---

*For simplicity, we won't consider the electron's position in the vertical direction—we focus solely on its position on a map of Manhattan. Also, let me re-emphasize that while this section will make clear that Schrödinger's equation doesn't allow waves to undergo an instantaneous collapse as in Figure 8.6, waves *can* be carefully prepared by the experimenter in a spiked shape (or, more precisely, very close to a spiked shape).

Figure 8.9 *An electron's probability wave, at a given moment, is spiked at Thirty-fourth Street and Broadway. A measurement of the electron's position, at that moment, confirms that it is located where its wave is spiked.*

measure the electron's position with a well-made piece of equipment, the result should be accurate—the device's readout should say "Thirty-fourth Street and Broadway." Indeed, if you do this experiment, that's just what happens, as in Figure 8.9.

It would be extraordinarily complicated to work out how Schrödinger's equation entwines the probability wave of the electron with that of the trillion trillion or so atoms that make up the measuring device, coaxing a collection of the latter to arrange themselves in the readout to spell "Thirty-fourth Street and Broadway," but whoever designed the device has done the heavy lifting for us. It's been engineered so that its interaction with such an electron causes the readout to display the single definite position where, at that moment, the electron is located. If the device did anything else in this situation, we'd be wise to exchange it for a new one that functions properly. And, of course, Macy's notwithstanding, there's nothing special about Thirty-fourth and Broadway; if we do the same experiment with the electron's probability wave spiked at the Hayden Planetarium near Eighty-first and Central

Park West, or at Bill Clinton's office on 125th near Lenox Avenue, the device's readout will return these locations.

Let's now consider a slightly more complicated wave shape, as in Figure 8.10. This probability wave indicates that, at the given moment, there are two places the electron might be found— Strawberry Fields, the John Lennon memorial in Central Park, and Grant's Tomb in Riverside Park. (The electron's in one of its dark moods.) If we measure the electron's position but, in opposition to Bohr and in keeping with the most refined experiments, assume that Schrödinger's equation continues to apply—to the electron, to the particles in the measuring device, to everything—what will the device's output read? Linearity is the key to the answer. We know what happens when we measure spiked waves individually. Schrödinger's equation causes the device's display to spell out the spike's location, as in Figure 8.9. Linearity then tells us that to find the answer for two spikes, we combine the results of measuring each spike separately.

Here's where things get weird. At first blush, the combined

Figure 8.10 *An electron's probability wave is spiked at two locations. The linearity of Schrödinger's equation suggests that a measurement of the electron's position would yield a confusing amalgam of both locations.*

results suggest that the display should simultaneously register the locations of both spikes. As in Figure 8.10, the words "Strawberry Fields" and "Grant's Tomb" should flash simultaneously, one location commingled with the other, like the confused monitor of a computer that's about to crash. Schrödinger's equation also dictates how the probability waves of the photons emitted by the measuring device's display entangle with those of the particles in your rods and cones, and subsequently those rushing through your neurons, creating a mental state reflecting what you see. Assuming unlimited Schrödinger hegemony, linearity applies here too, so not only will the device simultaneously display both locations but also your brain will be caught up in the confusion, thinking that the electron is simultaneously positioned at both.

For yet more complicated wave shapes, the confusion becomes yet wilder. A shape with four spikes doubles the dizziness. With six, it triples. Notice that if you keep on going, putting wave spikes of various heights at every location in the model Manhattan, their combined shape fills out an ordinary, more gradually varying quantum wave shape, as schematically illustrated in Figure 8.11. Lin-

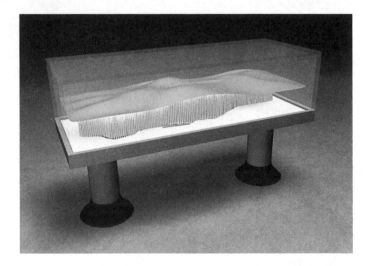

Figure 8.11 *A general probability wave is the union of many spiked waves, each representing a possible position of the electron.*

earity still holds, and this implies that the final device reading, as well as your final brain state and mental impression, are dictated by the union of the results for each spike individually. The device should simultaneously register the location of each and every spike—each and every location in Manhattan—as your mind becomes profoundly puzzled, being unable to settle on a single definite location for the electron.[5]

But, of course, this seems grossly at odds with experience. No properly functioning device, when taking a measurement, displays conflicting results. No properly functioning person, on performing a measurement, has the mental impression of a dizzying mélange of simultaneous yet distinct outcomes.

You can now see the appeal of Bohr's prescription. Hold the Dramamine, he'd declare. According to Bohr, we don't see ambiguous meter readings because they don't happen. He'd argue that we've come to an incorrect conclusion because we've overextended the reach of Schrödinger's equation into the domain of big things: laboratory equipment that takes measurements, and scientists who read the results. Although Schrödinger's equation and its feature of linearity dictate that we should combine the results from distinct possible outcomes—nothing collapses—Bohr tells us that this is wrong because the act of measurement tosses Schrödinger's math out the window. Instead, he'd pronounce, the measurement causes all but one of the spikes in Figure 8.10 or Figure 8.11 to collapse to zero; the probability that a particular spike will be the sole survivor is proportional to the spike's height. That unique remaining spike determines the device's unique reading, as well as your mind's recognition of a unique result. Dizziness done.

But for Everett, and later DeWitt, the cost of Bohr's approach was too high. Schrödinger's equation is meant to describe particles. All particles. Why would it somehow not apply to particular configurations of particles—those constituting the equipment that takes measurements, and those in the experimenters who monitor the equipment? This just didn't make sense. Everett therefore suggested that we not dispense with Schrödinger so quickly. Instead, he advocated that we analyze where Schrödinger's equation takes us from a decidedly different perspective.

Many Worlds

The challenge we've encountered is that it's bewildering to think of a measuring device or a mind as simultaneously experiencing distinct realities. We can have conflicting opinions on this or that issue, mixed emotions regarding this or that person, but when it comes to the facts that constitute reality, everything we know attests to there being an unambiguous objective description. Everything we know attests that one device and one measurement will yield one reading; one reading and one mind will yield one mental impression.

Everett's idea was that Schrödinger's math, the core of quantum mechanics, *is* compatible with such basic experiences. The source of the supposed ambiguity in device readings and mental impressions is the manner in which we've executed that math — the manner, that is, in which we've combined the results of the measurements illustrated in Figure 8.10 and Figure 8.11. Let's think it through.

When you measure a single spiked wave, such as that in Figure 8.9, the device registers the spike's location. If it's spiked at Strawberry Fields, that's what the device reads; if you look at the result, your brain registers that location and you become aware of it. If it's spiked at Grant's Tomb, that's what the device registers; if you look, your brain registers that location and you become aware of it. When you measure the double spiked wave in Figure 8.10, Schrödinger's math tells you to combine the two results you just found. But, says Everett, be careful and precise when you combine them. The combined result, he argued, does not yield a meter and a mind each simultaneously registering two locations. That's sloppy thinking.

Instead, proceeding slowly and literally, we find that the combined result is a device and a mind registering Strawberry Fields, and a device and a mind registering Grant's Tomb. And what does that mean? I'll use broad strokes in painting the general picture, which I'll refine shortly. To accommodate Everett's suggested outcome, the device and you and everything else must split upon measurement, yielding two devices, two yous, and two everything elses—the only difference between the two being that one device and one you registers Strawberry Fields, while the other device and the other you registers Grant's Tomb. As in Figure 8.12, this implies that we now have two parallel realities, two parallel worlds. To the you occupying each, the measurement and your mental impression of the result are sharp and unique and thus feel like life as usual. The peculiarity, of course, is that there are two of you who feel this way.

To keep the discussion accessible, I've focused on the position measurement of a single particle, and one that has a particularly simple probability wave. But Everett's proposal applies generally. If

Figure 8.12 *In Everett's approach, the measurement of a particle whose probability wave has two spikes yields both outcomes. In one world, the particle is found at the first location; in another world, it is found at the second.*

you measured the position of a particle whose probability wave has any number of spikes, say, five, the result, according to Everett, would be five parallel realities differing only by the location registered on each reality's device, and within the mind of each reality's you. If one of these yous then measured the position of another particle whose wave had seven spikes, that you and that world would split again, into seven more, one for each possible outcome. And if you measured a wave like that of Figure 8.11, which can be partitioned into a great many tightly packed spikes, the result would be a great many parallel realities in which each possible particle location would be recorded on a device and read by a copy of you. In Everett's approach, everything that is *possible*, quantum-mechanically speaking (that is, all those outcomes to which quantum mechanics assigns a nonzero probability), is *realized in its own separate world*. These are the "many worlds" of the *Many Worlds* approach to quantum mechanics.

If we apply the terminology we've been using in earlier chapters, these many worlds would properly be described as many universes, composing a multiverse, the sixth we've encountered. I'll call it the *Quantum Multiverse*.

A Tale of Two Tales

In describing how quantum mechanics may generate many realities, I used the word "split." Everett used it. So did DeWitt. Nevertheless, in this context it's a loaded verb with the potential to grossly mislead, and I'd intended not to invoke it. But I gave in to temptation. In my defense, it's sometimes more effective to use a sledgehammer to break down a barrier separating us from an unfamiliar proposal about the workings of reality, and to subsequently repair the damage, than it is to delicately carve a pristine window that directly reveals the new vista. I've been using that sledgehammer; in this and the next section I'll undertake the necessary repairs. Some of the ideas are a touch more difficult than those we've so far encountered, and the explanatory chains are a bit longer as well, but I encourage you to stay with me. I've found that all too often, people who learn about, or are even somewhat familiar with, the Many Worlds idea have the impression that it emerged from speculation of the most extravagant sort. But nothing could be further from the truth. As I will explain, the Many Worlds approach is, in some ways, the most conservative framework for defining quantum physics, and it's important to understand why.

The essential point is that physicists must always tell two kinds of stories. One is the mathematical story of how the universe evolves according to a given theory. The other, also essential, is the physical story, which translates the abstract mathematics into experiential language. This second story describes how the mathematical evolution will appear to observers like you and me, and more generally, what the theory's mathematical symbols tell us about the nature of reality.[6] In the time of Newton, the two stories were essentially identical, as I suggested with my remarks in Chapter 7 about Newtonian "architecture" being immediate and palpable. Every mathematical symbol in Newton's equations has a direct and transparent physical correlate. The symbol x? Oh, that's the ball's position. The symbol v? The ball's velocity. By the time we get to quantum mechanics, however, translation between the mathematical symbols and what we can see in the world around us becomes far more subtle. In turn, the language used and the concepts deemed relevant to each of the two stories become so different that you need both to acquire a full understanding. But it's important to keep straight which story is which: to understand fully which ideas and descriptions are invoked as part of the theory's fundamental mathematical structure and which are used to build a bridge to human experience.

Let's tell the two stories for the Many Worlds approach to quantum mechanics. Here's the first.

The mathematics of Many Worlds, unlike that of Copenhagen, is pure, simple, and constant. Schrödinger's equation determines how probability waves evolve over time, and it is never set aside; it is *always* in effect. Schrödinger's math guides the shape of probability waves, causing them to shift, morph, and undulate over time. Whether it's addressing the probability wave for a particle, or for a collection of particles, or for the various assemblages of particles that constitute you and your measuring equipment, Schrödinger's equation takes the particles' initial probability wave shape as input and then, like the graphics program driving an elaborate screen saver, provides the wave's shape at any future time as output. And that, according to this approach, is how the universe evolves. Period. End of story. Or, more precisely, end of first story.

Notice that in telling the first story I did not need the word "split" nor the terms "many worlds," "parallel universes," or "Quantum Multiverse." The Many Worlds approach does not hypothesize these features. They play no role in the theory's fundamental mathematical structure. Rather, as we will now see, these ideas are called upon in the theory's second story, when, following Everett and others who've since extended his pioneering work, we investigate what the mathematics tells us about our observations and measurements.

Let's start simply—or, as simply as we can. Consider measuring an electron that has a spiked probability wave, as in Figure 8.9. (Again, don't worry about how it got this wave shape; just take it as a given.) As noted earlier, to tell the first story of even this measurement process in detail is beyond what we can do. We'd need to use Schrödinger's math to figure out how the probability wave describing the positions of the huge number of particles that constitute you and your measuring device joins with the probability wave of the electron, and how their union evolves forward in time. My undergraduate students, many of whom are quite able, often struggle to solve Schrödinger's equation for even a single particle. Between you and the device, there are something like 10^{27} particles. Working out Schrödinger's math for that many constituents is virtually impossible. Even so, we understand qualitatively what the math entails. When we measure the electron's position, we cause a mass particle migration. Some 10^{24} or so particles in the device's display, like performers in a crisply choreographed halftime show, race to the appropriate spot so that they collectively spell out "Thirty-fourth Street and Broadway," while a similar number in my eyes and brain do whatever's required for me to develop a firm mental grasp of the result. Schrödinger's math—however impenetrable explicit analysis of it might be when faced with so many particles—describes such a particle shift.

To visualize this transformation at the level of a probability wave is also far beyond reach. In Figure 8.9 and others in that sequence, I used two axes, the north-south and east-west street grid of our model Manhattan, to denote the possible positions of a single particle. The probability wave's value at each location was denoted by the wave's height. This already simplifies things because I've left out the third axis, the particle's vertical position (whether it's on the second floor of Macy's, or the fifth). Including the vertical would have been awkward, because if I'd used it to denote position, I'd have no axis left for recording the size of the wave. Such are the limitations of a brain and a visual system that evolution has firmly rooted in three spatial dimensions. To properly visualize the probability wave for roughly 10^{27} particles, I'd need to include three axes for each, allowing me to account mathematically for every possible position each particle could occupy.* Adding even a single vertical axis to Figure 8.9 would have made it difficult to visualize; to contemplate adding a billion billion billion more is, well, silly.

But a mental image of the key ideas is important; so, however imperfect the result, let's give it a try. In sketching the probability wave for the particles making up you and your device, I'll abide by the two-axis flat-page limit but will use an unconventional interpretation of what the axes mean. Roughly speaking, I'll think of each axis as comprising an enormous bundle of axes, tightly grouped together, which will symbolically delineate the possible positions of a similarly enormous number of particles. A wave drawn using these bundled axes will therefore lay out the probabilities for the positions of a huge group of particles. To emphasize the distinction between the many-particle and single-particle situations, I'll use a glowing outline for the many-particle probability wave, as in Figure 8.13.

The many-particle and single-particle illustrations have some features in common. Just as the spiked wave shape in Figure 8.6 indicates probabilities that are sharply skewed (being almost 100 percent at the spike's location and almost 0 percent everywhere else), so the peaked wave in Figure 8.13 also denotes sharply skewed

---

*For a mathematical depiction, see note 4.

Mathematical Story   Physical Story

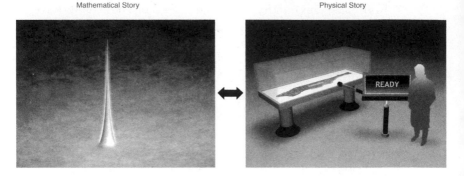

Figure 8.13 *A schematic depiction of the combined probability wave for all the particles making up you and your measuring device.*

probabilities. But you need to exercise care, because understanding based on the single-particle illustrations can take you only so far. For example, based on Figure 8.6 it is natural to think that Figure 8.13 represents particles that are all clustered around the same location. Yet, that's not right. The peaked shape in Figure 8.13 symbolizes that each of the particles making up you and each of the particles making up the device starts out in the ordinary, familiar state of having a position that is nearly 100 percent definite. But they are not all positioned at the same location. The particles constituting your hand, shoulder, and brain are, with near certainty, clustered within the location of your hand, shoulder, and brain; the particles constituting the measuring device are, with near certainty, clustered within the location of the device. The peaked wave shape in Figure 8.13 denotes that each of these particles has only the most remote chance of being found anywhere else.

If you now perform the measurement illustrated in Figure 8.14, the many-particle probability wave (for the particles inside you and the device), by virtue of the interaction with the electron, evolves (as illustrated schematically in Figure 8.14a). All the particles involved still have nearly definite positions (within you; within the device), which is why the wave in Figure 8.14a maintains a spiked shape. But a mass particle rearrangement occurs that results in the

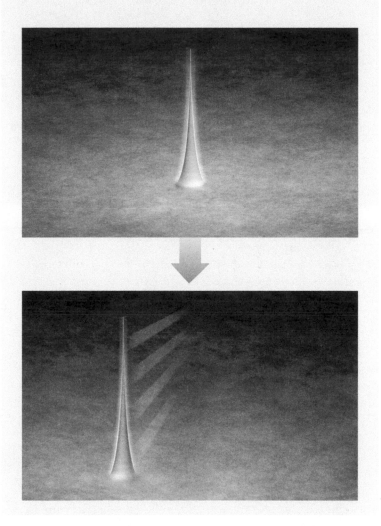

Figure 8.14 (a) *A schematic illustration of the evolution, dictated by Schrödinger's equation, of the combined probability wave for all the particles making up you and the measuring device, when you measure the position of an electron. The electron's own probability wave is spiked at Strawberry Fields.*

Physical Story

Figure 8.14 (b) *The corresponding physical, or experiential, story.*

words "Strawberry Fields" forming in the device's readout and also in your brain (as in Figure 8.14b). Figure 8.14a represents the mathematical transformation dictated by Schrödinger's equation, the first kind of story. Figure 8.14b illustrates the physical description of such mathematical evolution, the second kind of story. Similarly, if we perform the experiment in Figure 8.15, an analogous wave shift takes place (Figure 8.15a). This shift corresponds to a mass particle rearrangement that spells out "Grant's Tomb" in the display and generates within you the associated mental impression (Figure 8.15b).

Now use linearity to put the two together. If you measure the position of an electron whose probability wave is spiked at two locations, the probability wave for you and your device commingles with that of the electron, resulting in the evolution shown in Figure 8.16a—the combined evolutions depicted in Figure 8.14a and Figure 8.15a. So far, this is nothing but an illustrated and annotated version of the first type of quantum story. We start with a probability wave of a given shape, Schrödinger's equation evolves it forward in time, and we end up with a probability wave of a new shape. But the details we've overlaid now let us tell this mathematical story in more qualitative, type-two story language.

Physically, each spike in Figure 8.16a represents a configuration of an enormous number of particles that results in a device having a particular reading and your mind acquiring that information. In the left spike, the reading is Strawberry Fields; in the right, it's Grant's Tomb. Besides that difference, *nothing* distinguishes one spike from the other. I emphasize this because it's essential to realize that neither is somehow more real than the other. Nothing but the device's particular reading, and your reading of that reading, distinguishes the two multiparticle wave spikes.

Which means that our type-two story, as illustrated in Figure 8.16b, involves two realities.

In fact, the focus on the device and your mind is merely another simplification. I could also have included the particles that make up the laboratory and everything therein, as well as those of the earth, the sun, and so on, and the whole discussion would have been the same, essentially verbatim. The only difference would have been

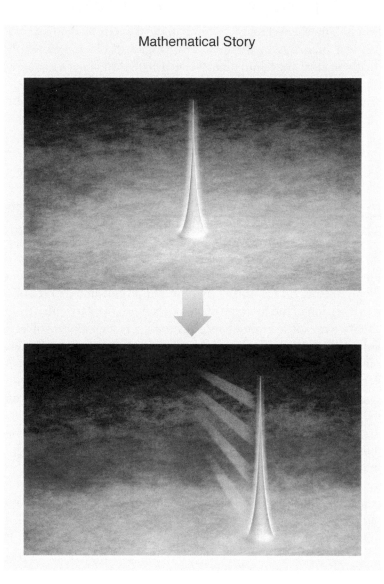

Figure 8.15 (a) *The same type of mathematical evolution as in Figure 8.14a, but with the electron's probability wave spiked at Grant's Tomb.*

Physical Story

Figure 8.15 (b) *The corresponding physical, or experiential, story.*

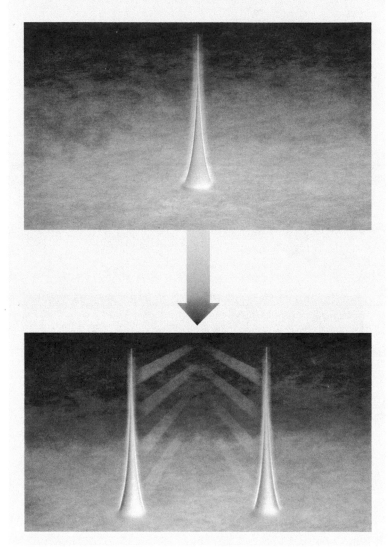

Figure 8.16 (a) *A schematic illustration of the evolution of the combined probability wave of all the particles making up you and your device, when measuring the position of an electron whose probability wave is spiked at two locations.*

Physical Story

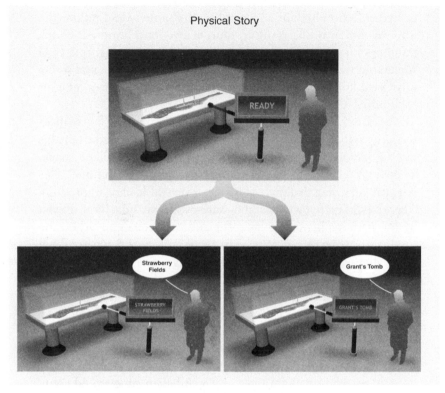

Figure 8.16 (b) *The corresponding physical, or experiential, story.*

that the glowing probability wave in Figure 8.16a would now have information about all those other particles, too. But because the measurement we're discussing has essentially no impact on them, they'd just come along for the ride. It's useful to include those particles, though, because our second story can now be augmented to comprise not only a copy of you examining a device that's undertaken a measurement, but also copies of the surrounding laboratory, the rest of the earth in orbit around the sun, and so on. This means that each spike, in story-two language, corresponds to what we'd traditionally call a bona fide universe. In one such universe, you see "Strawberry Fields" on the display's reading; in the other, "Grant's Tomb."

If the electron's original probability wave had, say, four spikes, or five, or a hundred, or any number, the same would follow: the wave evolution would result in four, or five, or a hundred, or any number of universes. In the most general case, as in Figure 8.11, a spread-out wave is composed of spikes at every location, and so the wave evolution would yield a vast collection of universes, one for each possible position.[7]

As advertised, though, the only thing that happens in any of these scenarios is that a probability wave enters Schrödinger's equation, his math goes to work, and out comes a wave with a modified shape. There's no "cloning machine." There's no "splitting machine." This is why I said earlier that such words can give a misleading impression. There's nothing but a probability-wave-evolution "machine" driven by the lean mathematical law of quantum mechanics. When the resulting waves have a *particular* shape, as in Figure 8.16a, we retell the mathematical story in type-two language, and conclude that in each spike there's a sentient being, situated within a normal-looking universe, certain he sees one and only one definite result for the given experiment, as in Figure 8.16b. If I could somehow interview all these sentient beings, I'd find each to be an exact replica of the others. Their only point of departure would be that each would attest to a different definite result.

And so, whereas Bohr and the Copenhagen gang would argue that only one of these universes would exist (because the act of measurement, which they claim lies outside of Schrödinger's purview, would collapse away all the others), and whereas a first-pass attempt to go beyond Bohr and extend Schrödinger's math to all particles, including those constituting equipment and brains, yielded dizzying confusion (because a given machine or mind seemed to internalize all possible outcomes simultaneously), Everett found that a more careful reading of Schrödinger's math leads somewhere else: to a plentiful reality populated by an ever-growing collection of universes.

Prior to the publication of Everett's 1957 paper, a preliminary version was circulated to a number of physicists around the world. Under Wheeler's guidance, the paper's language had been abbreviated so aggressively that many who read it were unsure as to whether Everett was arguing that all the universes in the mathematics were real. Everett became aware of this confusion and decided to clarify it. In a "note added in proof" that he seems to have slipped in just before publication, and apparently without Wheeler's notice, Everett sharply articulated his stance on the reality of the different outcomes: "From the viewpoint of the theory, all . . . are 'actual,' none any more 'real' than the rest."[8]

When Is an Alternative a Universe?

Besides the loaded words "splitting" and "cloning," we've freely invoked two other grand terms in our type-two stories—"world" and, interchangeably in this context, "universe." Are there guidelines for determining when this usage is appropriate? When we consider a probability wave for a single electron that has two (or more) spikes, we don't speak of two (or more) worlds. Instead, we speak of one world—ours—containing an electron whose position is ambiguous. Yet, in Everett's approach, when we measure or observe that electron, we speak in terms of multiple worlds. What is it that distinguishes the unmeasured and the measured particle, yielding descriptions that sound so radically different?

One quick answer is that for a single isolated electron, we don't tell a type-two story because without a measurement or an observation there's no link to human experience that's in need of articulation. The type-one story of a probability wave evolving via Schrödinger's math is all that's needed. And without a type-two story, there's no opportunity to invoke multiple realities. Although this explanation is adequate, it proves worthwhile to delve a little deeper, revealing a special feature of quantum waves that comes into play when many particles are involved.

To grasp the essential idea, it's easiest to look back at the double-slit experiment of Figures 8.2 and 8.4. Recall that an electron's probability wave encounters the barrier, and two wave fragments make it through the slits and travel onward to the detector screen. Inspired by our Many Worlds discussion, you might be tempted to think of the two racing waves as representing separate realities. In one, an electron whisks through the left slit; in the other, an electron whisks through the right slit. But you promptly realize that the intermingling of these supposedly "distinct realities" profoundly affects the experiment's outcome; the intermingling is why an interference pattern is produced. So it doesn't make much sense, nor does it yield any particular insight, to consider the two wave trajectories as existing in separate universes.

If we change the experiment, however, by placing a meter behind each slit that records whether or not an electron passes through it, the situation is radically different. Because macroscopic equipment is now involved, the two distinct trajectories of an electron generate differences in a huge number of particles—the huge number of particles in the meters' displays that register "electron passed through left slit" or "electron passed through right slit." And because of this, the respective probability waves for each possibility become so disparate that it's virtually impossible for them to have any subsequent influence on each other. Much as in Figure 8.16a, the differences between the billions and billions of particles in the meters cause the waves for the two outcomes to shift away from each other, leaving negligible overlap. With no overlap, the waves don't engage in any of the hallmark interference phenomena of quantum physics. Indeed, with the meters in place, the electrons no longer yield the striped pattern of Figure 8.2c; instead, they generate a simple, noninterfering amalgam of the results in Figure 8.2a and Figure 8.2b. Physicists say that the probability waves have *decohered* (something you can read about in more detail, for example, in Chapter 7 of *The Fabric of the Cosmos*).

The point, then, is that once decoherence sets in, the waves for each outcome evolve independently—there's no intermingling between the distinct possible outcomes—and each can thus be called a world or a universe of its own. For the case at hand, in one such universe the electron goes through the left slit, and the meter displays left; in another universe the electron goes through the right slit, and the meter records right.

In this sense, and only in this sense, there's resonance with Bohr. According to the Many Worlds approach, big things made of many particles do differ from small things made from one particle or a mere handful. Big things don't stand outside the basic mathematical law of quantum mechanics, as Bohr thought, but they do allow probability waves to acquire enough variations that their capacity to interfere with one another becomes negligible. And once two or more waves can't affect one another, they become mutually invisible; each "thinks" the others have disappeared. So, whereas Bohr argued away by fiat all but one outcome in a measurement, the Many Worlds approach, combined with decoherence, ensures that within each universe it *appears* as though the other outcomes have vanished. Within each universe, that is, it's *as if* the probability wave has collapsed. But, compared with the Copenhagen approach, the "as if" provides for a very different picture of the expanse of reality. In the Many Worlds view, all outcomes, not just one, are realized.

Uncertainty at the Cutting Edge

This might seem like a good place to end the chapter. We've seen how the bare-bones mathematical structure of quantum mechanics leads us by the nose to a new conception of parallel universes. Yet you'll note that the chapter still has a fair way to go. In those pages I'll explain why the Many Worlds approach to quantum physics remains controversial; we will see that the resistance goes well beyond the queasiness some feel about the conceptual leap into such an unfamiliar perspective on reality. But in case you've reached saturation and feel compelled to skip ahead to the next chapter, here is a short summary.

In day-to-day life, probability enters our thinking when we face a range of possible outcomes, but for one reason or another we're unable to figure out which will actually happen. Sometimes we have enough information to determine which outcomes are more or less likely to occur, and probability is the tool that makes such insights quantitative. Our confidence in a probabilistic approach grows when we find that the outcomes deemed likely happen often and those deemed unlikely happen rarely. The challenge facing the Many Worlds approach is that it needs to make sense of probability—quantum mechanics' probabilistic predictions—in a wholly different context, one that envisions *all* possible outcomes happening. The dilemma is simple to state: How can we speak of some outcomes being likely and others being unlikely when all take place?

In the remaining sections, I'll explain the issue more fully and discuss attempts to address it. Be warned: we are now deep into cutting-edge research, so opinions vary widely on where we currently stand.

A Probable Problem

A frequent criticism of the Many Worlds approach is that it's just too baroque to be true. The history of physics teaches us that successful theories are simple and elegant; they explain data with a minimum of assumptions and provide an understanding that's precise and economical. A theory that introduces an ever-growing cornucopia of universes falls way short of this ideal.

Proponents of the Many Worlds approach argue, credibly, that in assessing the complexity of a scientific proposal, you shouldn't focus on its *implications*. What matters is the fundamental features of the proposal itself. The Many Worlds approach assumes that a single equation—Schrödinger's—governs all probability waves all the time, so for simplicity of formulation and economy of assumptions, it's hard to beat. The Copenhagen approach is surely no simpler. It, too, invokes Schrödinger's equation, but it also includes a vague, ill-defined prescription for when Schrödinger's equation should be turned off, and then an even less detailed prescription regarding the process of wave collapse that is meant to take its place. That the Many Worlds approach leads to an exceptionally rich picture of reality is no more a black mark against it than the rich diversity of life on earth is a black mark against Darwinian natural selection. Mechanisms that are fundamentally simple can give rise to complicated consequences.

Nevertheless, while this establishes that Occam's razor isn't sharp enough to pare away the Many Worlds approach, the proposal's surfeit of universes does yield a potential problem. Earlier I said that in applying a theory, physicists need to tell two kinds of stories—the story describing how the world evolves mathematically and the story that links the math to our experiences. But there's actually a third story, related to these two, that the physicist must also tell. It's the story of how we've come to have confidence in a given theory. For quantum mechanics, the third story generally goes like this: our confidence in quantum mechanics comes from its phenomenal success in explaining data. If a quantum expert uses the theory to calculate that in repeating a given experiment we expect one outcome to happen, say, 9.62 times more often than another, that's what experimenters invariably see. Turning this around, had results not agreed with the quantum predictions, experimenters would have concluded that quantum mechanics wasn't right. Actually, being careful scientists, they would have been more cautious. They would have called it doubtful that quantum mechanics was right but would have noted that their results didn't rule out the theory definitively. Even a fair coin tossed 1,000 times can have surprising runs that defy the odds. But the larger the deviation, the more one suspects the coin is not fair; the larger the experimental deviations from those predicted by quantum mechanics, the more strongly the experimenters would have suspected that quantum theory was mistaken.

That confidence in quantum mechanics could have been undermined by data is essential; with any proposed scientific theory that has been suitably developed and understood, we should be able to say, at least in principle, that if upon doing such and such an experiment we don't find such and such results, our belief in the theory should diminish. And the more that observations deviate from predictions, the greater the loss of credibility should be.

The potential problem with the Many Worlds approach, and the reason it remains controversial, is that it may undercut this means for assessing the credibility of quantum mechanics. Here's why. When I flip a coin, I know there's a 50 percent chance that it will land heads and a 50 percent chance that it will land tails. But that conclusion rests on the usual assumption that a coin toss yields a unique result. If a coin toss yields heads in one world and tails in another, and moreover, if there's a copy of me in each world who witnesses the outcome, what sense can we make of the usual odds? There'll be someone who looks just like me, has all my memories, and emphatically claims to be me who sees heads, and another being, equally convinced that he's me, who sees tails. Since both outcomes happen—there's a Brian Greene who sees heads and a Brian Greene who sees tails—the familiar probability of there being an equal chance that Brian Greene will see either heads *or* tails seems nowhere to be found.

The same concern applies to an electron whose probability wave is hovering near Strawberry Fields and Grant's Tomb, as in Figure 8.16b. Traditional quantum reasoning says that you, the experimenter, have a 50 percent chance of finding the electron at either location. But in the Many Worlds approach, both outcomes happen. There's a you who will find the electron at Strawberry Fields and another you who will find the electron at Grant's Tomb. So, how can we make sense of the traditional probabilistic predictions, which in this case say that with equal odds you'll see one result *or* the other?

The natural inclination of many people when they first encounter this issue is to think that among the various yous in the Many Worlds approach, there's one who's somehow more real than the others. Even though each you in each world looks identical and has the same memories, the common thought is that only one of these beings is *really* you. And, this line of thought continues, it's *that* you, who sees one and only one outcome, to whom the probabilistic predictions apply. I appreciate this response. Years ago, when I first learned about these ideas, I had it too. But the reasoning runs completely counter to the Many Worlds approach. Many Worlds practices minimalist architecture. Probability waves simply evolve by Schrödinger's equation. That's it. To imagine that one of the copies of you is the "real" you is to slip in through the back door something closely akin to Copenhagen. Wave collapse in the Copenhagen approach is a brutish means for making one and only one of the possible outcomes real. If in the Many Worlds approach you imagine that one and only one of the yous is *really* you, you're doing the same thing, just a little more quietly. Such a move would erase the very reason for introducing the Many Worlds scheme. Many Worlds emerged from Everett's attempt to address the failings of Copenhagen, and his strategy was to invoke nothing beyond the battle-tested Schrödinger equation.

This realization shines an uncomfortable light on the Many Worlds approach. We have confidence in quantum mechanics because experiments confirm its probabilistic predictions. Yet, in the Many Worlds approach, it's hard to see how probability even plays a role. How, then, can we tell the third kind of story, the one that should provide the basis of our confidence in the Many Worlds scheme? That's the quandary.

On reflection, it's not surprising that we've bumped into this wall. There's nothing at all chancy in the Many Worlds approach. Waves simply evolve from one shape to another in a manner described fully and deterministically by Schrödinger's equation. No dice are thrown; no roulette wheels are spun. By contrast, in the Copenhagen approach, probability enters through the hazily defined measurement-induced wave collapse (again, the larger the wave's value at a given location, the larger the probability that the collapse will put the particle there). That's the point in the Copenhagen approach where "dice throwing" makes an appearance. But since the Many Worlds approach abandons collapse, it abandons the traditional entry point for probability.

So, is there a place for probability in the Many Worlds approach?

Probability and Many Worlds

Everett surely thought there was. The bulk of his 1956 draft dissertation, as well as the truncated 1957 version, was devoted to explaining how to incorporate probability in the Many Worlds approach. But a half century later, the debate still rages. Among those physicists and philosophers who spend their professional lives puzzling over the issue, there is a wide range of opinions on how, and whether, Many Worlds and probability come together. Some have argued that the problem is insoluble, and so the Many Worlds approach should be discarded. Others have argued that probability, or at least something that masquerades as probability, can indeed be incorporated.

Everett's original proposal provides a good example of the difficult points that arise. In everyday settings, we invoke probability because we generally have incomplete knowledge. If, when a coin is tossed, we know enough details (the coin's precise dimensions and weight, precisely how the coin was thrown, and so on), we'd be able to predict the outcome. But since we generally don't have that information, we resort to probability. Similar reasoning applies to the weather, the lottery, and every other familiar example where probability plays a role: we deem the outcomes chancy only because our knowledge of each situation is limited. Everett argued that probabilities find their way into the Many Worlds approach because an analogous ignorance, from a thoroughly different source, necessarily creeps in. Inhabitants of the Many Worlds only have access to their own single world; they do not experience the others. Everett argued that with such a limited perspective comes an infusion of probability.

To get a feel for how, leave quantum mechanics for a moment and consider an imperfect but helpful analogy. Imagine that aliens from the planet Zaxtar have succeeded in building a cloning machine that can make identical copies of you, me, or anyone. Were you to step into the cloning machine, and were two of you then to step out, both would be absolutely convinced that they were the real you, and both would be right. The Zaxtarians delight in subjecting less intelligent life-forms to existential dilemmas, so they swoop down to earth and make you the following offer. Tonight, when you go to sleep, you'll be carefully wheeled into the cloning machine; five minutes later two of you will be wheeled out. When one of you awakes, life will be normal—except that you will have been granted any wish of your choosing. When the other you awakes, life won't be normal; you will be escorted to a torture chamber back on Zaxtar, never to leave. And no, your lucky clone is not allowed to wish for your release. Do you accept the offer?

For most people, the answer is no. Since each of the clones really, truly *is* you, in accepting the offer you'd be guaranteeing that there will be a you who awakens to a lifetime of torment. Sure, there will also be a you who awakens to your usual life, augmented by the unlimited power of an arbitrary wish, but for the you on Zaxtar there'll be nothing but torture. The price is too high.

Anticipating your reluctance, the Zaxtarians up the ante. Same deal, but now they'll make a million and one copies of you. A million will wake up on a million identical-looking earths, with the power to fulfill any wish; one will get the Zaxtarian torture. Do you accept? At this point, you begin to waver. "Heck," you think, "the odds seem pretty good that *I* won't end up on Zaxtar but instead will wake up right here at home, wish in hand."

This last intuition is particularly relevant to the Many Worlds approach. If odds entered your thinking because you imagine that only one of the million and one clones is the "real" you, then you've not taken in the scenario fully. Each copy *is* you. There's a 100 percent certainty that one of you will wake up to an unbearable future. If this was indeed what led you to think in terms of odds, you need to let it go. However, probability may have entered your thinking in a more refined way. Imagine that you just agreed to the Zaxtarian offer and are now contemplating what it will be like to wake up tomorrow morning. Curled up under a warm duvet, just regaining consciousness but not yet having opened your eyes, you'll remember the Zaxtarian deal. At first it will seem like an unusually vivid nightmare, but as your heart starts to pound you'll recognize that it is real—that a million and one copies of you are in the process of waking up, with one of you destined for Zaxtar and the others about to be granted extraordinary power. "What are the odds," you'll ask yourself nervously, "that when I open my eyes I'll be shipping out to Zaxtar?"

Before the cloning there was no sensible way to speak of whether it was or wasn't likely that you'd be Zaxtar bound—it is absolutely certain that there will be such a you, so how could it be unlikely? But after the cloning, the situation seems different. Each clone experiences itself as the real you; indeed, each *is* the real you. But each copy is also a separate and distinct individual who can inquire about his or her own future. Each of the million and one copies can ask for the probability that *they* will go to Zaxtar. And since each knows that only one of the million and one will wake up to that outcome, each reckons that the odds of being that unlucky individual are low. Upon waking, a million will find their cheery expectation confirmed, and only one will not. So although there's nothing uncertain, nothing chancy, nothing probabilistic in the Zaxtarian scenario—again, no dice are rolled and no roulette wheels spun—probability nevertheless seems to enter. It does so through the subjective ignorance experienced by each individual clone regarding which outcome he or she will witness.

This suggests a tack for injecting probabilities into the Many Worlds approach. Before you undertake a given experiment, you are much like your precloned self. You contemplate all outcomes allowed by quantum mechanics and know that there's a 100 percent certainty that a copy of you will see each. Nothing at all chancy has made an appearance. You then undertake the experiment. At that point, as with the Zaxtarian scenario, a notion of probability presents itself. Each copy of you is an independent sentient being capable of wondering about which world he or she happens to inhabit—the likelihood, that is, that when the experiment's results are revealed, he or she will see this or that particular outcome. Probability enters through each inhabitant's subjective experience.

Everett's approach, which he described as "objectively deterministic" with probability "reappearing at the subjective level," resonated with this strategy. And he was thrilled by the direction. As he noted in the 1956 draft of his dissertation, the framework offered to bridge the position of Einstein (who famously believed that a fundamental theory of physics should not involve probability) and the position of Bohr (who was perfectly happy with a fundamental theory that did). According to Everett, the Many Worlds approach accommodated both positions, the difference between them merely being one of perspective. Einstein's perspective is the mathematical one in which the grand probability wave of all particles relentlessly evolves by the Schrödinger equation, with chance playing absolutely no role.* I like to picture Einstein soaring high

---

*This non-chancy perspective would argue strongly for abandoning the colloquial terminology that I've used, "probability wave," in favor of the technical name, "wavefunction."

above the many worlds of Many Worlds, watching as Schrödinger's equation fully dictates how the entire panorama unfolds, and happily concluding that even though quantum mechanics is correct, God *doesn't* play dice. Bohr's perspective is that of an inhabitant in one of the worlds, also happy, using probabilities to explain, with stupendous precision, those observations to which his limited perspective gives him access.

It's a captivating vision—Einstein and Bohr agreeing on quantum mechanics. But there are pesky details that for more than half a century have convinced many that it's still too early to sign on. Those who have studied Everett's thesis generally agree that while his intent was clear—a deterministic theory that to its inhabitants nevertheless appears probabilistic—he didn't convincingly spell out how to achieve it. For example, much in the spirit of material covered in Chapter 7, Everett sought to determine what a "typical" inhabitant of the many worlds would observe in any given experiment. But (unlike our focus in Chapter 7) in the Many Worlds approach, the inhabitants we need to contend with are all the same person; if you're the experimenter, they are all you, and collectively they will see a range of different outcomes. So who is the "typical" you?

Inspired by the Zaxtarian scenario, a natural suggestion is to count the number of yous who will see a given result; the outcome seen by the greatest number of yous would then qualify as typical. Or, more quantitatively, define the probability of a result to be proportional to the number of yous who see it. For simple examples, this works: in Figure 8.16, there's one of you who sees each outcome, and so you peg the odds at 50:50 for seeing one result or the other. That's good; the usual quantum mechanical prediction is also 50:50, because the probability wave heights at the two locations are equal.

However, consider a more general situation, such as that in Figure 8.17, in which the probability wave heights are unequal. If the wave is a hundred times larger at Strawberry Fields than at Grant's Tomb, then quantum mechanics predicts that you are a hundred times more likely to find the electron at Strawberry Fields. But in the Many Worlds approach, your measurement still generates one you who sees Strawberry Fields and another you who sees Grant's Tomb; the odds based on counting the number of yous is thus still

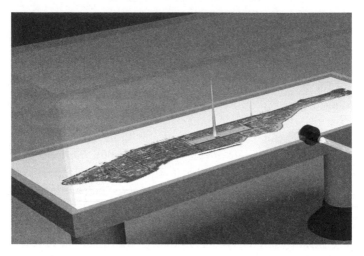

Figure 8.17 *The combined probability wave for you and your device encounters a probability wave that has multiple spikes of different magnitudes.*

50:50—the wrong result. The origin of the mismatch is clear. The number of yous who see one result or another is determined by the number of spikes in the probability wave. But the quantum mechanical probabilities are determined by something else—not by the number of spikes but by their relative heights. And it's these predictions, the quantum mechanical predictions, which have been convincingly confirmed by experiments.

Everett developed a mathematical argument that was meant to address this mismatch; many others have since pushed it further.[9] In broad strokes, the idea is that in calculating the odds of seeing one or another outcome, we should place ever-less weight on universes whose wave heights are ever smaller, as depicted symbolically in Figure 8.18. But this is perplexing. And controversial. Is the universe in which you find the electron at Strawberry Fields somehow a hundred times as genuine, or a hundred times as likely, or a hundred times as relevant as the one in which you find it at Grant's Tomb? These suggestions would surely create tension with the belief that every world is just as real as every other.

After more than fifty years, during which distinguished scientists

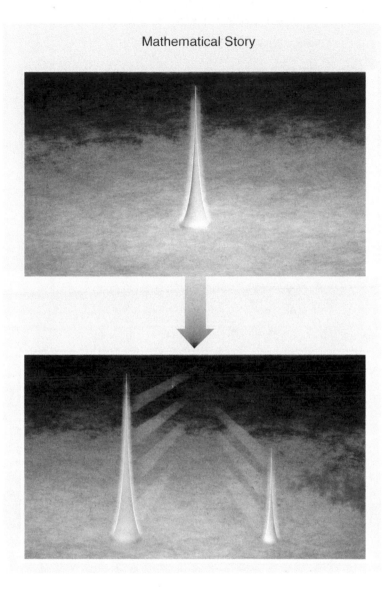

Figure 8.18 (a) *A schematic illustration of the evolution, dictated by Schrödinger's equation, of the combined probability wave for all the particles making up you and the measuring device, when you measure the position of an electron. The electron's own probability wave is spiked at two locations, but with unequal wave heights.*

Physical Story

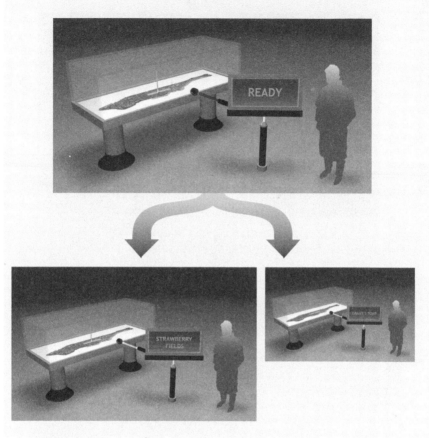

Figure 8.18 (b) *Some proposals suggest that in the Many Worlds approach, unequal wave heights imply that some worlds are less genuine, or less relevant, than others. There is controversy over what, if anything, this means.*

have revisited, revised, and extended Everett's arguments, many agree that the puzzles persist. Yet it remains seductive to imagine that the mathematically simple, totally bare-bones, profoundly revolutionary Many Worlds approach yields the probabilistic predictions that form the foundation of belief in quantum theory. This has inspired many other ideas, beyond the Zaxtarian-type reasoning, for joining probability and Many Worlds.[10]

A prominent proposal comes from a leading group of researchers at Oxford, including, among others, David Deutsch, Simon Saunders, David Wallace, and Hilary Greaves. They've developed a sophisticated line of attack that focuses on a seemingly boorish question. If you're a gambler, and you believe in the Many Worlds approach, what's the optimum strategy for placing bets on quantum mechanical experiments? Their answer, which they argue for mathematically, is that you'd bet just as Neils Bohr would. When speaking of maximizing your return, these authors have in mind something that would have sent Bohr into a tizzy—they're considering an average over the many inhabitants of the multiverse who claim to be you. But even so, their conclusion is that the numbers that Bohr and everyone since have been calculating and calling probabilities are the very numbers that should guide how you wager. That is, even though quantum theory is fully deterministic, you should treat the numbers *as if* they were probabilities.

Some are convinced that this completes Everett's program. Some are not.

The lack of consensus on the crucial question of how to treat probability in the Many Worlds approach is not all that unexpected. The analyses are highly technical and also deal with a topic—probability—that is notoriously tricky even outside its application to quantum theory. When you roll a die, we all agree that you have a 1 in 6 chance of getting a 3, and so we'd predict that over the course of, say, 1,200 rolls the number 3 will turn up about 200 times. But since it's possible, in fact likely, that the number of 3s will deviate from 200, what does the prediction mean? We want to say that it's highly probable that ⅙th of the outcomes will be 3s, but if we do that, then we've defined the probability of getting a 3 by invoking the concept of probability. We've gone circular. That's just a small taste of how the issues, beyond their intrinsic mathematical complexity, are conceptually slippery. Throw into the mix the added Many Worlds intricacy of "you" no longer referring to a single person, and it's no wonder researchers find ample points of contention. I have little doubt that full clarity will one day emerge, but not yet, and perhaps not for some time.

Predictions and Understanding

For all these controversies, quantum mechanics itself remains as successful as any theory in the history of ideas. The reason, as we've seen, is that for the kinds of experiments we can do in the laboratory, and for many of the observations we can make of astrophysical processes, we have a "quantum algorithm" that produces testable predictions. Use Schrödinger's equation to calculate the evolution of the relevant probability waves and use the results—the various wave heights—to predict the probability that you'll find one outcome or another. As far as predictions are concerned, why this algorithm works—whether the wave collapses upon measurement, whether all possibilities are realized in their own universes, whether some other process is at work—is secondary.

Some physicists argue that even calling the issue secondary accords it more status than it deserves. In their view, physics is *only* about making predictions, and as long as different approaches don't affect those predictions, why should we care which is ultimately correct? I offer three thoughts.

First, beyond making predictions, physical theories need to be mathematically coherent. The Copenhagen approach is a valiant effort, but it fails to meet this standard: at the critical moment of observation, it retreats into mathematical silence. That's a substantial gap. The Many Worlds approach attempts to fill it.[11]

Second, in some situations, the predictions of the Many Worlds approach *would* differ from those of the Copenhagen approach. In Copenhagen, the process of collapse would revise Figure 8.16a to have a single spike. So if you could cause the two waves depicted in the figure—representing macroscopically distinct situations—to interfere, generating a pattern similar to that in Figure 8.2c, it would establish that Copenhagen's hypothesized wave collapse didn't happen. Because of decoherence, as discussed earlier, it is an extraordinarily formidable task to do this, but, at least theoretically speaking, the Copenhagen and Many Worlds approaches yield different predictions.[12] It is an important point of principle. The Copenhagen and Many Worlds approaches are often referred to as different "interpretations" of quantum mechanics. This is an abuse of language. If two approaches can yield different predictions, you can't call them mere interpretations. Well, you can. And people do. But the terminology is off the mark.

Third, physics is not just about making predictions. If one day we were to find a black box that always and accurately predicted the outcome of our particle physics experiments and our astronomical observations, the existence of the box would not bring inquiry in these fields to a close. There's a difference between *making* predictions and *understanding* them. The beauty of physics, its raison d'être, is that it offers insights into *why* things in the universe behave the way they do. The ability to predict behavior is a big part of physics' power, but the heart of physics would be lost if it didn't give us a deep understanding of the hidden reality underlying what we observe. And should the Many Worlds approach be right, what a spectacular reality our unwavering commitment to understanding predictions will have uncovered.

I don't expect theoretical or experimental consensus to come in my lifetime concerning which version of reality—a single universe, a multiverse, something else entirely—quantum mechanics embodies. But I have little doubt that future generations will look back upon our work in the twentieth and twenty-first centuries as having nobly laid the basis for whatever picture finally emerges.

CHAPTER 9

Black Holes and Holograms

The Holographic Multiverse

Plato likened our view of the world to that of an ancient forebear watching shadows meander across a dimly lit cave wall. He imagined our perceptions to be but a faint inkling of a far richer reality that flickers beyond reach. Two millennia later, it seems that Plato's cave may be more than a metaphor. To turn his suggestion on its head, reality—not its mere shadow—may take place on a distant boundary surface, while everything we witness in the three common spatial dimensions is a projection of that faraway unfolding. Reality, that is, may be akin to a hologram. Or, really, a holographic movie.

Arguably the strangest parallel world entrant, the *holographic principle* envisions that all we experience may be fully and equivalently described as the comings and goings that take place at a thin and remote locus. It says that if we could understand the laws that govern physics on that distant surface, and the way phenomena there link to experience here, we would grasp all there is to know about reality. A version of Plato's shadow world—a parallel but thoroughly unfamiliar encapsulation of everyday phenomena— would *be* reality.

The journey to this peculiar possibility combines developments deep and far flung—insights from general relativity; from research on black holes; from thermodynamics; quantum mechanics; and, most recently, string theory. The thread linking these diverse areas is the nature of information in a quantum universe.

Information

Beyond John Wheeler's knack for finding and mentoring the world's most gifted young scientists (besides Hugh Everett, Wheeler's students included Richard Feynman, Kip Thorne, and, as we will shortly see, Jacob Bekenstein), he had an uncanny ability to identify issues whose exploration could change our fundamental paradigm of nature's workings. During a lunch we had at Princeton in 1998, I asked him what he thought the dominant theme in physics would be in the decades going forward. As he had already done frequently that day, he put his head down, as if his aging frame had grown weary of supporting such a massive intellect. But now the length of his silence left me wondering, briefly, whether he didn't want to answer or whether, perhaps, he had forgotten the question. He then slowly looked up and said a single word: "Information."

I wasn't surprised. For some time, Wheeler had been advocating a view of physical law quite unlike what a fledgling physicist learns in the standard academic curriculum. Traditionally, physics focuses on *things*—planets, rocks, atoms, particles, fields—and investigates the forces that affect their behavior and govern their interactions. Wheeler was suggesting that *things*—matter and radiation—should be viewed as secondary, as carriers of a more abstract and fundamental entity: information. It's not that Wheeler was claiming that matter and radiation were somehow illusory; rather, he argued that they should be viewed as the material manifestations of something more basic. He believed that information—where a particle is, whether it is spinning one way or another, whether its charge is positive or negative, and so on—forms an irreducible kernel at the heart of reality. That such information is instantiated in real particles, occupying real positions, having definite spins and charges, is something like an architect's drawings being realized as a skyscraper. The fundamental information is in the blueprints. The skyscraper is but a physical realization of the information contained in the architect's design.

From this perspective, the universe can be thought of as an information processor. It takes information regarding how things are now and produces information delineating how things will be at the next now, and the now after that. Our senses become aware of such processing by detecting how the physical environment changes over time. But the physical environment itself is emergent; it arises from the fundamental ingredient, information, and evolves according to the fundamental rules, the laws of physics.

I don't know whether such an information-theoretic stance will reach the dominance in physics that Wheeler envisioned. But recently, driven largely by the work of physicists Gerard 't Hooft and Leonard Susskind, a major shift in thinking has resulted from puzzling questions regarding information in one particularly exotic context: black holes.

Black Holes

Within a year of general relativity's publication, the German astronomer Karl Schwarzschild found the first exact solution to Einstein's equations, a result that determined the shape of space and time in the vicinity of a massive spherical object such as a star or a planet. Remarkably, not only had Schwarzschild found his solution while calculating artillery trajectories on the Russian front during World War I, but also he had beaten the master at his own game: to that point, Einstein had found only approximate solutions to the equations of general relativity. Impressed, Einstein publicized Schwarzschild's achievement, presenting the work before the Prussian Academy, but even so he failed to appreciate a point that would become Schwarzschild's most tantalizing legacy.

Schwarzschild's solution shows that familiar bodies like the sun and the earth produce a modest curvature, a gentle depression in the otherwise flat spacetime trampoline. This matched well the approximate results Einstein had managed to work out earlier, but by dispensing with approximations, Schwarzschild could go further. His exact solution revealed something startling: if enough mass were crammed into a small enough ball, a gravitational abyss would form. The spacetime curvature would become so extreme that anything venturing too close would be trapped. And because "anything" includes light, such regions would fade to black, a characteristic that inspired the early term "dark stars." The extreme warping would also bring time to a grinding halt at the star's edge; hence another early label, "frozen stars." Half a century later, Wheeler, who was nearly as adept at marketing as he was at physics, popularized such stars both within and beyond the scientific community with a new and more memorable name: black holes. It stuck.

When Einstein read Schwarzschild's paper, he agreed with the mathematics as applied to ordinary stars or planets. But as to what we now call black holes? Einstein scoffed. In those early days it was a challenge, even for Einstein, to fully understand the intricate mathematics of general relativity. While the modern understanding of black holes was still decades away, the intense folding of space and time already apparent in the equations was, in Einstein's view, too radical to be real. Much as he would resist cosmic expansion a few years later, Einstein refused to believe that such extreme configurations of matter were anything more than mathematical manipulations—based on his own equations—run amok.[1]

When you see the numbers that are involved, it's easy to come to a similar conclusion. For a star as massive as the sun to be a black hole, it would need to be squeezed into a ball about three kilometers across; a body as massive as the earth would become a black hole only if squeezed to a centimeter across. The idea that there might be such extreme arrangements of matter seems nothing short of ludicrous. Yet, in the decades since, astronomers have gathered overwhelming observational evidence that black holes are both real and plentiful. There is wide agreement that a great many galaxies are powered by an enormous black hole at their center; our very own Milky Way galaxy is believed to revolve around a black hole whose mass is about three million times that of the sun. There's even a chance, as discussed in Chapter 4, that the Large Hadron Collider may produce tiny black holes in the laboratory by packing the mass (and energy) of violently colliding protons into such a minuscule volume that Schwarzschild's result again applies, though on microscopic scales. Extraordinary emblems of math's ability to illuminate the dark corners of the cosmos, black holes have become the cynosures of modern physics.

Besides serving as a boon for observational astronomy, black holes have also been a fertile source of inspiration for theoretical research by providing a mathematical playground in which physicists can push ideas to their limits, conducting pen-and-paper explorations of one of nature's most extreme environments. As a weighty case in point, in the early 1970s Wheeler realized that when the venerable Second Law of Thermodynamics—a guiding light for over a century in understanding the interplay between energy, work, and heat—was considered in the vicinity of a black hole, it seemed to flounder. The fresh thinking of Wheeler's young graduate student Jacob Bekenstein came to the rescue, and in doing so planted the seeds of the holographic proposal.

The Second Law

The aphorism "less is more" takes many forms. "Let's have the executive summary." "Just the facts." "TMI." "You had me at hello." These idioms are so common because every moment of every day we're bombarded with information. Thankfully, in most cases our senses pare down the details to those that really matter. If I'm out on the savanna and encounter a lion, I don't care about the motion of every photon reflecting off his body. Way TMI. I just want particular overall features of those photons, the very ones our eyes have evolved to sense and our brains to rapidly decode. Is the lion coming toward me? Is he crouched and stalking? Provide me with a moment-to-moment catalog of every reflected photon and, sure, I'll be in possession of all the details. What I won't have is any understanding. Less would indeed be very much more.

Similar considerations play a central role in theoretical physics. Sometimes we want to know every microscopic detail of a system we're studying. At the locations along the Large Hadron Collider's seventeen-mile-long tunnel where particles are steered into head-on collisions, physicists have placed mammoth detectors capable of tracking, with extreme precision, the motion of the particle fragments produced. Essential for gaining insight into the fundamental laws of particle physics, the data are so detailed that a year's worth would fill a stack of DVDs about fifty times as tall as the Empire State Building. But, as in that impromptu meeting with a lion, there are other situations in physics where that level of detail would obscure, not clarify. A nineteenth-century branch of physics called *thermodynamics* or, in its more modern incarnation, *statistical mechanics*, focuses on such systems. The steam engine, the technological innovation that initially drove thermodynamics—as well as the Industrial Revolution—provides a good illustration.

The core of a steam engine is a vat of water vapor that expands when heated, driving the engine's piston forward, and contracts when cooled, returning the piston to its initial position, ready to drive forward once again. In the late nineteenth and early twentieth centuries, physicists worked out the molecular underpinnings of matter, which among other things provided a microscopic picture of the steam's action. As steam is heated, its H_2O molecules pick up increasing speed and career into the underside of the piston. The hotter they are, the faster they go and the bigger the push. A simple insight, but one essential to thermodynamics, is that to understand the steam's force we don't need the details of which particular molecules happen to have this or that velocity or which happen to hit the piston precisely here or there. Provide me with a list of billions and billions of molecular trajectories, and I'll look at you just as blankly as I would if you listed the photons bouncing off the lion. To figure out the piston's push, I need only the average number of molecules that will hit it in a given time interval, and the average speed they'll have when they do. These are much coarser data, but it's exactly such pared-down information that's useful.

In crafting mathematical methods for systematically sacrificing detail in favor of such higher-level aggregate understanding, physicists honed a wide range of techniques and developed a number of powerful concepts. One such concept, encountered briefly in earlier chapters, is *entropy*. Initially introduced in the mid-nineteenth century to quantify energy dissipation in combustion engines, the modern view, emerging from Ludwig Boltzmann's work in the 1870s, is that entropy provides a characterization of how finely arranged—or not—the constituents of a given system need to be for it to have the overall appearance that it does.

To get a feel for this, imagine that Felix is frantic because he believes the apartment he shares with Oscar has been broken into. "They've ransacked us!" he tells Oscar. Oscar brushes him off— surely Felix is having one of his moments. To make his point, Oscar throws open the door to his bedroom, revealing clothing, empty pizza boxes, and crushed beer cans strewn everywhere. "It looks just like it always does," Oscar barks. Felix isn't swayed. "Of course it looks the same—ransack a pigsty and you get a pigsty. But look at my room." And he throws open his own door. "Ransacked," mocks Oscar; "it's neater than a straight whiskey." "Neat, yes. But the intruders have left their mark. My vitamin bottles? Not lined up in order of size. My collected works of Shakespeare? Out of alphabet- ical order. And my sock drawer? Look at this—some black pairs are in the blue bin! Ransacked, I tell you. Obviously ransacked."

Putting Felix's hysteria aside, the scenario makes plain a simple but essential point. When something is highly disordered, like Oscar's room, a great many possible rearrangements of its con- stituents leave its overall appearance intact. Grab the twenty-six crumpled shirts that were scattered across the bed, floor, and dresser, and toss them this way and that, fling the forty-two crushed beer cans randomly here and there, and the room will look the same. But when something is highly ordered, like Felix's room, even small rearrangements are easily detected.

This distinction underlies Boltzmann's mathematical definition of entropy. Take any system and count the number of ways its constituents can be rearranged without affecting its gross, overall, macroscopic appearance. That number is the system's entropy.* If there's a large number of such rearrangements, then entropy is high: the system is highly disordered. If the number of such rearrange- ments is small, entropy is low: the system is highly ordered (or, equivalently, has low disorder).

For more conventional examples, consider a vat of steam and a cube of ice. Focus only on their overall macroscopic properties, those you can measure or observe without accessing the detailed state of either's molecular constituents. When you wave your hand

---

*This loose definition will suffice for now; in a moment, I'll be more precise.

through the steam, you rearrange the positions of billions upon billions of H_2O molecules, and yet the vat's uniform haze looks undisturbed. But randomly change the positions and speeds of that many molecules in a piece of ice, and you'll immediately see the impact—the ice's crystalline structure will be disrupted. Fissures and fractures will appear. The steam, with H_2O molecules randomly flitting through the container, is highly disordered; the ice, with H_2O molecules arranged in a regular, crystalline pattern, is highly ordered. The entropy of the steam is high (many rearrangements will leave it looking the same); the entropy of the ice is low (few rearrangements will leave it looking the same).

By assessing the sensitivity of a system's macroscopic appearance to its microscopic details, entropy is a natural concept in a mathematical formalism that focuses on aggregate physical properties. The Second Law of Thermodynamics developed this line of insight quantitatively. The law states that, over time, the total entropy of a system will increase.[2] Understanding why requires only the most elementary grasp of chance and statistics. By definition, a higher-entropy configuration can be realized through many more microscopic arrangements than a lower-entropy configuration. As a system evolves, it's overwhelmingly likely to pass through higher-entropy states since, simply put, there are more of them. *Many* more. When bread is baking, you smell it throughout the house because there are trillions more arrangements of the molecules streaming from the bread that are spread out, yielding a uniform aroma, than there are arrangements in which the molecules are all tightly packed in a corner of the kitchen. The random motions of the hot molecules will, with near certainty, drive them toward one of the numerous spread-out arrangements, and not toward one of the few clustered configurations. The collection of molecules evolves, that is, from lower to higher entropy, and that's the Second Law in action.

The idea is general. Glass shattering, a candle burning, ink spilling, perfume pervading: these are different processes, but the statistical considerations are the same. In each, order degrades to disorder and does so because there are so many ways to be disordered. The beauty of this kind of analysis—the insight provided one of the most potent "Aha!" moments in my physics education—is that, without getting lost in the microscopic details, we have a guiding principle to explain why a great many phenomena unfold the way they do.

Notice, too, that, being statistical, the Second Law does not say that entropy *can't* decrease, only that it is extremely unlikely to do so. The milk molecules you just poured into your coffee might, as a result of their random motions, coalesce into a floating figurine of Santa Claus. But don't hold your breath. A floating milk Santa has very low entropy. If you move around a few billion of his molecules, you'll notice the result—Santa will lose his head or an arm, or he'll disperse into abstract white tendrils. By comparison, a configuration in which the milk molecules are uniformly spread around has enormously more entropy: a vast number of rearrangements continue to look like ordinary coffee with milk. With a huge likelihood, then, the milk poured into your dark coffee will turn it a uniform tan, with nary a Santa in sight. Similar considerations hold for the vast majority of high-to-low-entropy evolutions, making the Second Law appear inviolable.

The Second Law and Black Holes

Now to Wheeler's point about black holes. Back in the early 1970s, Wheeler noticed that when black holes amble onto the scene, the Second Law appears compromised. A nearby black hole seems to provide a ready-made and reliable means for reducing overall entropy. Throw whatever system you're studying—smashed glass, burned candles, spilled ink—into the hole. Since nothing escapes from a black hole, the system's disorder would appear permanently gone. Crude the approach may be, but it seems easy to lower total entropy if you have a black hole to work with. The Second Law, many thought, had met its match.

Wheeler's student Bekenstein was not convinced. Perhaps, Bekenstein suggested, entropy is not lost to the black hole but merely transferred to it. After all, no one claimed that, in gorging themselves on dust and stars, black holes provide a mechanism for violating the First Law of Thermodynamics, the conservation of energy. Instead, Einstein's equations show that when a black hole gorges, it gets bigger and heftier. The energy in a region can be redistributed, with some falling into the hole and some remaining outside, but the total is preserved. Maybe, Bekenstein suggested, the same idea applies to entropy. Some entropy stays outside a given black hole and some entropy falls in, but none gets lost.

This sounds reasonable, but experts shot Bekenstein down. Schwarzschild's solution, and much work that followed, seemed to establish that black holes are the epitome of order. Infalling matter and radiation, however messy and disordered, are crushed to infinitesimal size at a black hole's center: a black hole is the ultimate in orderly trash compaction. True, no one knows exactly what happens during such powerful compression, because the extremes of curvature and density disrupt Einstein's equations; but there just doesn't seem to be any capacity for a black hole's center to harbor disorder. And outside the center, a black hole is nothing but an empty region of spacetime extending to the boundary of no return—the event horizon—as in Figure 9.1. With no atoms or molecules wafting this way and that, and thus no constituents to rearrange, a black hole would seem to be entropy-free.

In the 1970s, this view was reinforced by the so-called *no hair theorems*, which established mathematically that black holes, much like the bald performers of Blue Man Group, have a dearth of distinguishing characteristics. According to the theorems, any two black holes that have the same mass, charge, and angular momentum (rate of rotation) are *identical*. Lacking any other intrinsic traits—as the Blue Men lack bangs, mullets, or dreads—black holes seemed to lack the underlying differences that would harbor entropy.

By itself, this was a fairly convincing argument, but there was a yet more damning consideration that seemed to definitively undercut Bekenstein's idea. According to basic thermodynamics, there's a close association between entropy and temperature. Temperature is a measure of the average motion of an object's constituents: hot objects have fast-moving constituents, cold objects have slow-moving constituents. Entropy is a measure of the possible rearrangements of

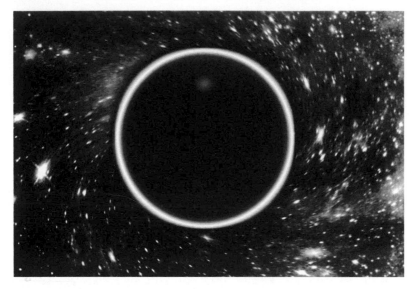

Figure 9.1 *A black hole comprises a region of spacetime surrounded by a surface of no return, the event horizon.*

these constituents that, from a macroscopic viewpoint, would go unnoticed. Both entropy and temperature thus depend on aggregate features of an object's constituents; they go hand in hand. When worked out mathematically, it became clear that if Bekenstein was right and black holes carried entropy, they should also have a temperature.[3] *That* idea set off alarm bells. Any object with a nonzero temperature radiates. Hot coal radiates visible light; we humans, typically, radiate in the infrared. If a black hole has a nonzero temperature, the very laws of thermodynamics that Bekenstein was seeking to preserve state that it too should radiate. But that conflicts blatantly with the established understanding that nothing can escape a black hole's gravitational grip. Most everyone concluded that Bekenstein was wrong. Black holes do not have a temperature. Black holes do not harbor entropy. Black holes are entropy sinkholes. In their presence, the Second Law of Thermodynamics fails.

Despite the evidence mounting against him, Bekenstein had one tantalizing result on his side. In 1971, Stephen Hawking realized that black holes obey a curious law. If you have a collection of black holes with various masses and sizes, some engaged in stately orbital waltzes, others pulling in nearby matter and radiation, and still others crashing into each other, *the total surface area of the black holes increases over time.* By "surface area," Hawking meant the area of each black hole's event horizon. Now, there are many results in physics that ensure quantities don't change over time (conservation of energy, conservation of charge, conservation of momentum, and so on), but there are very few that require quantities to increase. It was natural, then, to consider a possible relation between Hawking's result and the Second Law. If we envision that, somehow, the surface area of a black hole is a measure of the entropy it contains, then the increase in total surface area could be read as an increase in total entropy.

It was an enticing analogy, but no one bought it. The similarity between Hawking's area theorem and the Second Law was, in almost everyone's view, nothing more than a coincidence. Until, that is, a few years later, when Hawking completed one of the most influential calculations in modern theoretical physics.

Hawking Radiation

Because quantum mechanics plays no role in Einstein's general relativity, Schwarzschild's black hole solution is based purely in classical physics. But proper treatment of matter and radiation—of particles like photons, neutrinos, and electrons that can carry mass, energy, and entropy from one location to another—requires quantum physics. To fully assess the nature of black holes and understand how they interact with matter and radiation, we must update Schwarzschild's work to include quantum considerations. This isn't easy. Notwithstanding advances in string theory (as well as in other approaches we haven't discussed, such as loop quantum gravity, twistors, and topos theory), we are still at an early stage in our attempt to meld quantum physics and general relativity. Back in the 1970s, there was still less theoretical basis for understanding how quantum mechanics would affect gravity.

Even so, a number of early researchers developed a partial union of quantum mechanics and general relativity by considering quantum fields (the quantum part) evolving in a fixed but curved spacetime environment (the general relativity part). As I pointed out in Chapter 4, a full union would, at the very least, consider not only the quantum jitters of fields within spacetime but the jitters of spacetime itself. To facilitate progress, the early work steadfastly avoided this complication. Hawking embraced the partial union and studied how quantum fields would behave in a very particular spacetime arena: that created by the presence of a black hole. What he found knocked physicists clear off their seats.

A well-known feature of quantum fields in ordinary, empty, uncurved spacetime is that their jitters allow pairs of particles, for instance an electron and its antiparticle the positron, to momentarily erupt out of the nothingness, live briefly, and then smash into each other, with mutual annihilation the result. This process, *quantum pair production*, has been intensively studied both theoretically and experimentally, and is thoroughly understood.

A novel characteristic of quantum pair production is that while one member of the pair has positive energy, the law of energy conservation dictates that the other must have an equal amount of *negative* energy—a concept that would be meaningless in a classical universe.* But the uncertainty principle provides a window of weirdness whereby negative-energy particles are allowed as long as they don't overstay their welcome. If a particle exists only fleetingly, quantum uncertainty establishes that no experiment will have adequate time, even in principle, to determine the sign of its energy. This is the very reason why the particle pair is condemned by quantum laws to swift annihilation. So, over and over again, quantum jitters result in particle pairs being created and annihilated, created and annihilated, as the unavoidable rumbling of quantum uncertainty plays itself out in otherwise empty space.

*In Chapter 3, we discussed how the energy embodied by a gravitational field can be negative; this energy, however, is potential energy. The energy we're discussing here, kinetic energy, comes from the electron's mass and its motion. In classical physics this has to be positive.

Hawking reconsidered such ubiquitous quantum jitters not in the setting of empty space but near the event horizon of a black hole. He found that sometimes events look much as they ordinarily do. Pairs of particles are randomly created; they quickly find each other; they are destroyed. But every so often something new happens. If the particles are formed sufficiently close to the black hole's edge, one can get sucked in while the other careens into space. In the absence of a black hole this never happens, because if the particles failed to annihilate each other then the one with negative energy would outlive the protective haze of quantum uncertainty. Hawking realized that the black hole's radical twisting of space and time can cause particles that have negative energy, as determined by anyone outside the hole, to appear to have *positive* energy to any unfortunate observer inside the hole. In this way, a black hole provides the negative energy particles a safe haven, and so eliminates the need for a quantum cloak. The erupting particles can forgo mutual annihilation and blaze their own separate trails.[4]

The positive-energy particles shoot outward from just above the black hole's event horizon, so to someone watching from afar they look like radiation, a form since named *Hawking radiation*. The negative-energy particles are not directly seen, because they fall into the black hole, but they nevertheless have a detectable impact. Much as a black hole's mass increases when it absorbs anything that carries positive energy, so its mass decreases when it absorbs anything that carries negative energy. In tandem, these two processes make the black hole resemble a piece of burning coal: the black hole emits a steady outward stream of radiation as its mass gets ever smaller.[5] When quantum considerations are included, black holes are thus not completely black. This was Hawking's bolt from the blue.

Which is not to say that your average black hole is red hot, either. As particles stream from just outside the black hole, they fight an uphill battle to escape the strong gravitational pull. In doing so, they expend energy and, because of this, cool down substantially. Hawking calculated that an observer far from the black hole would find that the temperature for the resulting "tired" radiation was inversely proportional to the black hole's mass. A huge black hole, like the one at the center of our galaxy, has a temperature that's less than a trillionth of a degree above absolute zero. A black hole with the mass of the sun would have a temperature less than a millionth of a degree, minuscule even compared with the 2.7-degree cosmic background radiation left to us by the big bang. For a black hole's temperature to be high enough to barbecue the family dinner, its mass would need to be about a ten-thousandth of the earth's, extraordinarily small by astrophysical standards.

But the magnitude of a black hole's temperature is secondary. Although the radiation coming from distant astrophysical black holes won't light up the night sky, the fact that they *do* have a temperature, that they *do* emit radiation, suggests that the experts had too quickly rejected Bekenstein's suggestion that black holes *do* have entropy. Hawking then nailed the case. His theoretical calculations determining a given black hole's temperature and the radiation it emits gave him all the data he needed to determine the amount of entropy the black hole should contain, according to the standard laws of thermodynamics. And the answer he found is proportional to the surface area of the black hole, just as Bekenstein had proposed.

So by the end of 1974, the Second Law was law once again. The insights of Bekenstein and Hawking established that in any situation, total entropy increases, as long as you account for not only the entropy of ordinary matter and radiation but also that contained within black holes, as measured by their total surface area. Rather than being entropy sinks that subvert the Second Law, black holes play an active part in upholding the law's pronouncement of a universe with ever-increasing disorder.

The conclusion provided a welcome relief. To many physicists, the Second Law, emerging from seemingly unassailable statistical considerations, came as close to sacred as just about anything in science. Its restoration meant that, once again, all was right with the world. But, in time, a vital little detail in the entropy accounting made it clear that the Second Law's balance sheet was not the deepest issue in play. *That* honor went to identifying *where entropy is stored*, a matter whose importance becomes clear when we recognize the deep link between entropy and the central theme of this chapter: information.

Entropy and Hidden Information

So far, I've described entropy, loosely, as a measure of disorder and, more quantitatively, as the number of rearrangements of a system's microscopic constituents that leave its overall macroscopic features unchanged. I've left implicit, but will now make explicit, that you can think of entropy as measuring the *gap in information* between the data you have (those overall macroscopic features) and the data you don't (the system's particular microscopic arrangement). Entropy measures the additional information hidden within the microscopic details of the system, which, should you have access to it, would distinguish the configuration at a micro level from all the macro look-alikes.

To illustrate, imagine that Oscar has straightened up his room, except that the thousand silver dollars he won in last week's poker game remain scattered across the floor. Even after he gathers them in a neat cluster, Oscar sees only a haphazard assortment of dollar coins, some heads and others tails. Were you to randomly change some heads to tails and other tails to heads, he'd never notice—evidence that the thousand-dropped-silver-dollar system has high entropy. Indeed, this example is so explicit that we can do the entropy counting. If there were only two coins, there'd be four possible configurations: (heads, heads), (heads, tails), (tails, heads), and (tails, tails)—two possibilities for the first dollar, times two for the second. With three coins, there'd be eight possible arrangements: (heads, heads, heads), (heads, heads, tails), (heads, tails, heads), (heads, tails, tails), (tails, heads, heads), (tails, heads, tails), (tails, tails, heads), (tails, tails, tails), arising from two possibilities for the first, times two for the second, times two for the third. With a thousand coins, the number of possibilities follows exactly the same pattern—a factor of 2 for each coin—yielding a total of 2^{1000}, which is 10715086071862673209484250490600018105614048117055336074437503883703510511249361224931983788156958581275946729175531468251871452856923140435984577574698574803934567774824230985421074605062371141877954182153046474983581941267398767559165543946077062914571196477686542167660429831652624386837205668069376. The vast majority of these heads-tails arrangements would have no distinguishing features, so they would not stand out in any way. Some *would*, for instance, if all 1,000 coins were heads or all were tails, or if 999 were heads, or 999 tails. But the number of such unusual configurations is so extraordinarily small, compared with the huge total number of possibilities, that removing them from the count would hardly make a difference.*

From our earlier discussion, you'd deduce that the number 2^{1000} is the entropy of the coins. And, for some purposes, that conclusion would be fine. But to draw the strongest link between entropy and information, I need to sharpen up the description I gave earlier. The entropy of a system is *related to* the number of indistinguishable rearrangements of its constituents, but properly speaking is not equal to the number itself. The relationship is expressed by a mathematical operation called a *logarithm*; don't be put off if this brings back bad memories of high school math class. In our coin example, it simply means that you pick out the exponent in the number of rearrangements—that is, the entropy is defined as 1,000 rather than 2^{1000}.

Using logarithms has the advantage of allowing us to work with more manageable numbers, but there's a more important motivation. Imagine I ask you how much information you'd need to supply in order to describe one particular heads-tails arrangement of the 1,000 coins. The simplest response is that you'd need to provide the list—heads, heads, tails, heads, tails, tails . . .—that specifies the disposition of each of the 1,000 coins. Sure, I respond, that would tell me the details of the configuration, but that wasn't my question. I asked *how much information* is contained in that list.

So, you start to ponder. What actually *is* information, and what does it do? Your response is simple and direct. Information answers questions. Years of research by mathematicians, physicists, and computer scientists have made this precise. Their investigations have established that the most useful measure of information content is the *number of distinct yes-no questions the information can*

---

*Besides flipping the coins, you could also swap around their locations, but for the purpose of illustrating the main ideas, we can safely ignore this complication.

answer. The coins' information answers 1,000 such questions: Is the first dollar heads? Yes. Is the second dollar heads? Yes. Is the third dollar heads? No. Is the fourth dollar heads? No. And so on. A datum that can answer a single yes-no question is called a *bit*—a familiar computer-age term that is short for *binary digit*, meaning a 0 or a 1, which you can think of as a numerical representation of *yes* or *no*. The heads-tails arrangement of the 1,000 coins thus contains 1,000 bits' worth of information. Equivalently, if you take Oscar's macroscopic perspective and focus only on the coins' overall haphazard appearance while eschewing the "microscopic" details of the heads-tails arrangement, the coins' "hidden" information content is 1,000 bits.

Notice that the value of the entropy and the amount of hidden information are equal. That's no accident. The number of possible heads-tails rearrangements *is* the number of possible answers to the 1,000 questions—(yes, yes, no, no, yes, . . .) or (yes, no, yes, yes, no, . . .) or (no, yes, no, no, no, . . .), and so on—namely, 2^{1000}. With entropy defined as the logarithm of the number of such rearrangements—1,000 in this case—entropy *is* the number of yes-no questions any one such sequence answers.

I've focused on the 1,000 coins so as to offer a specific example, but the link between entropy and information is general. The microscopic details of any system contain information that's hidden when we take account of only macroscopic, overall features. For instance, you know the temperature, pressure, and volume of a vat of steam, but did an H_2O molecule just hit the upper right-hand corner of the box? Did another just hit the midpoint of the lower left edge? As with the dropped dollars, *a system's entropy is the number of yes-no questions that its microscopic details have the capacity to answer, and so the entropy is a measure of the system's hidden information content.*[6]

Entropy, Hidden Information, and Black Holes

How does this notion of entropy, and its relation to hidden information, apply to black holes? When Hawking worked out the detailed quantum mechanical argument linking a black hole's entropy to its

surface area, he not only brought quantitative precision to Bekenstein's original suggestion, he also provided an algorithm for calculating it. Take the event horizon of a black hole, Hawking
instructed, and divide it into a gridlike pattern in which the sides of
each cell are one Planck length (10^{-33} centimeters) long. Hawking
proved mathematically that the black hole's entropy is the number
of such cells needed to cover its event horizon—the black hole's
surface area, that is, as measured in square Planck units (10^{-66}
square centimeters per cell). In the language of hidden information, it's as if each such cell secretly carries a single bit, a 0 or a 1,
that provides the answer to a single yes-no question delineating
some aspect of the black hole's microscopic makeup.[7] This is
schematically illustrated in Figure 9.2.

Einstein's general relativity, as well as the black hole no-hair
theorems, ignores quantum mechanics and so completely misses
this information. Choose values for its mass, its charge, and its

Figure 9.2 *Stephen Hawking showed mathematically that the
entropy of a black hole equals the number of Planck-sized cells
that it takes to cover its event horizon. It's as if each cell carries
one bit, one basic unit of information.*

angular momentum, and you've uniquely specified a black hole, says general relativity. But the most straightforward reading of Bekenstein and Hawking tells us you haven't. Their work established that there must be many different black holes with the same macroscopic features that, nevertheless, differ microscopically. And much as is the case in more commonplace settings—coins on the floor, steam in a vat—the black hole's entropy reflects information hidden within the finer details.

Exotic as black holes may be, these developments suggested that, when it comes to entropy, black holes behave much like everything else. But the results also raised puzzles. Although Bekenstein and Hawking tell us how much information is hidden within a black hole, they don't tell us what that information is. They don't tell us the specific yes-no questions the information answers, nor do they even specify the microscopic constituents that the information is meant to describe. The mathematical analyses pinned down the *quantity* of information a given black hole contains, without providing insight into the information itself.[8]

These were—and remain—perplexing issues. But there's yet another puzzle, one that seems even more basic: Why would the amount of information be dictated by the area of the black hole's surface? I mean, if you asked me how much information was stored in the Library of Congress, I'd want to know about the available space *inside* the Library of Congress. I'd want to know the capacity, within the library's cavernous interior, for shelving books, filing microfiche, and stacking maps, photographs, and documents. The same goes for the information in my head, which seems tied to the volume of my brain, the available space for neural interconnections. And it goes for the information in a vat of steam, which is stored in the properties of the particles that fill the container. But, surprisingly, Bekenstein and Hawking established that for a black hole, the information storage capacity is determined not by the volume of its interior but by the area of its surface.

Prior to these results, physicists had reasoned that since the Planck length (10^{-33} centimeters) was apparently the shortest length for which the notion of "distance" continues to have meaning, the smallest meaningful volume would be a tiny cube whose edges were each one Planck length long (a volume of 10^{-99} cubic centimeters). A reasonable conjecture, widely believed, was that irrespective of future technological breakthroughs, the smallest possible volume could store no more than the smallest unit of information—one bit. And so the expectation was that a region of space would max out its information storage capacity when the number of bits it contained equaled the number of Planck cubes that could fit inside it. That Hawking's result involved the Planck length was therefore not surprising. The surprise was that the black hole's storehouse of hidden information was determined by the number of Planck-sized squares covering its surface and not by the number of Planck-sized cubes filling its volume.

This was the first hint of holography—information storage capacity determined by the area of a bounding surface and not by the volume interior to that surface. Through twists and turns across three subsequent decades, this hint would evolve into a dramatic new way of thinking about the laws of physics.

Locating a Black Hole's Hidden Information

The Planckian chessboard with 0s and 1s scattered across the event horizon, Figure 9.2, is a symbolic illustration of Hawking's result for the amount of information harbored by a black hole. But how literally can we take the imagery? When the math says that a black hole's store of information is measured by its surface area, does that merely reflect a numerical accounting, or does it mean that the black hole's surface is where the information is actually stored?

It's a deep issue and has been pursued for decades by some of the most renowned physicists.* The answer depends sensitively on whether you view the black hole from the outside or from the inside—and from the outside, there's good reason to believe that information is indeed stored at the horizon.

To anyone familiar with the finer details of how general relativ-

---

*If you're interested in the full story, I highly recommend Leonard Susskind's excellent book *The Black Hole Wars*.

ity depicts black holes, this is an astoundingly odd claim. General relativity makes clear that were you to fall through a black hole's event horizon, you would encounter nothing—no material surface, no signposts, no flashing lights—that would in any way mark your crossing the boundary of no return. It's a conclusion that derives from one of Einstein's simplest but most pivotal insights. Einstein realized that when you (or any object) assume free-fall motion, you become weightless; jump from a high diving board, and a scale strapped to your feet falls with you and so its reading drops to zero. In effect, you cancel gravity by giving in to it fully. From this, Einstein leaped to an immediate consequence. Based on what you experience in your immediate environment, there's no way for you to distinguish between freely falling toward a massive object and freely floating in the depths of empty space: in both situations you are perfectly weightless. Sure, if you look beyond your immediate environment and see, say, the earth's surface rapidly getting closer, that's a pretty good clue that it's time to pull your parachute cord. But if you are confined to a small, windowless capsule, the experiences of free fall and free float are indistinguishable.[9]

In the early years of the twentieth century, Einstein seized on this simple but profound interconnection between motion and gravity; after a decade of development, he leveraged it into his general theory of relativity. Our application here is more modest. Suppose you are in that capsule and are freely falling not toward the earth but toward a black hole. The very same reasoning ensures that there's no way for your experience to be any different from floating in empty space. And that means that nothing special or unusual will happen as you freely fall through the black hole's horizon. When you eventually hit the black hole's center, you'll no longer be in free fall, and that experience will certainly distinguish itself. And spectacularly so. But until then, you could just as well be aimlessly floating in the dark depths of outer space.

This realization renders the black hole's entropy all the more puzzling. If as you pass through the horizon of a black hole you find nothing there, nothing at all to distinguish it from empty space, how can it store information?

An answer that has gained traction over the last decade resonates with the duality theme encountered in early chapters. Recall that duality refers to a situation in which there are complementary perspectives that seem completely different, and yet are intimately connected through a shared physical anchor. The Albert-Marilyn image of Figure 5.2 provides a good visual metaphor; mathematical examples come from the mirror shapes of string theory's extra dimensions (Chapter 4) and the naïvely distinct yet dual string theories (Chapter 5). In recent years, researchers, led by Susskind, have realized that black holes present another context in which complementary yet widely divergent perspectives yield fundamental insight.

One essential perspective is yours, as you freely fall toward a black hole. Another is that of a distant observer, watching your journey through a powerful telescope. The remarkable thing is that as you pass uneventfully through a black hole's horizon, the distant observer perceives a very different sequence of events. The discrepancy has to do with the black hole's Hawking radiation.* When the distant observer measures the Hawking radiation's temperature, she finds it to be tiny; let's say it's 10^{-13} K, indicating that the black hole is roughly the size of the one at the center of our galaxy. But the distant observer knows that the radiation is cold only because the photons, traveling to her from just outside the horizon, have expended their energy valiantly fighting against the black hole's gravitational pull; in the description I gave earlier, the photons are tired. She deduces that as you get ever closer to the black hole's horizon, you'll encounter ever-fresher photons, ones that have only just begun their journey and so are ever more energetic and ever hotter. Indeed, as she watches you approach to within a hair's breadth of the horizon, she sees your body bombarded by increasingly intense Hawking radiation, until finally all that's left is your charred remains.

Happily, however, what you experience is much more pleasant.

---

*The reader familiar with black holes will note that even without the quantum considerations that lead to Hawking radiation, the two perspectives would differ with regards to the rate of time's passage. Hawking radiation makes the perspectives yet more distinct.

You don't see or feel or otherwise obtain any evidence of this hot radiation. Again, because your free-fall motion cancels the effects of gravity,[10] your experience is indistinguishable from that of floating in empty space. And one thing we know for sure is that when you float in empty space, you don't suddenly burst into flames. So the conclusion is that from your perspective, you pass seamlessly through the horizon and (less happily) hurtle on toward the black hole's singularity, while from the distant observer's perspective, you are immolated by a scorching corona that surrounds the horizon.

Which perspective is right? The claim advanced by Susskind and others is that both are. Granted, this is hard to square with ordinary logic—the logic by which you are either alive *or* not alive. But this is no ordinary situation. Most saliently, the wildly different perspectives can never confront each other. You can't climb out of the black hole and prove to the distant observer that you are alive. And, as it turns out, the distant observer can't jump into the black hole and confront you with evidence that you're not. When I said that the distant observer "sees" you immolated by the black hole's Hawking radiation, that was a simplification. The distant observer, by closely examining the tired radiation that reaches her, can piece together the story of your fiery demise. But for the information to reach her takes time. And the math shows that by the time she can conclude you've burned, she won't have enough time left to then hop into the black hole and catch up with you before you're destroyed by the singularity. Perspectives can differ, but physics has a built-in fail-safe against paradoxes.

What about information? From your perspective, all your information, stored in your body and brain and in the laptop you're holding, passes with you through the black hole's horizon. From the perspective of the distant observer, all the information you carry is absorbed by the layer of radiation incessantly bubbling just above the horizon. The bits contained in your body, brain, and laptop would be preserved, but would become thoroughly scrambled as they joined, jostled, and intermingled with the sizzling hot horizon. Which means that to the distant observer, the event horizon *is* a real place, populated by real things that give physical expression to the information symbolically depicted in the chessboard, Figure 9.2.

The conclusion is that the distant observer—us—infers that a black hole's entropy is determined by the area of its horizon because the horizon is where the entropy is stored. Said that way, it seems utterly sensible. But don't lose sight of how unexpected it is that the storage capacity isn't set by the black hole's volume. And, as we will now see, this result doesn't merely highlight a peculiar feature of black holes. Black holes don't just tell us about how black holes store information. Black holes inform us about information storage in any context. This paves a direct path to the holographic perspective.

Beyond Black Holes

Consider any object or collection of objects—the collections of the Library of Congress, all of Google's computers, the CIA's archives—situated in some region of space. For ease, imagine that we highlight the region by surrounding it with an imaginary sphere, as in Figure 9.3a. Assume further that the total mass of the objects, compared with the volume they fill, is of such an ordinary run-of-the-mill magnitude that it's nowhere near what it takes to

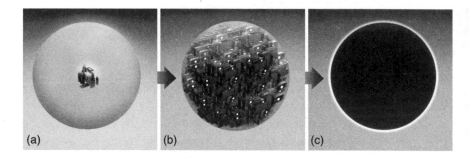

Figure 9.3 (a) *A variety of objects that store information, situated within a well-marked region of space.* (b) *We augment the region's capacity for storing information.* (c) *When the amount of matter crosses a threshold (whose value can be calculated from general relativity),*[11] *the region becomes a black hole.*

create a black hole. That's the setup. Now for the pivotal question: What is the maximum amount of information that can be stored within the region of space?

Those unlikely bedfellows, the Second Law and black holes, provide the answer. Imagine adding matter to the region, with the aim of augmenting its information storage capacity. You might insert high-capacity memory chips or voluminous hard drives into the bank of Google's computers; you might provide books or jam-packed Kindles to augment the Library of Congress collection. Since even raw matter carries information—Are the steam's mole-cules here or there? Are they moving at this speed or that?—you also cram every nook and cranny of the region with as much matter as you can get your hands on. Until you reach a critical juncture. At some point, the region will be so thoroughly stuffed that were you to add even a single grain of sand, the interior would go dark as the region turned into a black hole. When that happens, game over. A black hole's size is determined by its mass, so if you try to increase the information storage capacity by adding yet more matter, the black hole will respond by growing larger. And since we want to focus on the information that can inhabit a given *fixed* volume of space, this result falls afoul of the basic setup. You can't increase the black hole's information capacity without forcing the black hole to enlarge.[12]

Two observations take us across the finish line. The Second Law ensures that entropy increases throughout the entire process, and so the information hidden within the hard drives, Kindles, old-fashioned paper books, and everything else you packed into the region is less than that hidden in the black hole. From the results of Bekenstein and Hawking, we know that the black hole's hidden information content is given by the area of its event horizon. More-over, because you were careful not to overspill the original region of space, the black hole's event horizon coincides with the region's boundary, so the black hole's entropy equals the area of this sur-rounding surface. We thus learn an important lesson. *The amount of information contained within a region of space, stored in any objects of any design, is always less than the area of the surface that surrounds the region (measured in square Planck units).*

This is the conclusion we've been chasing. Notice that although black holes are central to the reasoning, the analysis applies to *any* region of space, whether or not a black hole is actually present. If you max out a region's storage capacity, you'll create a black hole, but as long as you stay under the limit, no black hole will form.

I hasten to add that in any practical sense, the information storage limit is of no concern. Compared with today's rudimentary storage devices, the potential storage capacity on the surface of a spatial region is humongous. A stack of five off-the-shelf terabyte hard drives fits comfortably within a sphere of radius 50 centimeters, whose surface is covered by about 10^{70} Planck cells. The surface's storage capacity is thus about 10^{70} bits, which is about a billion, trillion, trillion, trillion, trillion terabytes, and so enormously exceeds anything you can buy. No one in Silicon Valley cares much about these theoretical constraints.

Yet, as a guide to how the universe works, the storage limitations are telling. Think of any region of space, such as the room in which I'm writing or the one in which you're reading. Take a Wheelerian perspective and imagine that whatever happens in the region amounts to information processing—information regarding how things are right now is transformed by the laws of physics into information regarding how they will be in a second or a minute or an hour. Since the physical processes we witness, as well as those by which we're governed, seemingly take place within the region, it's natural to expect that the information those processes carry is also found within the region. But the results just derived suggest an alternative view. For black holes, we found that the link between information and surface area goes beyond mere numerical accounting; there's a concrete sense in which information is stored on their surfaces. Susskind and 't Hooft stressed that the lesson should be general: since the information required to describe physical phenomena within *any* given region of space can be fully encoded by data on a surface that surrounds the region, then there's reason to think that the surface is where the fundamental physical processes actually happen. Our familiar three-dimensional reality, these bold thinkers suggested, would then be likened to a holographic projection of those distant two-dimensional physical processes.

If this line of reasoning is correct, then there are physical processes taking place on some distant surface that, much like a puppeteer pulls strings, are fully linked to the processes taking place in my fingers, arms, and brain as I type these words at my desk. Our experiences here, and that distant reality there, would form the most interlocked of parallel worlds. Phenomena in the two—I'll call them *Holographic Parallel Universes*—would be so fully joined that their respective evolutions would be as connected as me and my shadow.

The Fine Print

That familiar reality may be mirrored, or perhaps even produced, by phenomena taking place on a faraway, lower-dimensional surface ranks among the most unexpected developments in all of theoretical physics. But how confident should we be that the holographic principle is right? We are navigating a realm deep in theoretical territory, and relying almost exclusively on developments that have not been experimentally tested, so there is surely grounds for skepticism. There are many places where the argument could be forced off course. Do black holes really have nonzero entropy and nonzero temperature, and, if so, do the values conform to theoretical predictions? Is the information capacity of a region of space really determined by the amount of information that can be stored on a surface that surrounds it? And on such a surface, is one bit per Planck area really the limit? We think the answer to each of these questions is yes because of the coherent, consistent, and carefully constructed theoretical edifice into which the conclusions perfectly fit. But since none of these ideas has been subject to the experimenter's scalpel, it is certainly possible (though in my view highly unlikely) that future advances will convince us that one or more of these essential intermediate steps are wrong. That could lay to waste the holographic idea.

Another important point is that throughout the discussion, we've spoken of a region of space, of a surface that surrounds it, and of the information content of each. But since our focus has been on entropy and the Second Law—both of which concern themselves primarily with the *quantity* of information in a given context—we've not elaborated on the details of *how* that information is physically realized or stored. When we talk about information residing on a sphere surrounding a region of space, what does that really mean? How does the information manifest itself? What form does it take? To what extent can we develop an explicit dictionary that translates from phenomena taking place on the boundary to those taking place in the interior?

Physicists have yet to articulate a general framework for addressing these questions. Given that gravity and quantum mechanics are both central to the reasoning, you might expect that string theory would provide a potent context for theoretical explorations. But when 't Hooft first formulated the holographic concept, he doubted that string theory would be able to advance the subject, noting, "Nature is much more crazy at the Planck scale than even string theorists could have imagined."[13] Less than a decade later, string theory proved 't Hooft wrong by proving him right. In a landmark paper, a young theorist showed that string theory provides an explicit realization of the holographic principle.

String Theory and Holography

When I was called to the stage at the University of California, Santa Barbara, to give my talk at the annual international string theory conference in 1998, I did something I'd never done before and suspect will never do again. I faced the audience, threw my right hand to my left shoulder and my left to my right shoulder, and then with both hands in succession grabbed the seat of my pants, bunny-hopped, and made a quarter turn, followed, thankfully, by audience laughter, which covered the three remaining steps necessary to reach the podium, where I began my talk. The crowd got the joke. At the banquet the night before, the conference participants had performed a song-and-dance celebrating—as only physicists can—a spectacular result of the Argentinian string theorist Juan Maldacena. With lyrics like "Black holes used to be a great mystery; /

Now we use D-branes to compute D-entropy," the crowd had reveled in a string theory version of the 1990s momentary dance craze, the Macarena—a touch more animated than Al Gore's version at the Democratic National Convention, a touch less mellifluous than Los del Rio's original one-hit wonder, but second to none in passion. I was one of the few at the conference whose talk was not focused on Maldacena's breakthrough, so when I took the stage the next morning I felt it only appropriate to preface my remarks with a personal gesture of appreciation.

Now, more than a decade later, many would agree that no work in string theory since is of comparable magnitude and influence. Of the numerous ramifications of Maldacena's result, one is directly relevant to the line we've been following. In a particular hypothetical setting, Maldacena's result *realized explicitly the holographic principle, and in doing so provided the first mathematical example of Holographic Parallel Universes.* Maldacena achieved this by considering string theory in a universe whose shape differs from ours but for the purpose at hand proves easier to analyze. In a precise mathematical sense, the shape has a boundary, an impenetrable surface that completely surrounds its interior. By zeroing in on this surface, Maldacena argued convincingly that everything taking place within the specified universe is a reflection of laws and processes acting themselves out on the boundary.

Although Maldacena's method may not seem directly applicable to a universe with the shape of ours, his results are decisive because they established a mathematical proving ground in which ideas regarding holographic universes could be made explicit and investigated quantitatively. The results of such studies won over a great many physicists who had previously eyed the holographic principle with much misgiving, and thus set off an avalanche of research that has yielded thousands of articles and considerably deeper understanding. Most exciting of all, there's now evidence that a link between these theoretical insights and physics in our universe *can* be forged. In the next few years, that link may very well allow the holographic ideas to be experimentally tested.

The rest of this and the next section will be devoted to explaining how Maldacena achieved this breakthrough; the material is the most difficult we will cover. I'll begin with a short summary, a CliffsNotes version that doubles as a guilt-free pass to jump to the last section should, at any point, the material overwhelm your appetite for detail.

Maldacena's inspired move was to invoke a new version of the duality arguments discussed in Chapter 5. Recall the branes—the "slice of bread" universes—introduced there. Maldacena considered, from two complementary perspectives, the properties of a tightly stacked collection of three-dimensional branes, as in Figure 9.4. One perspective, an "intrinsic" perspective, focused on strings that move, vibrate, and wiggle along the branes themselves. The other perspective, an "extrinsic" perspective, focused on how the branes influence their immediate environment gravitationally, much as the sun and the earth influence theirs. Maldacena argued that both perspectives describe one and the same physical situation, just from different vantage points. The intrinsic perspective involves strings moving on a stack of branes, while the extrinsic perspective involves strings moving through a region of curved spacetime that's bounded by the stack of branes. By equating the two, Maldacena found an explicit link between physics taking place in a region and physics taking place on that region's boundary; he found an explicit realization of holography. That's the basic idea.

With more color, the story goes like this.

Consider, Maldacena says, a stack of three-branes, so closely spaced that they appear as a single monolithic slab—Figure 9.4—and study the behavior of strings moving in this environment. You'll recall that there are two types of strings—open snippets and closed loops—and that the endpoints of open strings can move within and through branes but not off them, while closed strings have no ends and so can move freely through the entire spatial expanse. In the jargon of the field, we say that while open strings are confined to the branes, closed strings can move through the *bulk* of space.

Maldacena's first step was to confine his mathematical attention to strings that have low energy—that is, ones that vibrate relatively slowly. Here's why: the force of gravity between any two

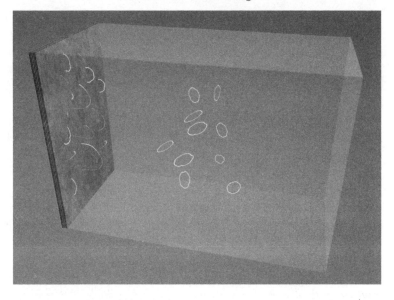

Figure 9.4 *A collection of closely spaced three-branes with open strings confined to the brane surfaces, and closed strings moving through the "bulk."*

objects is proportional to the mass of each; the same is true for the force of gravity acting between any two strings. Strings that have low energy have small mass, and so they hardly respond to gravity at all. By focusing on low energy strings, Maldacena was thus suppressing gravity's influence. That yielded a substantial simplification. In string theory, as we've seen (Chapter 5), gravity is transmitted from place to place by closed loops. Suppressing the force of gravity was therefore tantamount to suppressing the influence of closed strings on anything they might encounter—most notably, the open string snippets living on the brane stack. By ensuring that the two kinds of strings, open snippets and closed loops, wouldn't affect each other, Maldacena was ensuring that they could be analyzed independently.

Maldacena then changed gears and suggested thinking about the very same situation from a different perspective. Rather than treat the three-branes as a substrate that supports the motion of open strings, he encouraged viewing them as a single object, which has its own intrinsic mass and hence warps space and time in its vicinity. Maldacena was fortunate that previous research, by a number of physicists, had laid the groundwork for this alternative perspective. The earlier works had established that as you stack more and more branes together, their collective gravitational field grows ever stronger. Ultimately, the slab of branes behaves much like a black hole, but one that's brane-shaped, and so is called a *black brane*. As with a more ordinary black hole, if you get too close to a black brane, you can't escape. And, as is also the case with an ordinary black hole, if you stay far away but are watching something approach a black brane, the light you'll receive will be exhausted from its having fought against the black brane's gravity. This will make the object appear to have ever less energy and to be moving ever slower.[14]

From this second perspective, Maldacena again focused on the low-energy features of a universe containing such a black slab. Much as he had when working on the first perspective, he realized that the low-energy physics involved two components that could be analyzed independently. Slowly vibrating closed strings, moving anywhere in the bulk of space, are the most obvious low-energy carriers. The second component relies on the presence of the black brane. Imagine you are far from the black brane and have in your possession a closed string that's vibrating with an arbitrarily large amount of energy. Then, imagine lowering the string toward the event horizon while you maintain a safe distance. As recalled above, the black brane will make the string's energy appear ever lower; the light you'll receive will make the string look as though it's in a slow-motion movie. The second low-energy carriers are thus any and all vibrating strings that are sufficiently close to the black brane's event horizon.

Maldacena's final move was to compare the two perspectives. He noted that because they describe the same brane stack, only from different points of view, they must agree. Each description involves low-energy closed strings moving through the bulk of space, so this part of the agreement is manifest. But the remaining part of each description must also agree.

And that proves astonishing.

The remaining part of the first description consists of low-energy open strings moving on the three-branes. We recall from Chapter 4 that low-energy strings are well described by point particle quantum field theory, and that is the case here. The particular kind of quantum field theory involves a number of sophisticated mathematical ingredients (and it has an ungainly characterization: *conformally invariant supersymmetric quantum gauge field theory*), but two vital characteristics are readily understood. The absence of closed strings ensures the absence of the gravitational field. And, because the strings can move only on the tightly sandwiched three-dimensional branes, the quantum field theory lives in three spatial dimensions (in addition to the one dimension of time, for a total of four spacetime dimensions).

The remaining part of the second description consists of closed strings, executing any vibrational pattern, as long as they are close enough to the black branes' event horizon to appear lethargic— that is, to appear to have low energy. Such strings, although limited in how far they stray from the black stack, still vibrate and move through nine dimensions of space (in addition to one dimension of time, for a total of ten spacetime dimensions). And because this sector is built from closed strings, it contains the force of gravity.

However different the two perspectives might seem, they're describing one and the same physical situation, so they must agree. This leads to a thoroughly bizarre conclusion. A particular *non-gravitational, point particle* quantum field theory in *four* spacetime dimensions (the first perspective) describes the same physics as *strings, including gravity,* moving through a particular swath of *ten* spacetime dimensions (the second perspective). This would seem as far-fetched as claiming . . . Well, honestly, I've tried, and I can't come up with any two things in the real world more dissimilar than these two theories. But Maldacena followed the math, in the manner we've outlined, and ran smack into this conclusion.

The sheer strangeness of the result—and the audacity of the claim—isn't lessened by the fact that it takes but a moment to place it within the line of thought developed earlier in this chapter. As schematically illustrated in Figure 9.5, the gravity of the black

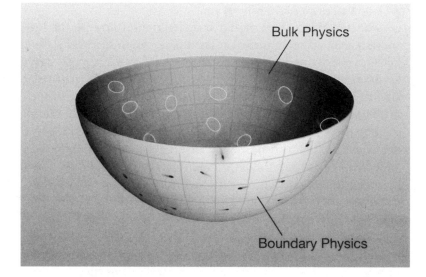

Figure 9.5 *A schematic illustration of the duality between string theory operating in the interior of a particular spacetime and quantum field theory operating on the boundary of that spacetime.*

brane slab imparts a curved shape to the ten-dimensional spacetime swath in its vicinity (the details are secondary, but the curved spacetime is called *anti–de Sitter five-space times the five sphere*); the black brane slab is itself the boundary of this space. And so, Maldacena's result is that string theory within the *bulk* of this spacetime shape is identical to a quantum field theory living on its *boundary*.[15]

This is holography come to life.

Maldacena had built a self-contained mathematical laboratory in which, among other things, physicists could explore in concrete detail a holographic realization of physical law. Within a few months, two papers, one by Edward Witten and one by Steven Gubser, Igor Klebanov, and Alexander Polyakov, supplied the next level of understanding. They established a precise mathematical dictionary for translating between the two perspectives: given a physical process on the brane boundary, the dictionary showed how it would appear in the bulk interior, and vice versa. In a hypothetical universe, then, the dictionary rendered the holographic principle explicit. On the boundary of this universe, information is embodied by quantum fields. When the information is translated by the mathematical dictionary, it reads as a story of stringy phenomena happening in the universe's interior.

The dictionary itself renders the holographic metaphor all the more appropriate. An everyday hologram bears no resemblance to the three-dimensional image it produces. On its surface appear only various lines, arcs, and swirls etched into the plastic. Yet a complex transformation, carried out operationally by shining a laser through the plastic, turns those markings into a recognizable three-dimensional image. Which means that the plastic hologram and the three-dimensional image embody the same data, even though the information in one is unrecognizable from the perspective of the other. Similarly, examination of the quantum field theory on the boundary of Maldacena's universe shows that it bears no obvious resemblance to the string theory inhabiting the interior. If a physicist were presented with both theories, not being told of the connections we've now laid out, he or she would more than likely conclude that they were unrelated. Nevertheless, the mathematical dictionary linking the two—functioning as a laser does for ordinary holograms—makes explicit that anything taking place in one has an incarnation in the other. At the same time, examination of the dictionary reveals that just as with a real hologram, the information in each appears scrambled on translation into the other's language.

As a particularly impressive example, Witten investigated what an ordinary black hole in the interior of Maldacena's universe would look like from the perspective of the boundary theory. Remember, the boundary theory does not include gravity, and so a black hole necessarily translates into something very unlike a black hole. Witten's result showed that much as the Wizard of Oz's frightening visage was produced by an ordinary man, a rapacious black hole is the holographic projection of something equally ordinary: a bath of hot particles in the boundary theory (Figure 9.6). Like a real hologram and the image it generates, the two theories—a black hole in the interior and a hot quantum field theory on the bound-

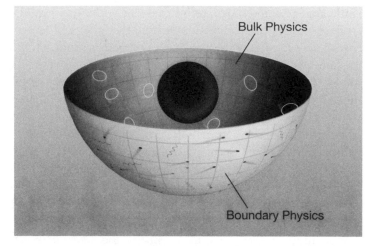

Figure 9.6 *The holographic equivalence applied to a black hole in the bulk of spacetime yields a hot bath of particles and radiation on the region's boundary.*

ary—bear no apparent resemblance to each other, and yet they embody identical information.*

In Plato's parable of the cave, our senses are privy only to a flattened, diminished version of the true, more richly textured, reality. Maldacena's flattened world is very different. Far from being diminished, it tells the full story. It's a profoundly different story from the one we're used to. But his flattened world may well be the primary narrator.

*There is a related story that I've not told in this chapter, to do with a long-standing debate regarding whether black holes require a modification of quantum mechanics—whether, by swallowing information, they upend the ability to fully evolve probability waves forward in time. A one-sentence summary is that Witten's result, by establishing an equivalence between a black hole and a physical situation that does *not* destroy information (a hot quantum field theory), supplied conclusive evidence that all information that falls into a black hole is ultimately available to the outside world. Quantum mechanics needs no modification. This application of Maldacena's discovery also establishes that the boundary theory provides a full description of the information (entropy) stored on a black hole's surface.

Parallel Universes or Parallel Mathematics?

Maldacena's result, and the many others it has spawned in the years since, is deemed conjectural. Because the mathematics is tremendously difficult, fashioning an airtight argument remains elusive. But the holographic ideas have been subject to a great many stringent mathematical tests; having come through unscathed, they've been propelled into mainstream thought among physicists searching for the deep roots of natural laws.

One factor contributing to the difficulty of rigorously proving that the boundary and bulk worlds are disguised versions of one another highlights why the result, if true, is so powerful. I described in Chapter 5 how physicists more often than not rely on approximation techniques, the perturbative methods that I outlined (recall the lottery example with Ralph and Alice). I also emphasized that such methods are accurate only if the relevant coupling constant is a small number. In analyzing the relationship between quantum field theory on the boundary and string theory in the bulk, Maldacena realized that when the coupling of one theory was small, that of the other was large, and vice versa. The natural test, and a possible means of proving that the two theories are secretly identical, is to perform independent calculations in each theory and then check for equality. But this is difficult to do, since when perturbative methods work for one, they fail for the other.[16]

However, if you accept Maldacena's more abstract argument, as outlined in the previous section, the perturbative vice becomes a calculational virtue. Much as we found with the string dualities in Chapter 5, the bulk-boundary dictionary translates daunting calculations, beset by a large coupling, in one framework into straightforward calculations, with a small coupling in the other. In recent years, this has been parlayed into results that may be experimentally testable.

At the Relativistic Heavy Ion Collider (RHIC) in Brookhaven, New York, gold nuclei are slammed into each other at just shy of light speed. Because the nuclei contain many protons and neutrons, the collisions create a commotion of particles that can be more than 200,000 times as hot as the sun's core. That's hot enough to melt the protons and neutrons into a fluid of quarks and the gluons that act between them. Physicists have exerted great effort to understand this fluidlike phase, called the *quark gluon plasma*, because it's likely that matter briefly assumed this form soon after the big bang.

The challenge is that the quantum field theory (*quantum chromodynamics*) describing the hot soup of quarks and gluons has a large value for its coupling constant, and that compromises the accuracy of perturbative methods. Ingenious techniques have been developed to skirt this hurdle, but experimental measurements continue to controvert some of the theoretical results. For example, as any fluid flows—be it water, molasses, or the quark gluon plasma—each layer of the fluid exerts a drag force on the layers flowing above and below. The drag force is known as *shear viscosity*. Experiments at RHIC measured the shear viscosity of the quark gluon plasma, and the results are far smaller than those predicted by the perturbative quantum field theory calculations.

Here's a possible way forward. In introducing the holographic principle, the perspective I've taken is to imagine that everything we experience lies in the interior of spacetime, with the unexpected twist being processes, mirroring those experiences, which take place on a distant boundary. Let's reverse that perspective. Imagine that our universe—or, more precisely, the quarks and gluons in our universe—lives on the boundary, and so that's where the RHIC experiments take place. Now invoke Maldacena. His result shows that the RHIC experiments (described by quantum field theory) have an alternative mathematical description in terms of strings moving in the bulk. The details are involved but the power of the rephrasing is immediate: difficult calculations in the boundary description (where the coupling is large) are translated into easier calculations in the bulk description (where the coupling is small).[17]

Pavel Kovtun, Andrei Starinets, and Dam Son did the math, and the results they found come impressively close to the experimental data. This pioneering work has motivated an army of theoreticians to undertake many other string theory calculations in an effort to make contact with RHIC observations, driving forward a vigorous interplay between theory and experiment—a welcome novelty for string theorists.

Bear in mind that the boundary theory doesn't model our universe fully since, for example, it doesn't contain the gravitational force. This doesn't compromise contact with RHIC data because in those experiments the particles have such small mass (even when traveling near light speed) that the gravitational force plays virtually no role. But it does make clear that in this application string theory is not being used as a "theory of everything"; instead, string theory provides a new calculational tool for breaking through obstacles that have impeded more traditional methods. Conservatively, analyzing quarks and gluons by using a higher dimensional theory of strings can be viewed as a potent string-based mathematical trick. Less conservatively, one can imagine that the higher dimensional string description is, in some yet to be understood way, physically real.

Regardless of perspective, conservative or not, the resulting confluence of mathematical results with experimental observations is extremely impressive. I am not a fan of hyperbole, but I view these developments as among the most exciting advances in decades. Mathematical manipulations that utilize strings moving through a particular ten-dimensional spacetime tell us something about quarks and gluons living in a four-dimensional spacetime—and the "something" the calculations tell us seems to be borne out by experiments.

Coda: The Future of String Theory

The developments we've covered in this chapter transcend evaluations of string theory. From Wheeler's emphasis on analyzing the universe in terms of information, to the recognition that entropy is a measure of hidden information, to the reconciliation between the Second Law of Thermodynamics and black holes, to the realization that black holes store entropy on their surface, to the understanding that black holes set a maximum for the amount of information that can occupy a given region of space, we've fol-

lowed a winding road across many decades and traversed an intricate web of results. The journey has been full of remarkable
insights, and has led us to a new unifying idea—the holographic
principle. The principle, as we've seen, suggests that the phenomena we witness are mirrored on a thin, distant bounding surface.
Looking to the future, I suspect that the holographic principle will
be a beacon for physicists well into the twenty-first century.

That string theory embraces the holographic principle, and provides concrete examples of holographic parallel worlds, is a testa-
ment to how cutting-edge developments are coming together in a
powerful synthesis. That these examples have provided the basis for
explicit calculations, some of whose results can be compared with
results from real-world experiments, is a gratifying step toward making contact with observable reality. But within string theory itself,
there's a broader frame within which these developments should
be seen.

For nearly thirty years after the initial discovery of string theory,
physicists lacked a full mathematical definition of the theory. Early
string theorists laid out the essential ideas of vibrating strings and
extra dimensions, but even after decades of further work, the mathematical foundations of the theory remained approximate and thus
incomplete. Maldacena's insight represents major progress. The
species of quantum field theory Maldacena identified as living on
the boundary is among the mathematically best understood of
those particle physicists have studied since the middle of the twentieth century. It does not include gravity, and that's a big plus since,
as we've seen, trying to bring general relativity directly into quantum field theory is like setting a campfire in a gunpowder factory.
We've now learned that this mathematically friendly, nongravitational quantum field theory *generates* string theory—a theory
that contains gravity—holographically. Operating way out on the
boundary of a universe with the specific shape schematically illustrated in Figure 9.5, this quantum field theory embodies all physical features, processes, and interactions of strings that move within
the interior, a link made explicit through the dictionary translating
phcnomena between the two. And since we have a sure-footed
mathematical definition of the boundary quantum field theory, *we*

can use it as a mathematical definition of string theory, at least for strings moving within this spacetime shape. The holographic parallel universes may thus be more than a potential outgrowth of fundamental laws; they may be part of the very definition of the fundamental laws.[18]

When I introduced string theory in Chapter 4, I noted that it fit the venerable pattern of providing a new approach to nature's laws that, nevertheless, did not erase past theories. The results we've now described take this observation to a whole different level. String theory doesn't just reduce to quantum field theory in certain circumstances. Maldacena's result suggests that string theory and quantum field theory are equivalent approaches expressed in different languages. The translation between them is complicated, which is why it took more than forty years for this connection to come to light. But if Maldacena's insights are fully valid, as all available evidence attests, string theory and quantum field theory may very well be two sides of the same coin.

Physicists are working hard to generalize the methods so they might apply to a universe with any shape; if string theory is right, that would include ours. But even with the current limitations, finally having a firm formulation of a theory we've worked on for many years is an essential foundation for future progress. It is surely enough to make many a physicist sing and dance.

CHAPTER 10

Universes, Computers, and Mathematical Reality

The Simulated and the Ultimate Multiverses

The parallel universe theories we considered in previous chapters emerged from mathematical laws developed by physicists in their pursuit of nature's deepest workings. The credence accorded one set of laws or another varies widely—quantum mechanics is viewed as established fact; inflationary cosmology has observational support; string theory is thoroughly speculative—as does the type and logical necessity of the parallel worlds associated with each. But the pattern is clear. When we hand over the steering wheel to the mathematical underpinnings of the major proposed physical laws, we're driven time and again to some version of parallel worlds.

Let's now change tack. What happens if we seize the wheel? Can we humans manipulate the cosmic unfolding to volitionally create universes parallel to our own? If you believe, as I do, that the behavior of living beings is dictated by nature's laws, then you may see this as no change of tack at all but simply as a narrowing of perspective, to the impact of physical law when funneled through human activity. This line of thought quickly engages thorny issues such as the age-old debate about determinism and free will, but that's not a direction in which I want to head. Rather, my question is this: With the same sense of intent and control you feel when you choose a movie or a meal, might you create a universe?

The question sounds outlandish. And it is. I'll tip you off now that in addressing it we will find ourselves in territory even more speculative than what we've already covered, and considering where we've been, that says a lot. But let's have a little fun and see where it takes us. Let me lay out the perspective I'll take. In contemplating universe creation, I'm less interested in practical constraints than in the possibilities made available by the laws of physics. So, when I speak of "you" creating a universe, what I really mean is you, or a distant descendant, or an army of such descendants possibly millennia down the road. These present or future humans will still be subject to the laws of physics, but I will imagine that they're in possession of arbitrarily advanced technologies. I will also consider the creation of two distinct types of universes. The first type comprises the usual universes, ones that encompass an expanse of space and are filled with various forms of matter and energy. The second kind is less tangible: virtual computer-generated universes. The discussion will also naturally forge a link to a third multiverse proposal. This variety does not originate from thinking about universe creation, per se, but instead addresses the question of whether mathematics is "real" or is instead created by the mind.

To Create a Universe

Despite uncertainties in delineating the composition of the universe—What is the dark energy? What is the full list of fundamental particulate ingredients?—scientists are confident that were you to weigh everything that's within our cosmic horizon, the tally would come in at about 10 billion billion billion billion billion billion grams. If the contents weighed significantly more or less than this, their gravitational influence on the cosmic microwave background radiation would cause the splotches in Figure 3.4 to be much larger or smaller, and that would conflict with refined measurements of their angular size. But the precise weight of the observable universe is secondary; my point is that it's huge. So huge that the notion of us humans creating another such realm seems utterly fatuous.

Using big bang cosmology as our blueprint for universe formation, we find no guidance on how to clear this hurdle. In the standard big bang theory, the observable universe was ever-smaller at ever-earlier times, but the stupendous quantities of matter and energy we now measure were always present; they were just squeezed into an ever-smaller volume. If you want a universe like the one we see today, you have to start with raw material whose mass and energy are those we see today. The big bang theory takes such raw material as an unexplained given.[1]

In broad strokes, then, the big bang's instructions for creating a universe like ours require that we gather a gargantuan amount of mass and compress it to a fantastically small size. But having achieved that, however improbable, we would face another challenge. How do we ignite the bang? It's an obstacle that becomes only more daunting when we recall that the big bang is not an explosion that takes place within a static region of space; the big bang propels the expansion of space itself.

If the big bang theory were the pinnacle of cosmological thought, the scientific pursuit of universe creation would stop here. But it's not. We've seen that the big bang theory has given way to the more robust inflationary cosmology, and inflation offers a strategy for going forward. With a powerful outward burst of spatial expansion being its trademark, the inflationary theory puts a bang in the big bang, and a big one at that; according to inflation, an anti-gravity blast is what set the outward expansion of space in motion. Of equal importance, as we'll now see, inflation establishes that vast amounts of matter can be *created* from the most modest of seeds.

Recall from Chapter 3 that in the inflationary approach, a universe like ours—a hole in the cosmic Swiss cheese—formed when the inflaton's value rolled down its potential energy curve, bringing to a close the phenomenal outward surge in our vicinity. As the inflaton's value dropped, the energy it contained was transformed into a bath of particles uniformly filling our bubble. That's where the matter we see originated. Progress, for sure, but the insight raises the next question: What's the source of the inflaton's energy?

It comes from gravity. Remember that inflationary expansion is much like viral replication: a high-valued inflaton field drives the region it inhabits to rapidly grow, and in doing so creates an increasingly large spatial volume that is itself infused with a high-valued inflaton field. And because a uniform inflaton field contributes a constant energy per unit volume, the larger the volume it fills, the more energy it embodies. The driving force behind the expansion is gravity—in its repulsive guise—and so gravity is the source of the ever-larger energy the region contains.

Inflationary cosmology can thus be thought of as creating a sustained energy flow from the gravitational field to the inflaton field. This might seem like one more passing of the energy buck—where does gravity get *its* energy?—but the situation is a good deal better than that. Gravity is different from the other forces because where there's gravity, there's a virtually unlimited reservoir of energy. It's a familiar idea expressed in unfamiliar language. When you jump off a cliff, your kinetic energy—the energy of your motion—gets ever larger. Gravity, the force driving your motion, is the energy's source. In any realistic situation, you will hit the ground, but in principle you could fall arbitrarily far, tumbling down an increasingly long rabbit hole, while your kinetic energy grows ever larger. The reason gravity can supply such unlimited quantities of energy is that, much like the U.S. Treasury, it has no fear of debt. As you fall and your energy gets ever more positive, gravity compensates by its energy becoming ever more negative. You know intuitively that the gravitational energy is negative because to climb out of the rabbit hole, you need to exert positive energy—pushing with your legs, pulling with your arms; that's how you repay the energy debt gravity incurred on your behalf.[2]

The essential conclusion is that as an inflaton-filled region rapidly grows, the inflaton extracts energy from the gravitational field's inexhaustible resources, resulting in the region's energy rapidly growing too. And because the inflaton field supplies the energy that's converted into ordinary matter, inflationary cosmology—unlike the big bang model—does not need to posit the raw material for generating planets, stars, and galaxies. Gravity is matter's sugar daddy.

The only independent energy budget required by inflationary cosmology is what's needed to create an initial inflationary seed, a small spherical nugget of space filled with a high-valued inflaton field that gets the inflationary expansion rolling in the first place. When you put in numbers, the equations show that the nugget need be only about 10^{-26} centimeters across and filled with an inflaton field whose energy, when converted to mass, would weigh less than ten grams.[3] Such a tiny seed would, faster than a flash, undergo spectacular expansion, growing far larger than the observable universe while harboring ever-increasing energy. The inflaton's total energy would quickly soar beyond what's necessary to generate all the stars in all the galaxies we observe. And so, with inflation in the cosmological driver's seat, the impossible starting point of the big bang's recipe—gather more than 10^{55} grams and squeeze the whole lot into an infinitesimally small speck—is radically transformed. Gather ten grams of inflaton field and squeeze it into a lump that's about 10^{-26} centimeters across. That's a lump you could put in your wallet.

This approach, nevertheless, presents daunting challenges. For one thing, the inflaton remains a purely hypothetical field. Cosmologists freely incorporate the inflaton field into their equations, but unlike with electron and quark fields, there is as yet no evidence that the inflaton field exists. For another, even if the inflaton proves real, and even if we one day develop the means to manipulate it much as we do the electromagnetic field, still the *density* of the requisite inflaton seed would be enormous: about 10^{67} times that of an atomic nucleus. Although the seed would weigh less than a handful of popcorn, the compressive force we would need to apply is trillions and trillions of times beyond what we can now muster.

But this is just the kind of technological hurdle that we're imagining an arbitrarily advanced civilization might one day overcome. So, if our distant descendants one day harness the inflaton field and develop extraordinary compressors capable of producing such dense nuggets, will we have attained the status of universe creators? And, as we contemplate such a step toward Olympus, should we worry that if we artificially set off new inflationary realms, our own corner of space may be swallowed by the ballooning expanse? Alan Guth and a number of collaborators investigated these questions in a series of papers, and found both good news and bad. Start with the last question, as that's where we'll find the good news.

Guth, together with Steven Blau and Eduardo Guendelman, showed that there's no need to be concerned about an artificial phase of inflationary expansion ripping through our existing environment. The reason has to do with pressure. If an inflationary seed were created in the laboratory, it would harbor the inflaton field's characteristic positive energy and negative pressure, but it would be surrounded by ordinary space in which the inflaton field's value, and its pressure, would be zero (or nearly so).

We usually don't ascribe much power to zero, but in this case zero makes all the difference. Zero pressure is larger than negative pressure, and so the pressure outside the seed would be larger than the pressure inside. This would subject the seed to a net external force pressing upon it, much like what your eardrums experience when deep-sea diving. The pressure differential is powerful enough to prevent the seed from expanding into the surrounding environment.

But this does not prevent the inflaton's drive to expand. If you blow air into a balloon while tightly clasping its surface, the balloon will bubble out from between your hands. The inflaton seed can behave similarly. The seed can generate a new expanding spatial realm that sprouts from the original spatial environment, as illustrated by the little growing sphere in Figure 10.1. The calculations show that once the new expanding realm reaches a critical size, its

Figure 10.1 *Because of the greater pressure in the ambient environment, an inflationary seed is forced to expand into newly formed space. As the bubble universe grows, it detaches from the parent environment, yielding a separate, expanding spatial domain. To someone in the ambient environment, the process looks like the formation of a black hole.*

umbilical cord to the parent space severs, as in the final image of Figure 10.1, and an independent inflating universe is born.

As enticing as the process might be—*the artificial creation of a new universe*—the view from the laboratory wouldn't live up to the advance billing. It's a relief that the inflationary bubble would not gobble up the surrounding environment, but the flip side is that there would be little evidence of the creation itself. A universe that expands by generating new space, which then detaches from ours, is a universe we can't see. Indeed, as the new universe pinches off, its sole residue would be a deep gravitational well—you can see this in the last image of Figure 10.1—which would appear to us as a black hole. And since we have no capacity to see beyond a black hole's edge, we wouldn't even be assured that our experiment had been a success; without access to the new universe, we would have no means of establishing observationally that the universe had been created at all.

Physics protects us, but the price for safety is total separation from our handiwork. And that's the good news.

The bad news for aspiring universe creators is a more sobering result derived by Guth and his MIT colleague Edward Farhi. Their careful mathematical treatment showed that the sequence depicted in Figure 10.1 requires an additional ingredient. Much as some balloons require that you give a strong initial burst of air, after which they more easily inflate, Guth and Farhi found that the nascent universe in Figure 10.1 needs a strong kick-start to get the inflationary expansion off and running. So strong that there's only one entity that can provide it: a white hole. A white hole, the opposite of a black hole, is a hypothetical object that spews matter out rather than drawing it in. This requires conditions so extreme that known mathematical methods break down (much as is the case at the center of a black hole); suffice it to say, no one anticipates generating white holes in the laboratory. Ever. Guth and Farhi found a fundamental wrench in the universe-creation works.

A number of research groups have since suggested possible ways of skirting the problem. Guth and Farhi, joined by Jemal Guven, found that by creating the inflationary seed through a quantum tunneling process (similar to what we discussed in the context of the Landscape Multiverse) the white hole singularity can be avoided; but the probability for the quantum tunneling process to happen is so fantastically small that there's essentially no chance of its happening over timescales that anyone would consider worth contemplating. A group of Japanese physicists, Nobuyuki Sakai, Ken-ichi Nakao, Hideki Ishihara, and Makoto Kobayashi, showed that a magnetic monopole—a hypothetical particle that has either the north pole or the south pole of a standard bar magnet—might catalyze inflationary expansion, also avoiding singularities; but after nearly forty years of intense searching, no one has yet found a single one of these particles.*

As of today, then, the summary is that the door to creating new universes remains open, but only barely. Given the proposals' heavy reliance on hypothetical elements, future developments may well shut this door permanently. But if they don't—or, perhaps, if subsequent work makes a stronger case for the possibility of universe creation—would there be motivation to proceed? Why create a universe if there's no way to see it, or interact with it, or even know for sure that it *was* created? Andrei Linde, famous not just for his deep cosmological insights but also for his flair for mock drama, has noted that the allure of playing god would simply prove irresistible.

I don't know that it would. Admittedly, it would be thrilling to have so thoroughly grasped nature's laws that we could reenact the most pivotal of all events. I suspect, however, that by the time we can seriously consider universe creation—if that time ever comes—our scientific and technical advancements would have made available so many other spectacular undertakings, whose results we could not just imagine but truly experience, that the intangible nature of universe creation would make it much less interesting.

---

*Ironically, an explanation for why magnetic monopoles have not been found (even though they are predicted by many approaches to unified theories) is that their population was diluted by the rapid expansion of space that takes place in inflationary cosmology. The suggestion now being made is that magnetic monopoles may themselves play a role in initiating future inflationary episodes.

The appeal would surely be stronger were we to learn how to manufacture universes that we could see or even interact with. For "real" universes, in the usual sense of a universe constituted from the standard ingredients of space, time, matter and energy, we don't yet have any strategy for doing so that's compatible with the laws of physics as we currently understand them.

But what if we set aside real universes and consider virtual ones?

The Stuff of Thought

A couple of years ago, I had a bout of feverish flu that came with hallucinations far more vivid than any ordinary dream or nightmare. In one that has stayed with me, I'd find myself with a group of people sitting in a sparse hotel room, locked in a hallucination within the hallucination. I was absolutely certain that days and weeks went by—until I was thrust back into the primary hallucination, where I'd learn, shockingly, that hardly any time had passed at all. Each time I felt myself drifting back to the room, I resisted strenuously, since I knew from previous iterations that once there I'd be swallowed whole, unable to recognize the realm as false until I found myself back in the primary hallucination, where I'd again be distraught to learn that what I'd thought real was illusory. Periodically, when the fever subsided, I'd pull out one level further, back to ordinary life, and realize that all those translocations had been taking place within my own swirling mind.

I don't usually learn much from having a fever. But this experience added immediacy to something which, to that point, I'd largely understood only in the abstract. Our grip on reality is more tenuous than day-to-day life can lead us to believe. Modify normal brain function just a bit, and the bedrock of reality may suddenly shift; though the outside world remains stable, our perception of it does not. This raises a classic philosophical question. Since all of our experiences are filtered and analyzed by our respective brains, how sure are we that our experiences reflect what's real? In the framing philosophers like to use: How do you know you're reading this sentence, and not floating in a vat on a distant planet, with alien scientists stimulating your brain to produce the thoughts and experiences you deem real?

These issues are central to epistemology, a philosophical subfield that asks what constitutes knowledge, how we acquire it, and how sure we are that we have it. Popular culture has brought these scholarly pursuits to a wide audience in films such as *The Matrix*, *The Thirteenth Floor*, and *Vanilla Sky*, tussling with them in entertaining and thought-provoking ways. So, in looser language, the question we're asking is: How do you know you're not hooked into the Matrix?

The bottom line is that you can't know for sure. You engage the world through your senses, which stimulate your brain in ways your neural circuitry has evolved to interpret. If someone artificially stimulates your brain so as to elicit electrical crackles exactly like those produced by eating pizza, reading this sentence, or skydiving, the experience will be indistinguishable from the real thing. Experience is dictated by brain processes, not by what activates those processes.

Going a step further, we can consider dispensing with the sloppiness of biological material altogether. Might all your thoughts and experiences be nothing more than a simulation that leverages software and circuitry sufficiently elaborate to mimic ordinary brain function? Are you convinced of the reality of flesh, blood, and the physical world, when actually your experience is only a crowd of electrical impulses firing through a hyper-advanced supercomputer?

An immediate challenge in considering such scenarios is that they easily set off a spiraling skeptical collapse; we wind up trusting nothing, not even our powers of deductive reasoning. My first response to questions like the ones just posed is to work out how much computer power you'd need to stand a chance of simulating a human brain. But if I am indeed part of such a simulation, why should I believe anything I read in neurobiology texts? The books would be simulations too, written by simulated biologists, whose findings would be dictated by the software running the simulation and thus could easily be irrelevant to the workings of "real" brains. The very notion of a "real" brain might itself be computer-generated artifice. Once you can't trust your knowledge base, reality quickly sails to sea.

We'll return to these concerns, but I don't want them to sink us—at least, not yet. So, for the time being, let's drop anchor. Imagine that you are real flesh and blood, and so am I, and that everything you and I take to be real, in the everyday sense of the term, *is* real. With all that assumed, let's take up the question of computers and brainpower. What, roughly, is the processing speed of the human brain, and how does it compare with the capacity of computers?

Even if we are not stuck in a skeptical morass, this is a difficult question. Brain function is largely an uncharted territory. But just to get a glimpse of the terrain, however foggy, consider some numbers. The human retina, a thin slab of 100 million neurons that's smaller than a dime and about as thick as a few sheets of paper, is one of the best-studied neuronal clusters. The robotics researcher Hans Moravec has estimated that for a computer-based retinal system to be on a par with that of humans, it would need to execute about a billion operations each second. To scale up from the retina's volume to that of the entire brain requires a factor of roughly 100,000; Moravec suggests that effectively simulating a brain would require a comparable increase in processing power, for a total of about 100 million million (10^{14}) operations per second.[4] Independent estimates based on the number of synapses in the brain and their typical firing rates yield processing speeds within a few orders of magnitude of this result, about 10^{17} operations per second. Although it's difficult to be more precise, this gives a sense of the numbers that come into play. The computer I'm now using has a speed that's about a billion operations per second; today's fastest supercomputers have a peak speed of about 10^{15} operations per second (a statistic that no doubt will quickly date this book). If we use the faster estimate for brain speed, we find that a hundred million laptops, or a hundred supercomputers, approach the processing power of a human brain.

Such comparisons are likely naïve: the mysteries of the brain are manifold, and speed is only one gross measure of function. But most everyone agrees that one day we will have raw computing capacity equal to, and likely far in excess of, what biology has provided. Futurists contend that such technological leaps will yield a world so far beyond familiar experience that we lack the capacity to imagine what it will be like. Invoking an analogy with phenomena that lie outside the bounds of our most refined physical theories, they call this visionary roadblock a singularity. One broad-brush prognosis holds that the surpassing of brainpower by computers will completely blur the boundary between humans and technology. Some anticipate a world run rampant with thinking and feeling machines, while those of us still based in old-fashioned biology routinely upload our brain content, safely storing knowledge and personalities *in silico*, complete with backup drives, for unlimited durations.

This vision may well be hyperbolic. There's little dispute regarding projections of computer power, but the obvious unknown is whether we will ever leverage such power into a radical fusion of mind and machine. It's a modern-day question with ancient roots; we've been thinking about thinking for thousands of years. How is it that the external world generates our internal responses? Is your sensation of color the same as mine? How about your sensations of sound and touch? What exactly is that voice we hear in our heads, the stream of internal chatter we call our conscious selves? Does it derive from purely physical processes? Or does consciousness arise from a layer of reality that transcends the physical? Penetrating thinkers through the ages, Plato and Aristotle, Hobbes and Descartes, Hume and Kant, Kierkegaard and Nietzsche, James and Freud, Wittgenstein and Turing, among countless others, have tried to illuminate (or debunk) processes that animate the mind and create the singular inner life available through introspection.

A great many theories of mind have emerged, differing in ways significant and subtle. We won't need the finer points, but just to get a feel for where the trails have led, here are a few: *dualist* theories, of which there are many varieties, maintain that there's an essential nonphysical component vital to mind. *Physicalist* theories of mind, of which there are also many varieties, deny this, emphasizing instead that underlying each unique subjective experience is a unique brain state. *Functionalist* theories go further in this direction, suggesting that what really matters to making a mind are the processes and functions—the circuits, their interconnections, their relationships—and not the particulars of the physical medium within which these processes take place.

Physicalists would largely agree that were you to faithfully replicate my brain by whatever means—molecule by molecule, atom by atom—the end product would indeed think and feel as I do. Functionalists would largely agree that were you to focus on higher-level structures—replicating all my brain connections, preserving all brain processes while changing only the physical substrate through which they occur—the same conclusion would hold. Dualists would largely disagree on both counts.

The possibility of artificial sentience clearly relies on a functionalist viewpoint. A central assumption of this perspective is that conscious thought is not overlaid on a brain but rather *is* the very sensation generated by a particular kind of information processing. Whether that processing happens within a three-pound biological mass or within the circuits of a computer is irrelevant. The assumption could be wrong. Maybe a bundle of connections needs a substrate of wrinkled wet matter if it's to gain self-awareness. Maybe you need the actual physical molecules that constitute a brain, not just the processes and connections those molecules facilitate, if conscious thought is to animate the inanimate. Maybe the kinds of information processing that computers carry out will always differ in some essential way from brain functioning, preventing the leap to sentience. Maybe conscious thought is fundamentally nonphysical, as claimed by various traditions, and so lies permanently beyond the reach of technological innovation.

With the rise of ever more sophisticated technologies, the questions have become sharper and the pathway toward answers more tangible. A number of research groups have already taken the initial steps toward simulating a biological brain on a computer. For example, the Blue Brain Project, a joint venture between IBM and the École Polytechnique Fédérale in Lausanne, Switzerland, is dedicated to modeling brain function on IBM's fastest supercomputer. Blue Gene, as the supercomputer is called, is a more powerful version of Deep Blue, the computer that triumphed in 1997

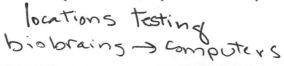

over the world chess champion Garry Kasparov. Blue Brain's approach is not all that different from the scenarios I just described. Through painstaking anatomical studies of real brains, researchers are gathering ever more precise insight into the cellular, genetic, and molecular structure of neurons and their interconnections. The project aims to encode such understanding, for now mostly at the cellular level, in digital models simulated by the Blue Gene computer. To date, researchers have drawn on results from tens of thousands of experiments focused on a pinhead-sized section of a rat brain, the neocortical column, to develop a three-dimensional computer simulation of roughly 10,000 neurons communicating through some 10 million interconnections. Comparisons between the response of a real rat's neocortical column and the computer simulation to the same stimuli show an encouraging fidelity of the synthetic model. This is far from the 100 billion neurons firing away in a typical human head, but the project's leader, the neuroscientist Henry Markram, anticipates that before 2020 the Blue Brain Project, leveraging processing speeds that are projected to increase by a factor of more than a million, will achieve a full simulated model of the human brain. Blue Brain's goal is not to produce artificial sentience, but rather to have a new investigative tool for developing treatments for various forms of mental illness; still, Markram has gone out on a limb to speculate that, when completed, Blue Brain may very well have the capacity to speak and to feel.

Regardless of the outcome, such hands-on explorations are pivotal to our theories of mind; I'm quite certain that the issue of which, if any, of the competing perspectives are on target cannot be settled through purely hypothetical speculation. In practice, too, challenges are immediately evident. Suppose a computer one day professes to be sentient—how would we know whether it really is? I can't even verify such claims of sentience when made by my wife. Nor she with me. That's a burden arising from consciousness being a private affair. But because our human interactions yield abundant circumstantial evidence supporting the sentience of others, solipsism quickly becomes absurd. Computer interactions may one day reach a similar point. Conversing with computers, consoling and cajoling them, may one day convince us that the simplest explanation for their apparent conscious self-awareness is that they are indeed conscious and self-aware. Let's take a functionalist viewpoint, and see where it leads.

Simulated Universes

If we ever create computer-based sentience, some would likely implant the thinking machines in artificial human bodies, creating a mechanical species—robots—that would be integrated into conventional reality. But my interest here is in those who would be drawn by the purity of electrical impulses to program simulated environments populated by simulated beings that would exist within a computer's hardware; instead of C-3PO or Data, think Sims or Second Life, but with inhabitants who have self-aware and responsive minds. The history of technological innovation suggests that iteration by iteration, the simulations would gain verisimilitude, allowing the physical and experiential characteristics of the artificial worlds to reach convincing levels of nuance and realism. Whoever was running a given simulation would decide whether the simulated beings knew that they existed within a computer; simulated humans who surmised that their world was an elaborate computer program might find themselves taken away by simulated technicians in white coats and confined to simulated locked wards. But probably the vast majority of simulated beings would consider the possibility that they're in a computer simulation too silly to warrant attention.

You may well be having that very reaction right now. Even if you accept the possibility of artificial sentience, you may be persuaded that the overwhelming complexity of simulating an entire civilization, or just a smaller community, renders such feats beyond computational reach. On this point, it's worth looking at some more numbers. Our distant descendants will likely fashion ever-larger quantities of matter into vast computing networks. So allow imagination free rein. Think big. Scientists have estimated that a present-day high-speed computer the size of the earth could perform anywhere from 10^{33} to 10^{42} operations per second. By comparison, if we assume that our earlier estimate of 10^{17} operations per second for a human brain is on target, then an average brain performs about 10^{24} total operations in a single hundred-year life span. Multiply that by the roughly 100 billion people who have ever walked the planet, and the total number of operations performed by every human brain since Lucy (my archaeology friends tell me I should say "Ardi") is about 10^{35}. Using the conservative estimate of 10^{33} operations per second, we see that the collective computational capacity of the human species could be achieved with a run of less than two minutes on an earth-sized computer.

And that's with today's technology. Quantum computing—harnessing all the distinct possibilities represented in a quantum probability wave so as to do many different calculations simultaneously—has the capacity to increase processing speeds by spectacular factors. Although we are still very far from mastering this application of quantum mechanics, researchers have estimated that a quantum computer no bigger than a laptop has the potential to perform the equivalent of all human thought since the dawn of our species in a tiny fraction of a second.

To simulate not just individual minds but also their interactions among themselves and with an evolving environment, the computational load would grow orders of magnitude larger. But a sophisticated simulation could cut computational corners with minimal impact on quality. Simulated humans on a simulated earth won't be bothered if the computer simulates only things lying within the cosmic horizon. We can't see beyond that range, so the computer can safely ignore it. More boldly, the simulation might simulate stars beyond the sun only during simulated nights, and then only when the simulated local weather resulted in clear skies. When no one's looking, the computer's celestial simulator routines could take a break from working out the appropriate stimulus to provide each and every person who *could* look skyward. A sufficiently well-structured program would keep track of the mental states and intentions of its simulated inhabitants, and so would anticipate, and appropriately respond to, any impending stargazing. The same goes for simulating cells, molecules, and atoms. For the most part, they'd be necessary only for simulated specialists of one scientific persuasion or another, and then only when such specialists were in the act of studying these exotic realms. A computationally cheaper replica of familiar reality that adjusts the simulation's degree of detail on an as-needed basis would be adequate.

Such simulated worlds would forcefully realize Wheeler's vision of information's primacy. Generate circuits that carry the right information and you've generated parallel realities that are as real to their inhabitants as this one is to us. These simulations constitute our eighth variety of multiverse, which I'll call the *Simulated Multiverse*.

Are You Living in a Simulation?

The idea that universes might be simulated on computers has a long history, dating as far back as suggestions made in the 1960s by the computer pioneer Konrad Zuse and the digital guru Edward Fredkin. I worked at IBM during five summers spanning college and graduate school; my boss, the late John Cocke, himself a revered computer specialist, spoke frequently of Fredkin's view that the universe was nothing but a giant computer chugging along, executing something akin to cosmic Fortran. The idea struck me as taking the digital paradigm to a ridiculous extreme. Through the years, I hardly gave it a thought—until I encountered, much more recently, a simple but curious conclusion by the Oxford philosopher Nick Bostrom.

To appreciate Bostrom's point (one that Moravec had also hinted at), begin with a straightforward comparison: the difficulty of creating a real universe versus the difficulty of creating a simulated universe. To create a real one, as we've discussed, presents enormous obstacles. And if we succeeded, the resulting universe would be beyond our ability to see, which invites the question of what motivated us to create it in the first place.

The creation of a simulated universe is a wholly different enterprise. The march toward increasingly powerful computers, running ever more sophisticated programs, is inexorable. Even with today's rudimentary technology, the fascination of creating simulated environments is strong; with more capability it's hard to imagine anything but more intense interest. The question is not whether our descendants will create simulated computer worlds. We're already doing it. The unknown is how realistic the worlds will become. Should there be an inherent obstacle to generating artificial sentience, all bets are off. But Bostrom, assuming that realistic simulations prove possible, makes a simple observation.

Our descendants are bound to create an immense number of simulated universes, filled with a great many self-aware, conscious inhabitants. If someone can come home at night, kick back, and fire up the create-a-universe software, it's easy to envision that they'll not only do so, but do so often. Think about what this scenario might entail. One future day, a cosmic census that takes account of all sentient beings might find that the number of flesh-and-blood humans pales in comparison with those made of chips and bytes, or their future equivalents. And, Bostrom reasons, if the ratio of simulated humans to real humans were colossal, then brute statistics suggests that we are *not* in a real universe. The odds would overwhelmingly favor the conclusion that you and I and everyone else are living within a simulation, perhaps one created by future historians with a fascination for what life was like back on twenty-first-century earth.

You may object that we have now run headlong into the skeptical quicksand we planned at the outset to avoid. Once we conclude that there's a high likelihood that we're living in a computer simulation, how do we trust anything, including the very reasoning that led to the conclusion? Well, our confidence in a great many things might diminish. Will the sun rise tomorrow? Maybe, as long as whoever is running the simulation doesn't pull the plug. Are all our memories trustworthy? They seem so, but whoever is at the keyboard may have a penchant for adjusting them from time to time.

Nevertheless, Bostrom notes, the conclusion that we're in a simulation does not fully sever our grasp on the true underlying reality. Even if we believe that we're in a simulation, we can still identify one feature that the underlying reality definitely possesses: it allows for realistic computer simulations. After all, according to our belief, we're in one. The unbridled skepticism generated by the suspicion that we're simulated aligns with that very knowledge and so fails to undermine it. While it was useful when we began to weigh anchor and declare the reality of all that seems real, it wasn't necessary. Logic alone can't ensure that we're not in a computer simulation.

The only way to dodge the conclusion that we're likely living in a simulation is to leverage intrinsic weaknesses in the reasoning. Maybe sentience can't be simulated, full stop. Or maybe, as Bostrom also suggests, civilizations en route to the technological mastery necessary to create sentient simulations will inevitably turn that technology inward and destroy themselves. Or maybe when our distant descendants gain the capacity to create simulated universes they choose not to do so, perhaps for moral reasons or simply because other currently inconceivable pursuits prove so much more interesting that, much as we noted with universe creation, universe simulation falls by the wayside.

These are among numerous loopholes, but whether they're large enough for the proverbial truck to drive through, who knows?* If not, you might want to spice up your life a bit, make your mark. Whoever is running the simulation is bound to get tired of wallflowers. Being a cynosure would seem a likely path toward longevity.[5]

Seeing Beyond a Simulation

If you were living in a simulation, could you figure that out? The answer depends in no small part on who is running your simulation—call him or her the Simulator—and the manner in which your simulation was programmed. The Simulator, for

---

*Another loophole arises from an incarnation of the measure problem from Chapter 7. If the number of real (nonvirtual) universes is infinite (if we're part of, say, the Quilted Multiverse), then there will be an infinite collection of worlds like ours in which descendants run simulations, yielding an infinite number of simulated worlds. Even though it would still seem that the number of simulated worlds would vastly outnumber the real ones, we saw in Chapter 7 that comparing infinities is a treacherous business.

instance, might choose to let you in on the secret. One day while taking a shower you might hear a gentle "ding-ding," and when you'd cleared the shampoo from your eyes you'd see a floating window in which your smiling Simulator would appear and introduce herself. Or maybe this revelation would happen on a worldwide scale, with giant windows and a booming voice surrounding the planet, announcing that there is in fact an All Powerful Programmer up in the heavens. But even if your Simulator shied away from exhibitionism, less obvious clues might turn up.

Simulations allowing for sentient beings would certainly have reached a minimum fidelity threshold, but as they do with designer clothes and cheap knockoffs, quality and consistency would likely vary. For example, one approach to programming simulations— call it the "emergent strategy"—would draw on the accumulated mass of human knowledge, judiciously invoking relevant perspectives as dictated by context. Collisions between protons in particle accelerators would be simulated using quantum field theory. The trajectory of a batted ball would be simulated using Newton's laws. The reactions of a mother watching her child's first steps would be simulated by melding insights from biochemistry, physiology, and psychology. The actions of governmental leaders would fold in political theory, history, and economics. Being a patchwork of approaches focused on different aspects of simulated reality, the emergent strategy would need to maintain internal consistency as processes nominally construed to lie in one realm spilled over into another. A psychiatrist needn't fully grasp the cellular, chemical, molecular, atomic, and subatomic processes underlying brain function—which is a good thing for psychiatry. But in simulating a person, the challenge for the emergent strategy would be to consistently meld coarse and fine levels of information, ensuring for example that emotional and cognitive functions interface sensibly with physiochemical data. This kind of cross-border meshing takes place in all phenomena and has always compelled science to seek deeper, more unified explanations.

Simulators employing emergent strategies would have to iron out mismatches arising from the disparate methods, and they'd need to ensure that the meshing was smooth. This would require fiddles and tweaks which, to an inhabitant, might appear as sudden, baffling changes to the environment with no apparent cause or explanation. And the meshing might fail to be fully effective; the resulting inconsistencies could build over time, perhaps becoming so severe that the world became incoherent, and the simulation crashed.

A possible way to obviate such challenges would be to use a different approach—call it the "ultra-reductionist strategy"—in which the simulation would proceed by a single set of fundamental equations, much as physicists imagine is the case for the real universe. Such simulations would take as input a mathematical theory of matter and the fundamental forces and a choice of "initial conditions" (how things were at the starting point of the simulation); the computer would then evolve everything forward in time, thereby avoiding the meshing issues of the emergent approach. But simulations of this kind would encounter their own computational problems, even beyond the staggering computational burden of simulating "everything," right down to the behavior of individual particles. If the equations our descendants have in their possession are similar to those we work with today—involving numbers that can vary continuously—then the simulations would necessarily invoke approximations. To *exactly* follow a number as it varies continuously, we would need to track its value to an infinite number of decimal places (for instance, as such a quantity varies, say, from .9 to 1, it would pass through numbers like .9, .95, .958, .9583, .95831, .958317, and on and on, with an arbitrarily large number of digits required for full accuracy). That's something a computer with finite resources can't manage: it will run out of time and memory. So, even if *the* deepest equations were used, it's still possible that computer-based calculations would inevitably be approximate, allowing errors to build up over time.*

Of course, by "error" I mean a deviation between what occurs

---

*A theory that allows for only a finite number of distinct states within a finite spatial volume (in accord, for example, with the entropy bounds discussed in the previous chapter) can still involve continuous quantities as part of its mathematical formalism. This is the case, for instance, with quantum mechanics: the probability wave's value can vary continuously even when only finitely many different outcomes are possible.

in the simulation and the description inherent in the most refined physical theories the simulator has at his or her disposal. But to those like you who are within the simulation, the mathematical rules driving the computer *would be* your laws of nature. The issue, then, is not how closely the mathematical laws used by the computer model the external world; we're imagining that you don't observe the external world from within the simulation. Rather, the problem for a simulated universe is that when a computer's necessary approximations permeate otherwise exact mathematical equations, calculations easily lose their stability. Round-off errors, when accumulated over a great many computations, can yield inconsistencies. You and other simulated scientists might witness anomalous results from experiments; cherished laws might start yielding inaccurate predictions; measurements that had long since converged on a single widely confirmed result might start producing different answers. For long stretches, you and your simulated colleagues would think that you'd encountered evidence, much as your forebears had throughout the previous centuries and millennia, that your final theory wasn't so final after all. Collectively, you'd closely reexamine the theory, perhaps coming up with new ideas, equations, and principles that better described the data. But, assuming the inaccuracies didn't result in contradictions that crashed the program, at some point you'd hit a wall.

After an exhaustive search through possible explanations, none of which was able to fully explain what was happening, an iconoclastic thinker might suggest a radically different idea. If the continuum laws that physicists had developed over many millennia were input to a powerful digital computer and used to generate a simulated universe, the errors built up from the inherent approximations would yield anomalies of the very kind being observed. "Are you suggesting that we're in a computer simulation?" you'd ask. "Yes," your colleague would answer. "Well, that's nutty," you'd say. "Really?" she'd reply. "Take a look." And she'd produce a monitor showing a simulated world, which she had programmed using those very same deep laws of physics, and—catching your breath after the shock of encountering a simulated world at all—you would see that the simulated scientists were indeed puzzling over the very same kind of strange data that troubled you.[6]

A Simulator who sought more assiduously to conceal herself could, of course, use more aggressive tactics. As inconsistencies started to build, she might reset the program and erase the inhabitants' memory of the anomalies. So it would seem a stretch to claim that a simulated reality would reveal its true nature through glitches and irregularities. And certainly I'd be hard pressed to argue that inconsistencies, anomalies, unanswered questions, and stalled progress would reflect anything more than our own scientific failings. The sensible interpretation of such evidence would be that we scientists need to work harder and be more creative in seeking explanations. However, there is one serious conclusion that emerges from the fanciful scenario I've told. If and when we do generate simulated worlds, with apparently sentient inhabitants, an essential question will arise: Is it reasonable to believe that we occupy a rarefied place in the history of scientific-technological development— that we have become the very first creators of sentient simulations? We *may* have—but if we're keen to go with the odds, we must consider alternative explanations that, in the grand scheme of things, don't require us to be so extraordinary. And there is a ready-made explanation that fits the bill. Once our own work convinces us that sentient simulations are possible, the guiding principle of "garden variety," discussed in Chapter 7, suggests that there's not just one such simulation out there but a swarming ocean of simulations, which constitute a Simulated Multiverse. While the simulation we've created might be a landmark feat in the limited domain to which we have access, within the context of the entire Simulated Multiverse it's nothing special, having been achieved a gazillion times over. Once we accept that idea, we're led to consider that we too may be in a simulation, since that's the status of the vast majority of sentient beings in a Simulated Multiverse.

Evidence for artificial sentience and for simulated worlds is grounds for rethinking the nature of your own reality.

The Library of Babel

During my first semester in college, I enrolled in an introductory philosophy course taught by the late Robert Nozick. From the very first lecture, it was a wild ride. Nozick was completing his voluminous *Philosophical Explanations*; he used the course as a dress rehearsal for many of the book's central arguments. Just about every class shook my grasp on the world, sometimes vigorously. This was an unexpected experience—I'd thought that upending reality would be the purview solely of my physics courses. Yet, there was an essential difference between the two. The physics lectures challenged comfortable views by exposing strange phenomena that arise in wholly unfamiliar realms where things move fast, are extremely heavy, or are fantastically tiny. The philosophy lectures shook comfortable views by challenging the foundations of *everyday* experience. How do we know there's a real world out there? Should we trust our perceptions? What thread binds our molecules and atoms to preserve our personal identity through time?

While I was hanging around after class one day, Nozick asked me what I was interested in, and I brazenly told him that I wanted to work on quantum gravity and unified theories. This was generally a conversation stopper, but for Nozick it presented a chance to educate a young mind by revealing a new perspective. "What drives your interest?" he asked. I told him that I wanted to find eternal truths, to help understand why things are the way they are. Naïve and blustery, for sure. But Nozick listened graciously and then took the idea further. "Let's say you find the unified theory," he said. "Would that really provide the answers you're looking for? Wouldn't you still be left asking why that particular theory, and not another, was the correct theory of the universe?" He was right, of course, but I replied that in the search for explanations there might come a point when we would just have to accept certain things as given. That was just where Nozick wanted me to go; in writing *Philosophical Explanations* he had developed an alternative to this view. It's based on what he called the principle of fecundity and is an attempt to frame explanations without "accepting certain things as given"; without, as Nozick explains it, accepting anything as brute-force truth.

The philosophical maneuver behind this trick is straightforward: defang the question. If you want to avoid explaining why one particular theory should be singled out over another, then don't single it out. Nozick suggests that we imagine we're part of a multiverse that comprises *every possible universe*.[7] The multiverse would include not only the alternative evolutions emerging from the Quantum Multiverse, or the many bubble universes of the Inflationary Multiverse, or the possible stringy worlds of the Brane or Landscape Multiverse. These multiverses wouldn't, on their own, fulfill Nozick's proposal, because you'd still be left wondering: Why quantum mechanics? Or why inflation? Or why string theory? Instead, come up with *any* possible universe whatsoever—it could be made of the usual atomic species, but a universe made solely of melted mozzarella would serve just as well—and it has a place in Nozick's scheme.

This is the last multiverse we will consider, since it's the most expansive of all—the most expansive possible. Any multiverse that's ever been or ever will be proposed is itself composed of possible universes, and will therefore be part of this mega-conglomerate, which I'll call the *Ultimate Multiverse*. Within this framework, if you ask why our universe is governed by the laws our research reveals, the answer harks back to anthropics: there are other universes out there, all possible universes in fact, and we inhabit the one we do because it's among those that support our form of life. In the other universes where we could live—of which there are many since, among other things, we can certainly survive sufficiently tiny changes to the various fundamental parameters of physics—there are people, much like us, asking the same question. And the same answer applies equally well to them. The point is that the attribute of existence affords a universe no special status, because in the Ultimate Multiverse all possible universes *do* exist. The question of why one set of laws describes a real universe—ours—while all others are sterile abstractions evaporates. There are no sterile laws. All sets of laws describe real universes.

Curiously, Nozick noted that within his multiverse there'd be a universe that consists of nothing. Absolutely nothing. Not empty space, but the nothing that Gottfried Leibniz referred to in his famous query "Why is there something rather than nothing?" Not that Nozick could have known, but for me this was an observation of particular resonance. When I was ten or eleven, I came upon Leibniz's question and found it deeply troubling. I'd pace my room, trying to grasp what nothing would be, often with one hand hovering behind the back of my head, thinking that the struggle to do the impossible—see my hand—would help me grasp the meaning of total absence. Even now, to focus on absolute true nothingness makes my heart sink. Total nothingness, from our familiar vantage point of somethingness, entails the most profound loss. But because nothing also seems so vastly simpler than something—no laws at work, no matter at play, no space to inhabit, no time to unfurl—Leibniz's question strikes many as right on the mark. *Why isn't there nothingness?* Nothingness would have been decidedly elegant.

In the Ultimate Multiverse, a universe consisting of nothing *does* exist. As far as we can tell, nothingness is a perfectly logical possibility and so must be included in a multiverse that embraces all universes. Nozick's answer to Leibniz, then, is that in the Ultimate Multiverse there is no imbalance between something and nothing that calls out for explanation. Universes of both types are part of this multiverse. A nothing universe is nothing to get exercised about. It's only because we humans are something that the nothing universe eludes us.

A theoretician, trained to speak in mathematics, understands Nozick's all-encompassing multiverse as one where all possible mathematical equations are realized physically. It's a version of Jorge Luis Borges' story "La Biblioteca de Babel," in which the books of Babel are written in the language of mathematics, and so contain all possible sensible, non-self-contradictory strings of mathematical symbols.* Some of the books would spell out familiar formulae, such as the equations of general relativity and those of quantum mechanics, as applied to the known particles of nature. But such recognizable strings of mathematical characters would be extremely rare. Most books would contain equations no one has previously written down, equations that would normally be deemed pure abstractions. The idea of the Ultimate Multiverse is to shed this familiar perspective. No longer do most equations lie dormant,

---

*Borges allows for books with all possible character strings, without regard to meaning.

with only the lucky few mysteriously coaxed to life through physical instantiation. Instead, every book in the Library of Mathematical Babel *is* a real universe.

Nozick's suggestion, in this mathematical framing, provides a concrete answer to a long-debated question. For centuries, mathematicians and philosophers have wondered whether mathematics is discovered or invented. Are mathematical concepts and truths "out there," waiting for an intrepid explorer to stumble upon them? Or, since that explorer is more than likely sitting at a desk, pencil in hand, scribbling arcane symbols furiously across a page, are the resulting mathematical concepts and truths invented as part of the mind's search for order and pattern?

At first sight, the uncanny way that a great many mathematical insights find application to physical phenomena provides compelling evidence that math is real. Examples abound. From general relativity to quantum mechanics, physicists have found that various mathematical discoveries are tailor-made for physical applications. Paul Dirac's prediction of the positron (the anti-particle of the electron) provides a simple but impressive case in point. In 1931, upon solving his quantum equations for the motion of electrons, Dirac found that the math provided an "extraneous" solution—apparently describing the motion of a particle just like the electron except that it carried a positive electric charge (whereas the electron's charge is negative). In 1932, that very particle was discovered by Carl Anderson through a close study of cosmic rays bombarding earth from space. What began as Dirac's manipulation of mathematical symbols in his notebooks concluded in the laboratory with the experimental discovery of the first species of antimatter.

The skeptic can counter, however, that mathematics still emanates from us. We were shaped by evolution to find patterns in the environment; the better we could do that, the better we could predict how to find the next meal. Mathematics, the language of pattern, emerged from our biological fitness. And with that language, we've been able to systematize the search for new patterns, going well beyond those relevant for mere survival. But mathematics, like any of the tools we developed and utilized through the ages, is a human invention.

My view on mathematics periodically changes. When I'm in the throes of a mathematical investigation that's going well, I often feel that the process is one of discovery, not invention. I know of no more exciting experience than watching the disparate pieces of a mathematical puzzle suddenly coalesce into a single coherent picture. When it happens, there's a feeling that the picture was there all along, like a grand vista hidden by the morning fog. On the other hand, when I more objectively survey mathematics, I'm less convinced. Mathematical knowledge is the literary output of humans conversant in the unusually precise language of mathematics. And as is surely the case with literature produced in one of the world's natural languages, mathematical literature is the product of human ingenuity and creativity. That's not to say that other intelligent life-forms wouldn't come upon the same mathematical results we've found; they very well might. But that could easily reflect similarities in our experiences (such as the need to count, the need to trade, the need to survey, and so on) and so would provide minimal evidence that math has a transcendent existence.

A number of years ago, in a public debate on the subject, I said that I could imagine an alien encounter during which, in response to learning of our scientific theories, the aliens remark, "Oh, math. Yeah, we tried that for a while. At first it seemed promising, but ultimately it was a dead end. Here, let us show you how it really works." But, to continue with my own vacillation, I don't know how the aliens would actually finish the sentence, and with a broad enough definition of mathematics (e.g., logical deductions following from a set of assumptions), I'm not even sure what kind of answers *wouldn't* amount to math.

The Ultimate Multiverse is unequivocal on the issue. All math is real in the sense that all math describes some real universe. Across the multiverse, all math gets its due. A universe governed by Newton's equations and populated solely by solid billiard balls (with no additional internal structure) is a real universe; an empty universe with 666 spatial dimensions governed by a higher-dimensional version of Einstein's equations is a universe too. If the aliens happened to be right, there would also be universes whose description would stand outside mathematics. But let's hold that possibility off to the side. A multiverse realizing all mathematical equations will be enough to keep us occupied; that's what the Ultimate Multiverse gives us.

Multiverse Rationalization

Where the Ultimate Multiverse differs from the other parallel universe proposals we've encountered is in the reasoning that leads to its consideration. The multiverse theories in previous chapters were not dreamed up to solve a problem or answer a question. Some of them do, or at least claim to, but they weren't developed for that purpose. We've seen that some theorists believe the Quantum Multiverse resolves the quantum measurement problem; some believe the Cyclic Multiverse addresses the question of time's beginning; some believe the Brane Multiverse clarifies why gravity is so much weaker than the other forces; some believe the Landscape Multiverse gives insight into the observed value of dark energy; some believe the Holographic Multiverse explains data emerging from the collision of heavy atomic nuclei. But such applications are secondary. Quantum mechanics was developed to describe the micro-realm; inflationary cosmology was developed to make sense of observed properties of the cosmos; string theory was developed to mediate between quantum mechanics and general relativity. The possibility that these theories generate various multiverses is a by-product.

The Ultimate Multiverse, by contrast, carries no explanatory weight apart from its assumption of a multiverse. It achieves precisely one goal: cleaving from our to-do list the project of finding an explanation for why our universe adheres to one set of mathematical laws and not another, and it accomplishes this singular feat precisely by introducing a multiverse. Cooked up specifically to address one issue, the Ultimate Multiverse lacks the independent rationale characterizing the multiverses discussed in earlier chapters.

That's my view, but not everyone agrees. There's a philosophical perspective (coming from the *structural realist* school of thought) that suggests physicists may have fallen prey to a false dichotomy between mathematics and physics. It's common for theoretical physicists to speak of mathematics providing a quantitative language for describing physical reality; I've done so on most every page of this book. But maybe, this perspective suggests, math is more than just a description of reality. Maybe math *is* reality.

It's a peculiar idea. We are not used to thinking of solid reality as being constructed from intangible mathematics. The simulated universes of the previous section provide a concrete and enlightening way to think about it. Consider that most celebrated of knee-jerk reactions, in which Samuel Johnson responded to Bishop Berkeley's claim that matter is a figment of the mind's conjuring by kicking a large stone. Imagine, however, that unbeknownst to Dr. Johnson, his kick happened within a hypothetical, high-fidelity computer simulation. In that simulated world, Dr. Johnson's experience of the stone would be just as convincing as in the historical version. Yet, the computer simulation is nothing but a chain of mathematical manipulations that take the computer's state at one moment—a complex arrangement of bits—and, according to specified mathematical rules, evolve those bits through subsequent arrangements.

Which means that were you to intently study the mathematical transformations the computer carried out during Dr. Johnson's demonstration, you'd see, right there in the math, the kick and the rebound of his foot, as well as the thought and the famous articulation "I refute it thus." Hook the computer to a monitor (or some futuristic interface), and you would see that the mathematically choreographed dancing bits yield Dr. Johnson and his kick. But don't let the simulation's bells and whistles—the computer's hardware, the fancy interface, and so on—obscure the essential fact: underneath the hood, there'd be nothing but math. Change the mathematical rules, and the dancing bits would tap out a different reality.

Now, why stop there? I put Dr. Johnson in a simulation only because that context provides an instructive bridge between mathematics and Dr. Johnson's reality. But the deeper point of this perspective is that the computer simulation is an inessential intermediate step, a mere mental stepping-stone between the experience of a tangible world and the abstraction of mathematical equations. The mathematics itself—through the relationships it creates, the connections it establishes, and the transformations it embodies—contains Dr. Johnson, both his actions and his thoughts. You don't need the computer. You don't need the dancing bits. *Dr. Johnson is in the mathematics.*[8]

And once you take on board the idea that mathematics itself can, through its inherent structure, embody any and all aspects of reality—sentient minds, heavy rocks, vigorous kicks, stubbed toes—you're led to envision that *our* reality is nothing but math. In this way of thinking, everything you're aware of—the sensation of holding this book, the thoughts you're now having, the plans you're making for dinner—is the experience of mathematics. Reality is how math feels.

To be sure, this perspective requires a conceptual leap not everyone will be persuaded to take; personally, it's a leap I've not taken. But for those who do, the worldview sees math as not just "out there," but as the only thing that's "out there." A body of mathematics, be it Newton's equations, those of Einstein, or any others, doesn't become real when physical entities arise that instantiate it. Mathematics—all mathematics—already is real; it doesn't require instantiation. Different collections of mathematical equations are different universes. The Ultimate Multiverse is thus the by-product of this perspective on mathematics.

Max Tegmark of the Massachusetts Institute of Technology, who has been a strong promoter of the Ultimate Multiverse (which he has called the Mathematical Universe Hypothesis), justifies this view through a related consideration. The deepest description of the universe should not require concepts whose meaning relies on human experience or interpretation. Reality transcends our existence and so shouldn't, in any fundamental way, depend on ideas of our making. Tegmark's view is that mathematics—thought of as collections of operations (like addition) that act on abstract sets of objects (like the integers), yielding various relations between them (like $1 + 2 = 3$)—is precisely the language for expressing statements that shed human contagion. But what, then, could possibly distinguish a body of mathematics from the universe it depicts? Tegmark argues that the answer is nothing. Were there some feature that did distinguish math from the universe, it would have to be non-mathematical; otherwise it could be absorbed into the mathematical depiction, erasing the purported distinction. But, according to this line of thought, if the feature were non-mathematical, it must bear a human imprint, and so can't be fundamental. Thus, there's no distinguishing what we conventionally call the mathematical description of reality from its physical embodiment. They are the same. There's no switch that turns math "on." Mathematical existence is synonymous with physical existence. And since this would be true for any and all math, this provides another road leading us to the Ultimate Multiverse.

While all these arguments are curious to contemplate, I remain skeptical. In evaluating a given multiverse proposal, I'm partial to there being a process, however tentative—a fluctuating inflaton field, collisions between braneworlds, quantum tunneling through the string theory landscape, a wave evolving via the Schrödinger equation—that we can imagine generating the multiverse. I prefer to ground my thinking in a sequence of events that can, at least in principle, result in the given multiverse unfolding. For the Ultimate Multiverse, it's hard to imagine what such a process could be; the process would need to yield different mathematical laws in different domains. In the Inflationary and Landscape Multiverses, we've seen that the details of how the laws of physics manifest themselves can vary from universe to universe, but this is because of environmental differences, such as the values of certain Higgs fields or the shape of the extra dimensions. The underlying mathematical equations, operating across all the universes, are the same. So what process, operating within a given set of mathematical laws, can change those mathematical laws? Like the number five desperately trying to be six, it seems plainly impossible.

However, before settling on that conclusion, consider this: there can be domains that *appear* as though they are governed by different mathematical rules. Think again about simulated worlds. In discussing Dr. Johnson above, I invoked a computer simulation as a pedagogical device to explain how mathematics may embody the essence of experience. But if we consider such simulations in their own right, as we do in the Simulated Multiverse, we see that they offer just the process we need: although the computer hardware on which a simulation is run is subject to the usual laws of physics, the simulated world itself will be founded on the mathematical equations the user happens to choose. From simulation to simulation, the mathematical laws can and generally will vary.

As we will now see, this provides a mechanism for generating a particular privileged part of the Ultimate Multiverse.

Simulating Babel

Earlier, I noted that for the kinds of equations we typically study in physics, computer simulations yield only approximations to the mathematics. Such is generally the case when continuous numbers confront a digital computer. For example, in classical physics (assuming, as we do in classical physics, that spacetime is continuous) a batted ball passes through an infinite number of different points as it travels from home plate to left field.[9] Keeping track of a ball through an infinity of locations, and of an infinity of possible speeds at those locations, will always remain beyond reach. At best, computers can perform highly refined but still approximate calculations, tracking a ball every millionth or billionth or trillionth of a centimeter, for instance. That's fine for many purposes, but it's still an approximation. Quantum mechanics and quantum field theory, by introducing various forms of discreteness, help in some ways. But both make extensive use of continuously varying numbers (values of probability waves, values of fields, and so on). The same reasoning holds for all the other standard equations of physics. A computer can approximate the math, but it can't simulate the equations exactly.*

---

*When we discussed the Quilted Multiverse (Chapter 2), I stressed that quantum physics assures us that in any finite region of space there are only finitely many different ways in which matter can arrange itself. Nevertheless, the mathematical formalism of quantum mechanics involves features that are continuous and that hence can assume infinitely many values. These features are things we can't directly observe (such as the height of a probability wave at a given point); it's with respect to the distinct results that measurements can acquire that there are only finitely many possibilities.

There are other types of mathematical functions, however, for which a computer simulation can be absolutely precise. They're part of a class called *computable functions*, which are functions that can be evaluated by a computer running through a finite set of discrete instructions. The computer may need to cycle through the collection of steps repeatedly but sooner or later it will produce the exact answer. No originality or novelty is needed at any step; it's just a matter of grinding out the result. In practice, then, to simulate the motion of a batted ball, computers are programmed with equations that are *computable approximations* to the laws of physics that you learned in high school. (Typically, continuous space and time are approximated on a computer by a fine grid.)

By contrast, a computer trying to calculate a noncomputable function will churn away indefinitely without coming to an answer, regardless of its speed or memory capacity. Such would be the case for a computer seeking the exact continuous trajectory of that batted ball. For a more qualitative example, imagine a simulated universe in which a computer is programmed to provide a wonderfully efficient simulated chef who provides meals for all those simulated inhabitants—and only those simulated inhabitants—who don't cook for themselves. As the chef furiously bakes, fries, and broils, he works up quite an appetite. The question is: Whom does the computer charge with feeding the chef?[10] Think about it, and it makes your head hurt. The chef can't cook for himself as he *only* cooks for those who don't cook for themselves, but if the chef doesn't cook for himself, he is among those for whom he is meant to cook. Rest assured, the computer's head would hardly fare better than yours. Noncomputable functions are much like this example: they stymie a computer's ability to complete its calculations, and so the simulation being run by the computer would hang. The successful universes constituting the Simulated Multiverse would therefore be based on computable functions.

The discussion suggests an overlap between the Simulated and Ultimate Multiverses. Consider a scaled-down version of the Ultimate Multiverse that includes only universes arising from computable functions. Then, rather than merely being posited as a resolution to one particular question—Why is this universe real, while other possible universes are not? — the scaled-down version of the Ultimate Multiverse can emerge from a process. An army of future computer users, perhaps not much different in temperament from today's Second Life enthusiasts, could spawn this multiverse through their insatiable fascination with running simulations based on ever-different equations. These users wouldn't generate all universes contained in the Mathematical Library of Babel, because the ones based on noncomputable functions wouldn't get off the ground. But the users would continually work their way through the library's computable wing.

The computer scientist Jürgen Schmidhuber, extending earlier ideas of Zuse, has come to a similar conclusion from a different angle. Schmidhuber realized that it's actually easier to program a computer to generate all possible computable universes than it is to program individual computers to generate them one by one. To see why, imagine programming a computer to simulate baseball games. For each game, the amount of information you'd need to supply is vast: every detail about every player, physical and mental, every detail about the stadium, the umpires, the weather, and so on. And each new game you simulate requires you to specify yet another mountain of data. However, if you decide to simulate not one or a few games, but *every* game imaginable, your programming job would be far easier. You'd just need to set up one master program that systematically makes its way through every possible variable — those that affect players, the environment, and all other relevant features — and let the program run. Finding any one particular game in the resulting voluminous output would be a challenge, but you'd be assured that sooner or later every possible game would appear.

The point is that whereas specifying one member of a large collection requires a great deal of information, specifying the entire collection can often be much easier. Schmidhuber found that this conclusion applies to simulated universes. A programmer hired to simulate a collection of universes based on specific sets of mathematical equations could take the easy way out: much like the baseball enthusiast, he could opt to write a single, relatively short program that would generate *all* computable universes, and turn the computer loose. Somewhere among the resulting gargantuan collection of simulated universes, the programmer would find those he'd been hired to simulate. I wouldn't want to be paying for computer usage by the hour as the turnaround time for generating these simulations would similarly be gargantuan. But I'd happily pay the programmer by the hour since the instruction set to generate all computable universes would be much less intensive than that required to yield any one universe in particular.[11]

Either of these scenarios—a great many users simulating a great many universes, or a master program that simulates them all—is how the Simulated Multiverse might be generated. And because the resulting universes would be based on a wide variety of different mathematical laws, we can equivalently think of these scenarios as generating part of the Ultimate Multiverse: the part encompassing universes based on computable mathematical functions.*

The drawback of generating only part of the Ultimate Multiverse is that this downsized version less effectively addresses the issue that inspired Nozick's principle of fecundity in the first place. If all possible universes don't exist, if the entire Ultimate Multiverse is not generated, the question resurfaces of why some equations come to life and others don't. Specifically, we're left wondering why universes based on computable equations hog the spotlight.

To continue along this chapter's highly speculative path, maybe the computable/noncomputable division is telling us something. Computable mathematical equations avoid the prickly issues raised in the middle of the last century by penetrating thinkers like Kurt Gödel, Alan Turing, and Alonzo Church. Gödel's famous *incompleteness theorem* shows that certain mathematical systems necessarily admit true statements that can't be proved within the mathematical system itself. Physicists have long wondered about

---

*Max Tegmark has noted that the entirety of a simulation, run from start to finish, is itself a collection of mathematical relations. Thus, if one believes that all mathematics is real, so is this collection. In turn, from this perspective there's no need to actually run any computer simulations since the mathematical relations each would produce are already real. Also, note that the focus on evolving a simulation forward in time, however intuitive, is overly restrictive. The computability of a universe should be evaluated by examining the computability of the mathematical relations that define its entire history, whether or not these relations describe the unfolding of the simulation through time.

the possible implications of Gödel's insights for their own work. Might physics, too, necessarily be incomplete, in the sense that some features of the natural world would forever elude our mathematical descriptions? In the context of the downsized Ultimate Multiverse, the answer is no. Computable mathematical functions, by definition, lie squarely within the bounds of calculation. They are the very functions that admit a procedure by which a computer can successfully evaluate them. And so, if all the universes in a multiverse were based on computable functions, they all would also do an end run around Gödel's theorem; this wing of the Library of Mathematical Babel, this version of the Ultimate Multiverse, would be free of Gödel's ghost. Maybe that's what singles out computable functions.

Would *our* universe find a place in this multiverse? That is, if and when we put our hands on the final laws of physics, will those laws describe the cosmos using mathematical functions that are computable? Not just approximately computable functions, as is the case with the physical laws we work with today. But exactly computable? No one knows. If so, developments in physics should drive us toward theories in which the continuum plays no role. Discreteness, the core of the computational paradigm, should prevail. Space surely seems continuous, but we've only probed it down to a billionth of a billionth of a meter. It's possible that with more refined probes we will one day establish that space is fundamentally discrete; for now, the question is open. A similar limited understanding applies to intervals of time. The discoveries recounted in Chapter 9, which yield information capacity of one bit per Planck area in any region of space, constitute a major step in the direction of discreteness. But the issue of how far the digital paradigm can be taken remains far from settled.[12] My guess is that whether or not sentient simulations ever come to be, we will indeed find that the world is fundamentally discrete.

The Roots of Reality

In the Simulated Multiverse, there's no ambiguity regarding which universe is "real"—that is, which universe lies at the root of the branching tree of simulated worlds. It's the one that houses those computers which, should they crash, would bring down the entire multiverse. A simulated inhabitant might simulate his or her own set of universes on simulated computers, as might the inhabitants of those simulations, but there are still real computers on which all these layered simulations appear as an avalanche of electrical impulses. There's no uncertainty about what facts, patterns, and laws are, in the traditional sense, real: they're the ones at work in the root universe.

However, typical simulated scientists across the Simulated Multiverse may have a different perspective. If these scientists are allowed sufficient autonomy—if the simulants rarely if ever tinker with inhabitants' memories or disrupt the natural flow of events—then, to judge by our own experiences, we can anticipate that they will make great progress in uncovering the mathematical code that propels their world. And they will treat that code as their laws of nature. Nevertheless, their laws won't necessarily be identical to the laws governing the real universe. Their laws merely need to be good enough, in the sense that when they're simulated on a computer they yield a universe with sentient inhabitants. If there are many distinct sets of mathematical laws that qualify as good enough, there could well be an ever-growing population of simulated scientists convinced of mathematical laws that, far from being fundamental, were simply chosen by whoever has programmed their simulation. If we are typical inhabitants in such a multiverse, this reasoning suggests that what we normally think of as science, a discipline charged with revealing fundamental truths about reality—the root reality operating at the base of the tree—would be undermined.

It's an uncomfortable possibility, but not one that keeps me up at night. Until I get my breath taken away by seeing a sentient simulation, I won't consider seriously the proposition that I am now in one. And, taking the long view, even if sentient simulations are achieved one day—itself a big if—I can well imagine that when a civilization's technical capabilities first enable such simulations, their appeal would be tremendous. But would that appeal be long-lived? I suspect the novelty of creating artificial worlds whose inhabitants are kept unaware of their simulated status would wear thin; there's just so much reality TV you can watch.

Instead, if I allow my imagination to run free within this speculative territory, my sense is that staying power would reside with applications that developed interactions between the simulated and the real worlds. Perhaps simulated inhabitants would be able to migrate into the real world or be joined in the simulated world by their real biological counterparts. In time, the distinction between real and simulated beings might become anachronistic. Such seamless unions strike me as a more probable outcome. In that case, the Simulated Multiverse would contribute to the expanse of reality—our expanse of reality, our real reality—in the most tangible way. It would become an intrinsic part of what we mean by "reality."

CHAPTER 11

The Limits of Inquiry

Multiverses and the Future

Isaac Newton cracked the scientific enterprise wide open. He discovered that a few mathematical equations could describe the way things move, both here on earth and up in space. Considering the power and simplicity of his results, one could easily have imagined that Newton's equations reflected eternal truths etched into the bedrock of the cosmos. But Newton himself didn't think so. He believed that the universe was far more rich and mysterious than his laws implied; later in life he famously reflected, "I do not know what I may appear to the world, but to myself I seem to have only been a boy playing on the seashore, diverting myself in now and then finding a smoother pebble or prettier shell than ordinary, whilst the great ocean of truth lay before me all undiscovered." The centuries since have abundantly affirmed this.

I'm glad. Had Newton's equations enjoyed unlimited reach, accurately describing phenomena in any context however big or small, heavy or light, fast or slow, the subsequent scientific odyssey would have taken on a distinctly different character. Newton's equations teach us much about the world, but their unlimited validity would have meant that the cosmic flavor was vanilla through and through. Once you understood physics on everyday scales, you'd be done. The same story would have held all the way up and all the way down.

In continuing Newton's explorations, scientists have ventured into realms far beyond the reach of his equations. What we've learned has required sweeping changes in our understanding of the nature of reality. Such changes are not made lightly. They are closely examined by the community of scientists, and they are often sharply resisted; only when the evidence reaches a critical abundance is the new view embraced. Which is just as it should be. There's no need to rush to judgment. Reality will wait.

The central fact, most forcefully emphasized by the last hundred years of theoretical and experimental progress, is that common experience fails to be a trustworthy guide for excursions that wander beyond everyday circumstances. But for all the radically new physics encountered in extreme conditions—described by general relativity, quantum mechanics, and, should it prove correct, string theory—the fact that radically new ideas would be required is not surprising. The basic assumption of science is that regularities and patterns exist on all scales, but as Newton himself anticipated, there's no reason to expect the patterns we directly encounter to be recapitulated on all scales.

The surprise would have been to find no surprises.

The same is undoubtedly true regarding what physics will reveal in the future. A given generation of scientists can never know whether the long view of history will judge their work as a diversion, as passing fascination, as a stepping-stone, or as having revealed insights that will stand the test of time. Such local uncertainty is balanced by one of physics' most gratifying features—global stability—that is, new theories generally do not erase those they supplant. As we've discussed, while new theories may require acclimation to new perspectives on the nature of reality, they almost never render past discoveries irrelevant. Instead, they incorporate and extend them. Because of this, the story of physics has maintained an impressive coherence.

In this book we've explored a candidate for the next major development in this story: the possibility that our universe is part of a multiverse. The journey has taken us through nine variations on the multiverse theme, which are summarized in Table 11.1. Although the various proposals differ widely in detail, they all suggest that our commonsense picture of reality is only part of a

PARALLEL UNIVERSE PROPOSAL	DESCRIPTION
Quilted Multiverse	Conditions in an infinite universe repeat across space, yielding pa___ ___ ___.
Inflationary Multiverse	Eternal cosmological inflation yields an enormous network of bubble universes, of which our universe would be one.
Brane Multiverse	In string/M-theory's braneworld scenario, our universe exists on one three-dimensional brane, which floats in a higher-dimensional expanse potentially populated by other branes—other parallel universes.
Cyclic Multiverse	Collisions between braneworlds can manifest as big bang–like beginnings, yielding universes that are parallel in time.
Landscape Multiverse	By combing inflationary cosmology and string theory, the many different shapes for string theory's extra dimensions give rise to many different bubble universes.
Quantum Multiverse	Quantum mechanics suggests that every possibility embodied in its probability waves is realized in one of a vast ensemble of parallel universes.
Holographic Multiverse	The holographic principle asserts that our universe is exactly mirrored by phenomena taking place on a distant bounding surface, a physically equivalent parallel universe.
Simulated Multiverse	Technological leaps suggest that simulated universes may one day be possible.
Ultimate Multiverse	The principle of fecundity asserts that every possible universe is a real universe, thereby obviating the question of why one possibility—ours—is special. These universes instantiate all possible mathematical equations.

Table 11.1 *Summary of Various Versions of Parallel Universes*

grander whole. And they all bear the indelible mark of human ingenuity and creativity. But determining whether any of these ideas goes beyond mathematical musings of the human mind will require more insight, knowledge, calculation, experiment, and observation than we've so far achieved. A final reckoning on whether parallel universes will be written into the next chapter of physics' story must therefore also await the perspective that only the future can bring.

As with the metaphorical book of nature, so with the book you're reading. In this last chapter, I'd be delighted to pull all the pieces together and answer the subject's most essential question: Universe or multiverse? But I can't. That's the nature of explorations that brush the edge of knowledge. Instead, to catch a glimpse of where the multiverse concept might be headed, as well as to emphasize the essential highlights of where it now stands, here are five central questions with which physicists will continue to grapple in the years ahead.

Is the Copernican Pattern Fundamental?

Regularities and patterns, evident in observations and in mathematics, are essential to formulating physical laws. Patterns of a different sort, in the nature of the physical laws accepted by each successive generation, are also revealing. Such patterns reflect how scientific discovery has shifted humankind's perspective on its place in the cosmic order. Over the course of nearly five centuries, the Copernican progression has been a dominant theme. From the rising and setting of the sun to the motion of constellations across the night sky to the leading role we each play in our mind's inner world, experience abounds with clues suggesting that we're a central hub around which the cosmos revolves. But the objective methods of scientific discovery have steadily corrected this perspective. At nearly every turn, we've found that were we not here, the cosmic order would hardly differ. We've had to give up our belief in earth's centrality among our cosmic neighbors, the sun's centrality in the galaxy, the Milky Way's centrality among the galaxies, and even the centrality of protons, neutrons, and electrons—the stuff of which we're made—in the cosmic recipe. There was a time when evidence contrary to long-held collective delusions of grandeur was viewed as a frontal assault on human worth. With practice, we've gotten better at valuing enlightenment.

The trek in this book has been toward what may be the capstone Copernican correction. Our universe itself may not be central to any cosmic order. Much as with our planet, star, and galaxy, our universe may merely be one among a great many. The idea that reality based on a multiverse extends the Copernican pattern and perhaps completes it is cause for curiosity. But what elevates the multiverse concept above idle speculation is a key fact that we've now repeatedly encountered. Scientists have not been on a hunt for ways to extend the Copernican revolution. They've not been plotting in darkened laboratories for ways to complete the Copernican pattern. Instead, scientists have been doing what they always do: using data and observations as a guide, they've been formulating mathematical theories to describe the fundamental constituents of matter and the forces that govern how those constituents behave, interact, and evolve. Remarkably, when diligently following the trail these theories blaze, scientists have run smack into one potential multiverse after another. Take a trip along a great many of the most traveled scientific highways, stay moderately attentive, and you'll encounter a diverse assortment of multiverse candidates. They're harder to avoid than they are to find.

Perhaps future discoveries will cast a different light on the series of Copernican corrections. But from our current vantage point, the more we understand, the less central we appear. Should the scientific considerations we've discussed in earlier chapters continue to push us toward multiverse-based explanations, it would be the natural step toward completing the Copernican revolution, five hundred years in the making.

Can Scientific Theories that Invoke a Multiverse Be Tested?

Although the multiverse concept fits snugly within the Copernican template, it differs qualitatively from our earlier migrations from center stage. By invoking realms that may be forever beyond our ability to examine—either with any degree of precision or, in some cases, even at all—multiverses seemingly erect substantial barriers to scientific knowledge. Regardless of one's view of humanity's place in the cosmic arrangement, a widely held assumption has been that through conscientious experimentation, observation, and mathematical calculation, the capacity for gaining deeper understanding is boundless. But if we're part of a multiverse, a reasonable expectation is that at best we can learn about our universe, our little corner of the cosmos. More distressing is the worry that by invoking a multiverse, we enter the domain of theories that can't be tested—theories that rely on "just so" stories, relegating everything we observe to "the way things just happen to be here."

As I've argued, however, the multiverse concept is more nuanced. We've seen various ways in which a theory that involves a multiverse might offer testable predictions. For instance, while the particular universes constituting a given multiverse may differ considerably, because they emerge from a common theory there may be features they all share. Failure to find those features, through measurements we undertake here in the one universe to which we have access, would prove that multiverse proposal wrong. Confirmation of those features, especially if they're novel, would build confidence that the proposal was right.

Or, if there aren't features common to all universes, correlations between physical features can provide another class of testable predictions. For example, we've seen that if all universes whose particle roster includes an electron also include an as-yet-undetected particle species, failure to find the particle through experiments undertaken here in our universe would rule out the multiverse proposal. Confirmation would build confidence. More complicated correlations—such as, those universes whose particle roster includes, say, all the known particles (electrons, muons, up-quarks, down-quarks, etc.) necessarily contain a new particle species—would similarly yield testable, falsifiable predictions.

In the absence of such tight correlations, the manner in which physical features vary from universe to universe can also provide predictions. Across a given multiverse, for example, the cosmological constant might take on a wide range of values. But if the vast majority of universes have a cosmological constant whose value agrees with what measurements have found here (as illustrated in Figure 7.1), confidence in that multiverse would deservedly grow.

Finally, even if most universes in a given multiverse have properties that differ from ours, there's one more diagnostic we can bring into play. We can invoke anthropic reasoning by considering only those universes in the multiverse hospitable to our form of life. If the vast majority of this subclass of universes has properties that agree with ours—if our universe is typical among those in which the conditions allow us to live—confidence in the multiverse would build. If we're atypical, we can't rule the theory out, but that's a familiar limitation of statistical reasoning. Unlikely outcomes can and sometimes do happen. Even so, the less typical we are, the less compelling the given multiverse proposal would be. If among all life-supporting universes in a given multiverse our universe would stick out like a sore thumb, that would provide a strong argument to deem that multiverse proposal irrelevant.

To probe a multiverse proposal quantitatively, therefore, we must determine the demographics of the universes that populate it. It's not enough to know the possible universes the multiverse proposal allows; we must determine the detailed features of the actual universes to which the proposal gives rise. This requires understanding the cosmological processes that bring the various universes of a given multiverse proposal into existence. Testable predictions can then emerge from the way physical features vary from universe to universe across the multiverse.

Whether this sequence of evaluations yields sharp results is something that can only be assessed multiverse by multiverse. But the conclusion is that theories that involve other universes—realms we can't access now or perhaps ever—can still provide testable, and hence falsifiable, predictions.

Can We Test the Multiverse Theories We've Encountered?

In the course of theoretical research, physical intuition is vital. Theorists need to navigate a bewildering array of possibilities.

Should I try this equation or that, invoke that pattern or this? The best physicists have sharp and wonderfully accurate hunches or gut feelings about which directions are promising and which are likely to be fruitless. But that happens behind the scenes. When scientific proposals are brought forward, they are not judged by hunches or gut feelings. Only one standard is relevant: a proposal's ability to explain or predict experimental data and astronomical observations.

Therein lies the singular beauty of science. As we struggle toward deeper understanding, we must give our creative imagination ample room to explore. We must be willing to step outside conventional ideas and established frameworks. But unlike the wealth of other human activities through which the creative impulse is channeled, science supplies a final reckoning, a built-in assessment of what's right and what's not.

A complication of scientific life in the late twentieth and early twenty-first centuries is that some of our theoretical ideas have soared past our ability to test or observe. String theory has for some time been the poster child for this situation; the possibility that we're part of a multiverse provides an even more sprawling example. I've laid out a general prescription for how a multiverse proposal might be testable, but at our current level of understanding none of the multiverse theories we've encountered yet meet the criteria. With ongoing research, this situation could greatly improve.

Our investigations of the Landscape Multiverse, for example, are in their earliest stages. The collection of possible string theory universes—the string landscape—is schematically illustrated in Figure 6.4, but detailed maps of this mountainous terrain have yet to be drawn. Like ancient seafarers, we have a rough sense of what's out there, but it will require extensive mathematical explorations to map the lay of the land. With such knowledge in hand, the next step will be to determine how these potential universes are distributed across the corresponding Landscape Multiverse. The essential physical process, the creation of bubble universes through quantum tunneling (illustrated in Figure 6.6 and Figure 6.7), is well understood in principle but has yet to be examined with quantitative depth in string theory. Various research groups (including my own) have undertaken initial reconnaissance, but there is vast terrain yet to scout. As we've seen in earlier chapters, a variety of similar uncertainties afflict the other multiverse proposals too.

No one knows whether it will take years, decades, or even longer for observational and theoretical progress to extract detailed predictions from any given multiverse. Should the current situation persist, we'll face a choice. Do we define science—"respectable science"—as including only those ideas, realms, and possibilities that fall within the capacity of contemporary human beings on Planet Earth to test or observe? Or do we take a more expansive view and consider as "scientific" ideas that might be testable with technological advances we can imagine achieving in the next hundred years? The next two hundred years? Longer? Or do we take a still more expansive view? Do we allow science to follow any and all paths it reveals, to travel in directions that radiate from experimentally confirmed concepts but that may lead our theorizing into hidden realms that lie, perhaps permanently, beyond human reach?

There's no clear-cut answer. It is here that personal scientific taste comes to the fore. I understand well the impulse to tether scientific investigations to those propositions that can be tested now, or in the near future; this is, after all, how we built the scientific edifice. But I find it parochial to bound our thinking by the arbitrary limits imposed by where we are, when we are, and who we are. Reality transcends these limits, so it's to be expected that sooner or later the search for deep truths will too.

My taste is for the expansive. But I draw the line at ideas that have no possibility of being confronted meaningfully by experiment or observation, not because of human frailty or technological hurdles, but because of the proposals' inherent nature. Of the multiverses we've considered, only the full-blown version of the Ultimate Multiverse falls into this netherland. If absolutely every possible universe is included, then no matter what we measure or observe, the Ultimate Multiverse will nod and embrace our result. The other eight multiverses, as summarized in Table 11.1, avoid this pitfall. Each emerges from a well-motivated, logical chain of reasoning, and each is open to judgment. Should observations provide convincing evidence that the spatial expanse is finite, the

Quilted Multiverse would drop from consideration. Should confidence in inflationary cosmology erode, perhaps because more precise cosmic microwave background data can be explained only by assuming contorted (and hence unconvincing) inflaton potential energy curves, the prominence of the Inflationary Multiverse would diminish too.* Should string theory suffer a theoretical setback, perhaps through the discovery of a subtle mathematical flaw showing that the theory is inconsistent (as early researchers initially thought was the case), the motivation for its various multiverses would evaporate. Conversely, observations of patterns in the microwave background radiation expected from bubble collisions could provide direct supporting evidence for the Inflationary Multiverse. Accelerator experiments searching for supersymmetric particles, missing energy signatures, and mini black holes could bolster the case for string theory and the Brane Multiverse, while evidence for bubble collisions could also provide support for the Landscape variety. Detection of gravitational wave imprints from the early universe, or lack thereof, could distinguish between cosmology based on the inflationary paradigm and that of the Cyclic Multiverse.

Quantum mechanics, in its Many Worlds guise, gives rise to the Quantum Multiverse. Should future research show that the equations of quantum mechanics, however reliable they've been so far, require small modifications to match more refined data, this type of multiverse could be ruled out. A modification of quantum theory that compromises the property of linearity (on which we relied extensively in Chapter 8) would do just that. We've noted as well that there are in-principle tests of the Quantum Multiverse, experiments whose results depend on whether or not Everett's Many Worlds picture is correct. The experiments are beyond what we can

---

*Note, as in Chapter 7, that an airtight observational refutation of inflation would require the theory's commitment to a procedure for comparing infinite classes of universes—something it has not yet achieved. However, most practitioners would agree that if, say, the microwave background data had looked different from Figure 3.4, their confidence in inflation would have plummeted, even though, according to the theory, there's a bubble universe in the Inflationary Multiverse in which those data would hold.

carry out now and perhaps always, but that's because they're fantastically difficult, not because some inherent feature of the Quantum Multiverse itself renders them fundamentally undoable.

The Holographic Multiverse emerges from considerations of established theories—general relativity and quantum mechanics—and receives its strongest theoretical support from string theory. Calculations based on holography are already making tentative contact with experimental results at the Relativistic Heavy Ion Collider, and all indications are that such experimental links will grow more robust in the future. Whether one views the Holographic Multiverse merely as a useful mathematical device or as evidence for holographic reality is a matter of opinion. We must await future work, theoretical and experimental, in order to build a stronger case for the physical interpretation.

The Simulated Multiverse rests not on any one theoretical structure but rather on the relentless rise of computational power. The linchpin assumption is that sentience is not fundamentally tied to a particular substrate—the brain—but is an emergent characteristic of a certain variety of information processing. It's a highly debatable proposition, with passionate arguments advanced on both sides. Maybe future research on the brain and on the nature of consciousness will undermine the idea of self-aware thinking machines. And maybe not. One means for judging this multiverse proposal, though, is clear. Should our descendants one day observe, or interact with, or virtually visit, or become part of a convincing simulated world, the issue would for all practical purposes be settled.

The Simulated Multiverse, at least in theory, might also be linked to a pared-down version of the Ultimate Multiverse that includes only universes based on computable mathematical structures. Unlike the full-blown version of the Ultimate Multiverse, this more limited incarnation has a genesis story that lifts it beyond mere assertion. The users, real and simulated, who are behind the Simulated Multiverse will, by definition, be simulating computable mathematical structures and thus will have the capacity to generate this part of the Ultimate Multiverse.

Gaining experimental or observational insight into the validity of any of the multiverse proposals is surely a long shot. But it's not an impossibility. And with the immensity of the potential payoff, if the exploration of multiverses is where the natural course of theoretical research takes us, we must follow the trail to see where it leads.

How Does a Multiverse Affect the Nature of Scientific Explanation?

Sometimes science focuses on details. It tells us why planets travel in elliptical orbits, why the sky is blue, why water is transparent, why my desk is solid. However familiar these facts may be, it is wondrous that we've been able to explain them. Sometimes science takes a larger view. It reveals that we live within a galaxy containing a few hundred billion stars, it establishes that ours is but one of hundreds of billions of galaxies, and it provides evidence for unseen dark energy permeating every nook and cranny of this vast arena. Looking back just a hundred years, to a time when the universe was thought to be static and populated solely by the Milky Way galaxy, we can rightly celebrate the magnificent picture science has since painted.

Sometimes science does something else. Sometimes it challenges us to reexamine our views of science itself. The usual centuries-old scientific framework envisions that when describing a physical system, a physicist needs to specify three things. We've seen all three in various contexts, but it's useful to gather them together here. First are the mathematical equations describing the relevant physical laws (for example, these might be Newton's laws of motion, Maxwell's equations of electricity and magnetism, or Schrödinger's equation of quantum mechanics). Second are the numerical values of all constants of nature that appear in the mathematical equations (for example, the constants determining the intrinsic strength of gravity and the electromagnetic forces or those determining the masses of the fundamental particles). Third, the physicist must specify the system's "initial conditions" (such as a baseball being hit from home plate at a particular speed in a particular direction, or an electron starting out with a 50 percent probability of being found at Grant's Tomb and an equal probability of being found at Strawberry Fields). The equations then determine what things will be like at any subsequent time. Both classical and quantum physics subscribe to this framework; they differ only in that classical physics purports to tell us how things will definitely be at a given moment, while quantum physics provides the probability that things will be one way or another.

When it comes to predicting where a batted ball will land, or how an electron will move through a computer chip (or a model Manhattan), this three-step process is demonstrably powerful. Yet, when it comes to describing the totality of reality, the three steps invite us to ask deeper questions: Can we explain the initial conditions—how things were at some purportedly earliest moment? Can we explain the values of the constants—the particle masses, force strengths, and so on—on which those laws depend? Can we explain why a particular set of mathematical equations describes one or another aspect of the physical universe?

The various multiverse proposals we've discussed have the potential to profoundly shift our thinking on these questions. In the Quilted Multiverse, the physical laws across the constituent universes are the same, but the particle arrangements differ; different particle arrangements now reflect different initial conditions in the past. In this multiverse, therefore, our perspective on the question of why the initial conditions in our universe were one way or another shifts. Initial conditions can and generally will vary from universe to universe, so there is no fundamental explanation for any particular arrangement. Asking for such an explanation is asking the wrong kind of question; it's invoking single-universe mentality in a multiverse setting. Instead, the question we should ask is whether somewhere in the multiverse is a universe whose particle arrangement, and hence initial conditions, agrees with what we see here. Better still, can we show that such universes abound? If so, the deep question of initial conditions would be explained with a shrug of the shoulders; in such a multiverse, the initial conditions of our universe would be in no more need of an explanation than the fact that somewhere in New York is a shoe store that carries your size.

In the inflationary multiverse, the "constants" of nature can and generally will vary from bubble universe to bubble universe. Recall from Chapter 3 that environmental differences—the different Higgs field values permeating each bubble—give rise to different particle masses and force properties. The same holds true in the Brane Multiverse, the Cyclic Multiverse, and the Landscape Multiverse, where the form of string theory's extra dimensions, together with various differences in fields and fluxes, result in universes with different features—from the electron's mass to whether there even is an electron to the strength of electromagnetism to whether there is an electromagnetic force to the value of the cosmological constant, and so on. In the context of these multiverses, asking for an explanation of the particle and force properties we measure is once again asking the wrong kind of question; it's a question borne of single-universe thinking. Instead, we should ask whether in any of these multiverses there's a universe with the physical properties we measure. Better would be to show that universes with our physical features are abundant, or at least are abundant among all those universes that support life as we know it. But as much as it's meaningless to ask for *the* word with which Shakespeare wrote *Macbeth*, so it's meaningless to ask the equations to pick out *the* values of the particular physical features we see here.

The Simulated and Ultimate Multiverses are horses of a different color; they don't emerge from particular physical theories. Yet, they too have the potential to shift the nature of our questions. In these multiverses, the mathematical laws governing the individual universes vary. Thus, much as with varying initial conditions and constants of nature, varying laws suggest that it's as misguided to ask for an explanation of the particular laws in operation here. Different universes have different laws; we experience the ones we do because these are among the laws compatible with our existence.

Collectively, we see that the multiverse proposals summarized in Table 11.1 render prosaic three primary aspects of the standard scientific framework that in a single-universe setting are deeply mysterious. In various multiverses, the initial conditions, the constants of nature, and even the mathematical laws are no longer in need of explanation.

Should We Believe Mathematics?

Nobel laureate Steven Weinberg once wrote, "Our mistake is not that we take our theories too seriously, but that we do not take them seriously enough. It is always hard to realize that these numbers and equations we play with at our desks have something to do with the real world."[1] Weinberg was referring to the pioneering results of Ralph Alpher, Robert Herman, and George Gamow on the cosmic microwave background radiation, which I described in Chapter 3. Although the predicted radiation is a direct consequence of general relativity combined with basic cosmological physics, it rose to prominence only after being discovered theoretically twice, a dozen years apart, and then being observed through a benevolent act of serendipity.

To be sure, Weinberg's remark has to be applied with care. Although *his* desk has played host to an inordinate amount of mathematics that has proved relevant to the real world, far from every equation with which we theorists tinker rises to that level. In the absence of compelling experimental or observational results, deciding which mathematics should be taken seriously is as much art as it is science.

Indeed, this issue is central to all we've discussed in this book; it has also informed the book's title. The breadth of multiverse proposals in Table 11.1 might suggest a panorama of hidden realities. But I've titled this book in the singular to reflect the unique and uniquely powerful theme that underlies them all: the capacity of mathematics to reveal secreted truths about the workings of the world. Centuries of discovery have made this abundantly evident; monumental upheavals in physics have emerged time and again from vigorously following mathematics' lead. Einstein's own complex dance with mathematics provides a revealing case study.

In the late 1800s when James Clerk Maxwell realized that light was an electromagnetic wave, his equations showed that light's speed should be about 300,000 kilometers per second—close to the value experimenters had measured. A nagging loose end was that his equations left unanswered the question: 300,000 kilometers per second relative to what? Scientists pursued the makeshift resolution that an invisible substance permeating space, the "aether," provided the unseen standard of rest. But in the early twentieth century, Einstein argued that scientists needed to take Maxwell's equations more seriously. If Maxwell's equations didn't refer to a standard of rest, then there was no need for a standard of rest; light's speed, Einstein forcefully declared, is 300,000 kilometers per second relative to *anything*. Although the details are of historical interest, I'm describing this episode for the larger point: everyone had access to Maxwell's mathematics, but it took the genius of Einstein to embrace the mathematics fully. And with that move, Einstein broke through to the special theory of relativity, overturning centuries of thought regarding space, time, matter, and energy.

During the next decade, in the course of developing the general theory of relativity, Einstein became intimately familiar with vast areas of mathematics that most physicists of his day knew little or nothing about. As he groped toward general relativity's final equations, Einstein displayed a master's skill in molding these mathematical constructs with the firm hand of physical intuition. A few years later, when he received the good news that observations of the 1919 solar eclipse confirmed general relativity's prediction that star light should travel along curved trajectories, Einstein confidently noted that had the results been different, "he would have been sorry for the dear Lord, since the theory is correct." I'm sure that convincing data contravening general relativity would have changed Einstein's tune, but the remark captures well how a set of mathematical equations, through their sleek internal logic, their intrinsic beauty, and their potential for wide-ranging applicability, can seemingly radiate reality.

Nevertheless, there was a limit to how far Einstein was willing to follow his own mathematics. Einstein did not take the general theory of relativity "seriously enough" to believe its prediction of black holes, or its prediction that the universe was expanding. As we've seen, others, including Friedmann, Lemaître, and Schwarzschild, embraced Einstein's equations more fully than he, and their achievements have set the course of cosmological understanding for nearly a century. By contrast, during the last twenty or so years of his life, Einstein threw himself into mathematical investigations, passionately striving for the prized achievement of a unified theory of physics. In assessing this work based on what we know now, one can't help but conclude that during those years Einstein was *too* heavily guided—some might say blinded—by the thicket of equations with which he was constantly surrounded. And so, even Einstein, at various times in his life, made the wrong decision regarding which equations to take seriously and which to not.

The third revolution in modern theoretical physics, quantum mechanics, provides another case study, one of direct relevance to the story I've told in this book. Schrödinger wrote down his equation for how quantum waves evolve in 1926. For decades, the equation was viewed as relevant only to the domain of small things: molecules, atoms, and particles. But in 1957, Hugh Everett echoed Einstein's Maxwellian charge of a half century earlier: *take the math seriously.* Everett argued that Schrödinger's equation should apply to everything because all things material, regardless of size, are made from molecules, atoms, and subatomic particles. And as we've seen, this led Everett to the Many Worlds approach to quantum mechanics and to the Quantum Multiverse. More than fifty years later, we still don't know if Everett's approach is right. But by taking the mathematics underlying quantum theory seriously—fully seriously—he may have discovered one of the most profound revelations of scientific exploration.

The other multiverse proposals similarly rely on a belief that mathematics is tightly stitched into the fabric of reality. The Ultimate Multiverse takes this perspective to its furthermost incarnation; mathematics, according to the Ultimate Multiverse, *is* reality. But even with their less panoptic view on the connection between mathematics and reality, the other multiverse theories in Table 11.1 owe their genesis to numbers and equations played with by theorists sitting at desks—and scribbling in notebooks, and writing on chalkboards, and programming computers. Whether invoking general relativity, quantum mechanics, string theory, or mathematical insight more broadly, the entries in Table 11.1 arise only because we assume that mathematical theorizing can guide us toward hidden truths. Only time will tell if this assumption takes the underlying mathematical theories too seriously, or perhaps not seriously enough.

If some or all of the mathematics that's compelled us to think about parallel worlds proves relevant to reality, Einstein's famous query, asking whether the universe has the properties it does simply because no other universe is possible, would have a definitive answer: no. Our universe is not the only one possible. Its properties could have been different. And in many of the multiverse proposals, the properties of the other member universes *would* be different. In turn, seeking a fundamental explanation for why certain things are the way they are would be pointless. Instead, statistical likelihood or plain happenstance would be firmly inserted in our understanding of a cosmos that would be profoundly vast.

I don't know if this is how things will turn out. No one does. But it's only through fearless engagement that we can learn our own limits. It's only through the rational pursuit of theories, even those that whisk us into strange and unfamiliar domains, that we stand a chance of revealing the expanse of reality.

Notes

Chapter 1: The Bounds of Reality

1. The possibility that our universe is a slab floating in a higher dimensional realm goes back to a paper by two renowned Russian physicists—"Do We Live Inside a Domain Wall?," V. A. Rubakov and M. E. Shaposhnikov, *Physics Letters B* 125 (May 26, 1983): 136—and does not involve string theory. The version I'll focus on in Chapter 5 emerges from advances in string theory in the mid-1990s.

Chapter 2: Endless Doppelgängers

1. The quote comes from the March 1933 issue of *The Literary Digest*. It is worth noting that the precision of this quote has recently been questioned by the Danish historian of science Helge Kragh (see his *Cosmology and Controversy*, Princeton: Princeton University Press, 1999), who suggests it may be a reinterpretation of a *Newsweek* report from earlier that year in which Einstein was referring to the origin of cosmic rays. What is certain, however, is that by this year Einstein had given up his belief that the universe was static and accepted the dynamic cosmology that emerged from his original equations of general relativity.

2. This law tells us the force of gravitational attraction, F, between two objects, given the masses, m_1 and m_2, of each, and the distance, r, between

them. Mathematically, the law reads: $F = Gm_1m_2/r^2$, where G stands for Newton's constant—an experimentally measured number that specifies the intrinsic strength of the gravitational force.

3. For the mathematically inclined reader, Einstein's equations are $R_{uv} - \frac{1}{2}g_{uv}R = 8\pi GT_{uv}$ where g_{uv} is the metric on spacetime, R_{uv} is the Ricci curvature tensor, R is the scalar curvature, G is Newton's constant, and T_{uv} is the energy-momentum tensor.

4. In the decades since this famous confirmation of general relativity, questions have been raised regarding the reliability of the results. For distant starlight grazing the sun to be visible, the observations had to be carried out during a solar eclipse; unfortunately, bad weather made it a challenge to take clear photographs of the solar eclipse of 1919. The question is whether Eddington and his collaborators might have been biased by foreknowledge of the result they were seeking, and so when they culled photographs deemed unreliable because of weather interference, they eliminated a disproportionate number containing data that appeared not to fit Einstein's theory. A recent and thorough study by Daniel Kennefick (see www.arxiv.org, paper arXiv:0709.0685, which, among other considerations, takes account of a modern reevaluation of the photograph plates taken in 1919) convincingly argues that the 1919 confirmation of general relativity is, indeed, reliable.

5. For the mathematically inclined reader, Einstein's equations of general relativity in this context reduce to $(\frac{da/dt}{a})^2 = \frac{8\pi G\rho}{3} - \frac{k}{a^2}$. The variable $a(t)$ is the scale factor of the universe—a number whose value, as the name indicates, sets the distance scale between objects (if the value of $a(t)$ at two different times differs, say, by a factor of 2, then the distance between any two particular galaxies would differ between those times by a factor of 2 as well), G is Newton's constant, ρ is the density of matter/energy, and k is a parameter whose value can be 1, 0, or –1 according to whether the shape of space is spherical, Euclidean ("flat"), or hyperbolic. The form of this equation is usually credited to Alexander Friedmann and, as such, is called the Friedmann equation.

6. The mathematically inclined reader should note two things. First, in general relativity we typically define coordinates that are themselves dependent on the matter space contains: we use galaxies as the coordinate carriers (acting as if each galaxy has a particular set of coordinates "painted" on it—so-called co-moving coordinates). So, to even identify a specific region of space, we usually make reference to the matter that occupies it. A more precise

rephrasing of the text, then, would be: The region of space containing a particular group of N galaxies at time t_1 will have a larger volume at a later time t_2. Second, the intuitively sensible statement regarding the density of matter and energy changing when space expands or contracts makes an implicit assumption regarding the equation of state for matter and energy. There are situations, and we will encounter one shortly, where space can expand or contract while the density of a particular energy contribution—the energy density of the so-called cosmological constant—remains unchanged. Indeed, there are even more exotic scenarios in which space can expand while the density of energy *increases*. This can happen because, in certain circumstances, gravity can provide a source of energy. The important point of the paragraph is that in their original form the equations of general relativity are not compatible with a static universe.

7. Shortly we will see that Einstein abandoned his static universe when confronted by astronomical data showing that the universe is expanding. It is worth noting, though, that his misgivings about the static universe predated the data. The physicist Willem de Sitter pointed out to Einstein that his static universe was unstable: nudge it a bit bigger, and it would grow; nudge it a bit smaller, and it would shrink. Physicists shy away from solutions that require perfect, undisturbed conditions for them to persist.

8. In the big bang model, the outward expansion of space is viewed much like the upward motion of a tossed ball: attractive gravity pulls on the upward-moving ball and so slows its motion; similarly, attractive gravity pulls on the outward-moving galaxies and so slows their motion. In neither case does the ongoing motion require a repulsive force. However, you can still ask: Your arm launched the ball skyward, so what "launched" the spatial universe on its outward expansion? We will return to this question in Chapter 3, where we will see that modern theory posits a short burst of repulsive gravity, operating during the earliest moments of cosmic history. We will also see that more refined data has provided evidence that the expansion of space is *not* slowing over time, which has resulted in a surprising—and as later chapters will make clear—potentially profound resurrection of the cosmological constant.

The discovery of the spatial expansion was a turning point in modern cosmology. In addition to Hubble's contributions, the achievement relied on the work and insights of many others, including Vesto Slipher, Harlow Shapley, and Milton Humason.

9. A two-dimensional torus is usually depicted as a hollow doughnut. A

two-step process shows that this picture agrees with the description provided in the text. When we declare that crossing the right edge of the screen brings you back to the left edge, that's tantamount to identifying the entire right edge with the left edge. Were the screen flexible (made of thin plastic, say) this identification could be made explicit by rolling the screen into a cylindrical shape and taping the right and left edges together. When we declare that crossing the upper edge brings you to the lower edge, that too is tantamount to identifying those edges. We can make this explicit by a second manipulation in which we bend the cylinder and tape the upper and lower circular edges together. The resulting shape has the usual doughnutlike appearance. A misleading aspect of these manipulations is that the surface of the doughnut looks curved; were it coated with reflective paint, your reflection would be distorted. This is an artifact of representing the torus as an object sitting within an ambient three-dimensional environment. Intrinsically, as a two-dimensional surface, the torus is not curved. It is flat, as is clear when it's represented as a flat video-game screen. That's why, in the text, I focus on the more fundamental description as a shape whose edges are identified in pairs.

10. The mathematically inclined reader will note that by "judicious slicing and paring" I am referring to taking quotients of simply connected covering spaces by various discrete isometry groups.

11. The quoted amount is for the current era. In the early universe, the critical density was higher.

12. If the universe were static, light that had been traveling for the last 13.7 billion years and has only just reached us would indeed have been emitted from a distance of 13.7 billion light-years. In an expanding universe, the object that emitted the light has continued to recede during the billions of years the light was in transit. When we receive the light, the object is thus farther away—much farther—than 13.7 billion light-years. A straightforward calculation using general relativity shows that the object (assuming it still exists and has been continually riding the swell of space) would now be about 41 billion light-years away. This means that when we look out into space we can, in principle, see light from sources that are now as far as roughly 41 billion light-years. In this sense, the observable universe has a diameter of about 82 billion light-years. The light from objects farther than this distance would not yet have had enough time to reach us and so are beyond our cosmic horizon.

13. In loose language, you can envision that because of quantum mechanics, particles always experience what I like to call "quantum jitter": a kind of

inescapable random quantum vibration that renders the very notion of the particle having a definite position and speed (momentum) approximate. In this sense, changes to position/speed that are so small that they're on par with the quantum jitters are within the "noise" of quantum mechanics and hence are not meaningful.

In more precise language, if you multiply the imprecision in the measurement of position by the imprecision in the measurement of momentum, the result—the uncertainty—is always larger than a number called *Planck's constant*, named after Max Planck, one of the pioneers of quantum physics. In particular, this implies that fine resolutions in measuring the position of a particle (small imprecision in position measurement) necessarily entail large uncertainty in the measurement of its momentum and, by association, its energy. Since energy is always limited, the resolution in position measurements is thus limited too.

Also note that we will always apply these concepts in a finite spatial domain—generally in regions the size of today's cosmic horizon (as in the next section). A finite-sized region, however large, implies a maximum uncertainty in position measurements. If a particle is assumed to be in a given region, the uncertainty of its position is surely no larger than the size of the region. Such a maximum uncertainty in position then entails, from the uncertainty principle, a minimum amount of uncertainty in momentum measurements—that is, limited resolution in momentum measurements. Together with the limited resolution in position measurements, we see the reduction from an infinite to a finite number of possible distinct configurations of a particle's position and speed.

You might still wonder about the barrier to building a device capable of measuring a particle's position with ever greater precision. It too is a matter of energy. As in the text, if you want to measure a particle's position with ever greater precision, you need to use an ever more refined probe. To determine whether a fly is in a room, you can turn on an ordinary, diffuse overhead light. To determine if an electron is in a cavity, you need to illuminate it with the sharp beam of a powerful laser. And to determine the electron's position with ever greater accuracy you need to make that laser ever more powerful. Now, when an ever more powerful laser zaps an electron, it imparts an ever greater disturbance to its velocity. Thus, the bottom line is that precision in determining particles' positions comes at the cost of huge changes in the particles' velocities—and hence huge changes in particle energies. If there's a limit to how much energy particles can have, as there always will be, there's a limit to how finely their positions can be resolved.

Limited energy in a limited spatial domain thus gives finite resolution on both position and velocity measurements.

14. The most direct way to make this calculation is by invoking a result I will describe in nontechnical terms in Chapter 9: the entropy of a black hole—the logarithm of the number of distinct quantum states—is proportional to its surface area measured in square Planck units. A black hole that fills our cosmic horizon would have a radius of about 10^{28} centimeters, or roughly 10^{61} Planck lengths. Its entropy would therefore be about 10^{122} in square Planck units. Hence the total number of distinct states is roughly 10 raised to the power 10^{122}, or $10^{10^{122}}$.

15. You might be wondering why I'm not also incorporating fields. As we will see, particles and fields are complementary languages—a field can be described in terms of the particles of which it's composed, much like an ocean wave can be described in terms of its constituent water molecules. The choice of using particle or field language is largely one of convenience.

16. The distance that light can travel in a given time interval depends sensitively on the rate at which space expands. In later chapters we will encounter evidence that the rate of spatial expansion is accelerating. If so, there is a limit to how far light can travel through space, even if we wait an arbitrarily long time. Distant regions of space would be receding from us so quickly that light we emit could not reach them; similarly, light they emit could not reach us. This would mean that cosmic horizons—the portion of space with which we can exchange light signals—would not grow in size indefinitely. (For the mathematically inclined reader, the essential formulae are in Chapter 6, note 7.)

17. G. Ellis and G. Bundrit studied duplicate realms in an infinite classical universe; J. Garriga and A. Vilenkin studied such realms in the quantum context.

Chapter 3: Eternity and Infinity

1. One point of departure from the earlier work was Dicke's perspective, which focused on the possibility of an oscillating universe that would repeatedly go through a series of cycles—big bang, expansion, contraction, big

crunch, big bang again. In any given cycle there would be remnant radiation suffusing space.

2. It is worth noting that even though they don't have jet engines, galaxies generally do exhibit some motion above and beyond that arising from the expansion of space—typically the result of large-scale intergalactic gravitational forces as well as the intrinsic motion of the swirling gas cloud from which stars in the galaxies formed. Such motion is called *peculiar velocity* and is generally small enough that it can be safely ignored for cosmological purposes.

3. The horizon problem is subtle, and my description of inflationary cosmology's solution slightly nonstandard, so for the interested reader let me elaborate here in a little more detail. First the problem, again: Consider two regions in the night sky that are so distant from one another that they have never communicated. And to be concrete, let's say each region has an observer who controls a thermostat that sets his or her region's temperature. The observers want the two regions to have the same temperature, but because the observers have been unable to communicate, they don't know how to set their respective thermostats. The natural thought is that since billions of years ago the observers were much closer, it would have been easy for them, way back then, to have communicated and thus to have ensured the two regions had equal temperatures. However, as noted in the main text, in the standard big bang theory this reasoning fails. Here's more detail on why. In the standard big bang theory, the universe is expanding, but because of gravity's attractive pull, the *rate* of expansion slows over time. It's much like what happens when you toss a ball in the air. During its ascent it first moves away from you quickly, but because of the tug of earth's gravity, it steadily slows. The slowing down of spatial expansion has a profound effect. I'll use the tossed ball analogy to explain the essential idea. Imagine a ball that undergoes, say, a six second ascent. Since it initially travels quickly (as it leaves your hand), it might cover the first half of the journey in only two seconds, but due to its diminishing speed it takes four more seconds to cover the second half of the journey. At the halfway point in time, three seconds, it was thus *beyond* the halfway mark in distance. Similarly, with spatial expansion that slows over time: at the halfway point in cosmic history, our two observers would be separated by *more* than half their current distance. Think about what this means. The two observers would be closer together, but they would find it harder—not easier—to have communicated. Signals one observer sends would have half the time to reach

the other, but the distance the signals would need to traverse is *more* than half of what it is today. Being allotted half the time to communicate across more than half their current separation only makes communication more difficult.

The distance between objects is thus only one consideration when analyzing their ability to influence each other. The other essential consideration is the amount of time that's elapsed since the big bang, as this constrains how far any purported influence could have traveled. In the standard big bang, although everything was indeed closer in the past, the universe was also expanding more quickly, resulting in less time, proportionally speaking, for influences to be exerted.

The resolution offered by inflationary cosmology is to insert a phase in the earliest moments of cosmic history in which the expansion rate of space doesn't decrease like the speed of the ball tossed upwards; instead, the spatial expansion starts out slow and then continually picks up speed: the expansion accelerates. By the same reasoning we just followed, at the halfway point of such an inflationary phase our two observers will be separated by *less* than half their distance at the end of that phase. And being allotted half the time to communicate across less than half the distance means it is easier at earlier times for them to communicate. More generally, at ever earlier times, accelerated expansion means there is more time, proportionally speaking—not less—for influences to be exerted. This would have allowed today's distant regions to have easily communicated in the early universe, explaining the common temperature they now have.

Because the accelerated expansion results in a much greater total spatial expansion of space than in the standard big bang theory, the two regions would have been *much* closer together at the onset of inflation than at a comparable moment in the standard big bang theory. This size disparity in the very early universe is an equivalent way of understanding why communication between the regions, which would have proved impossible in the standard big bang, can be easily accomplished in the inflationary theory. If at a given moment after the beginning, the distance between two regions is less, it is easier for them to exchange signals.

Taking the expansion equations seriously to arbitrarily early times (and for definiteness, imagine that space is spherically shaped), we also see that the two regions would have initially separated more quickly in the standard big bang than in the inflationary model: that's how they became so much farther apart in the standard big bang compared with their separation in the inflationary theory. In this sense, the inflationary framework involves a period of time during which the rate of separation between these regions is slower than in the usual big bang framework.

Often, in describing inflationary cosmology, the focus is solely on the fantastic increase in expansion speed over the conventional framework, not on a decrease in speed. The difference in description derives from which physical features between the two frameworks one compares. If one compares the trajectories of two regions of a given distance apart in the very early universe, then in the inflationary theory those regions separate much faster than in the standard big bang theory; by today they are also much farther apart in the inflationary theory than in the conventional big bang. But if one considers two regions of a given distance apart today (like the two regions on opposite sides of the night sky upon which we've been focused), the description I've given is relevant. Namely, at a given moment in time in the very early universe, those regions were much closer together, and had been moving apart much more slowly, in a theory that invokes inflationary expansion as compared with one that doesn't. The role of inflationary expansion is to make up for the slower start by then propelling those regions apart ever more quickly, ensuring that they arrive at the same location in the sky that they would have in the standard big bang theory.

A fuller treatment of the horizon problem would include a more detailed specification of the conditions from which the inflationary expansion emerges as well as the subsequent processes by which, for example, the cosmic microwave background radiation is produced. But this discussion highlights the essential distinction between accelerated and decelerated expansion.

4. Note that by squeezing the bag, you inject energy into it, and since both mass and energy give rise to the resulting gravitational warpage, the increase in weight will be partially due to the increase in energy. The point, however, is that the increase in pressure itself also contributes to the increase in weight. (Also note that to be precise, we should imagine doing this "experiment" in a vacuum chamber, so we don't need to consider the buoyant forces due to the air surrounding the bag.) For everyday examples the increase is tiny. However, in astrophysical settings the increase can be significant. In fact, it plays a role in understanding why, in certain situations, stars necessarily collapse to form black holes. Stars generally maintain their equilibrium through a balance between outward-pushing pressure, generated by nuclear processes in the star's core, and inward-pulling gravity, generated by the star's mass. As the star

exhausts its nuclear fuel, the positive pressure decreases, causing the star to contract. This brings all its constituents closer together and so increases their gravitational attraction. To avoid further contraction, additional outward pressure (what is labeled positive pressure, as in the next paragraph in the text) is needed. But the additional positive pressure itself generates additional attractive gravity and thus makes the need for additional positive pressure all the more urgent. In certain situations, this leads to a spiraling instability and the very thing that the star usually relies upon to counteract the inward pull of gravity—positive pressure—contributes so strongly to that very inward pull that a complete gravitational collapse becomes unavoidable. The star will implode and form a black hole.

5. In the approach to inflation I have just described, there is no fundamental explanation for why the inflaton field's value would begin high up on the potential energy curve, nor why the potential energy curve would have the particular shape it has. These are assumptions the theory makes. Subsequent versions of inflation, most notably one developed by Andrei Linde called *chaotic inflation*, find that a more "ordinary" potential energy curve (a parabolic shape with no flat section that emerges from the simplest mathematical equations for the potential energy) can also yield inflationary expansion. To initiate the inflationary expansion, the inflaton field's value needs to be high up on this potential energy curve too, but the enormously hot conditions expected in the early universe would naturally cause this to happen.

6. For the diligent reader, let me note one additional detail. The rapid expansion of space in inflationary cosmology entails significant cooling (much as a rapid compression of space, or of most anything, causes a surge in temperature). But as inflation comes to a close, the inflaton field oscillates around the minimum of its potential energy curve, transferring its energy to a bath of particles. The process is called "re-heating" because the particles so produced will have kinetic energy and thus can be characterized by a temperature. As space then continues to undergo more ordinary (non-inflationary) big bang expansion, the temperature of the particle bath steadily decreases. The important point, though, is that the uniformity set down by inflation provides uniform conditions for these processes, and so results in uniform outcomes.

7. Alan Guth was aware of the eternal nature of inflation; Paul Steinhardt wrote about its mathematical realization in certain contexts; Andrei Linde Alexander Vilenkin brought it to light in the most general terms.

8. The value of the inflaton field determines the amount of energy and

negative pressure it suffuses through space. The larger the energy, the greater the expansion rate of space. The rapid expansion of space, in turn, has a back reaction on the inflaton field itself: the faster the expansion of space, the more violently the inflaton field's value jitters.

9. Let me address a question that may have occurred to you, one we will return to in Chapter 10. As space undergoes inflationary expansion, its overall energy increases: the greater the volume of space filled with an inflaton field, the greater the total energy (if space is infinitely large, energy is infinite too — in this case we should speak of the energy contained in a finite region of space as the region grows larger). Which naturally leads one to ask: What is the source of this energy? For the analogous situation with the champagne bottle, the source of additional energy in the bottle came from the force exerted by your muscles. What plays the role of your muscles in the expanding cosmos? The answer is gravity. Whereas your muscles were the agent that allowed the available space inside the bottle to expand (by pulling out the cork), gravity is the agent that allows the available space in the cosmos to expand. What's vital to realize is that the gravitational field's energy can be arbitrarily negative. Consider two particles falling toward each other under their mutual gravitational attraction. Gravity coaxes the particles to approach each other faster and faster, and as they do, their kinetic energy gets ever more positive. The gravitational field can supply the particles with such positive energy because gravity can draw down its own energy reserve, which becomes arbitrarily negative in the process: the closer the particles approach each other, the more negative the gravitational energy becomes (equivalently, the more positive the energy you'd need to inject to overcome the force of gravity and separate the particles once again). Gravity is thus like a bank that has a bottomless credit line and so can lend endless amounts of money; the gravitational field can supply endless amounts of energy because its own energy can become ever more negative. And that's the energy source that inflationary expansion taps.

10. I will use the term "bubble universe," although the imagery of a "pocket universe" that opens up within the ambient inflaton-filled environment is a good one too (that term was coined by Alan Guth).

11. For the mathematically inclined reader, a more precise description of the horizontal axis in Figure 3.5 is as follows: consider the two-dimensional sphere comprising the points in space at the time the cosmic microwave background photons began to stream freely. As with any two-sphere, a convenient set of coordinates on this locus are the angular coordinates from a spherical

polar coordinate system. The temperature of the cosmic microwave background radiation can then be viewed as a function of these angular coordinates and, as such, can be decomposed in a Fourier series using as a basis the standard spherical harmonics, $Y_l^m(\theta, \phi)$. The vertical axis in Figure 3.5 is related to the size of the coefficients for each mode in this expansion—farther to the right on the horizontal axis corresponds to smaller angular separation. For technical details, see for example Scott Dodelson's excellent book *Modern Cosmology* (San Diego, Calif.: Academic Press, 2003).

12. A little more precisely, it is not the strength of the gravitational field, per se, that determines the slowing of time, but rather the strength of the gravitational potential. For instance, if you were to hang out inside a spherical cavity at the center of a massive star, you wouldn't feel a gravitational force at all, but because you were deep inside a gravitational-potential well, time for you would run slower than time for someone far outside the star.

13. This result (and closely related ideas) was found by a number of researchers in different contexts, and was most explicitly articulated by Alexander Vilenkin and also by Sidney Coleman and Frank De Luccia.

14. In our discussion of the Quilted Multiverse, you may recall that we assumed particle arrangements would vary randomly from patch to patch. The connection between the Quilted and Inflationary Multiverses also allows us to make good on that assumption. A bubble universe forms in a given region when the inflaton field's value drops; as it does, the energy the inflaton contained is converted into particles. The precise arrangement of these particles at any moment is determined by the precise value of the inflaton during the conversion process. But because the inflaton field is subject to quantum jitters, as its value drops it will be subject to random variations—the same random variations that give rise to the pattern of slightly hotter and slightly colder spots in Figure 3.4. When considered across the patches in a bubble universe, these jitters thus imply that the inflaton's value will display random quantum variations. And this randomness ensures randomness of the resulting particle distributions. That's why we expect any particle arrangement, such as the one responsible for all we see right now, to be replicated as often as any other.

Chapter 4: Unifying Nature's Laws

1. I thank Walter Isaacson for personal communications on this and a number of other historical issues related to Einstein.

2. In a little more detail, the insights of Glashow, Salam, and Weinberg suggested that the electromagnetic and weak forces were aspects of a combined *electroweak force*, a theory that was confirmed by accelerator experiments in the late 1970s and early 1980s. Glashow and Georgi went a step further and suggested that the electroweak and the strong forces were aspects of a yet more fundamental force, an approach that's called *grand unification*. The simplest version of grand unification, however, was ruled out when scientists failed to observe one of its predictions—that protons should, every so often, decay. Nevertheless, there are many other versions of grand unification that remain experimentally viable since, for example, the rate of proton decay they predict is so slow that existing experiments would not yet have the sensitivity to detect it. However, even if grand unification is not borne out by data, it is already beyond doubt that the three nongravitational forces can be described using the same mathematical language of quantum field theory.

3. The discovery of superstring theory spawned other, closely related, theoretical approaches seeking a unified theory of nature's forces. In particular, *supersymmetric quantum field theory*, and its gravitational extension *supergravity*, have been vigorously pursued since the mid-1970s. Supersymmetric quantum field theory and supergravity are based on the new principle of *supersymmetry*, which was discovered within superstring theory, but these approaches incorporate supersymmetry in conventional point-particle theories. We will briefly discuss supersymmetry later in the chapter, but for the mathematically inclined reader, I'll note here that supersymmetry is the last available symmetry (beyond rotational symmetry, translational symmetry, Lorentz symmetry, and, more generally, Poincaré symmetry) of a nontrivial theory of elementary particles. It relates particles of different quantum mechanical spin, establishing a deep mathematical kinship between particles that communicate forces and the particles making up matter. Supergravity is an extension of supersymmetry that includes the gravitational force. In the early days of string theory research, scientists realized that the frameworks of supersymmetry and supergravity emerged from a low-energy analysis of string theory. At low energies, the extended nature of a string generally cannot be discerned, so it appears to be a point particle. Correspondingly, as we will discuss in this chapter, when applied to low energy processes, the mathematics of string theory transforms into that of quantum field theory. Scientists found that because both supersymmetry and gravity survive the transformation, low energy string theory gives rise to supersymmetric quantum field theory and to supergravity. In more recent times, as we will discuss in Chapter 9, the link

between supersymmetric field theory and string theory has grown yet more profound.

4. The informed reader may take exception to my statement that every field is associated to a particle. So, more precisely, the small fluctuations of a field about a local minimum of its potential are generally interpretable as particle excitations. That's all we need for the discussion at hand. Additionally, the informed reader will note that localizing a particle at a point is itself an idealization, because it would take—from the uncertainty principle—infinite momentum and energy to do so. Again, the essence is that in quantum field theory there is, in principle, no limit to how finally localized a particle can be.

5. Historically speaking, a mathematical technique known as *renormalization* was developed to grapple with the quantitative implications of severe, small-scale (high-energy) quantum field jitters. When applied to the quantum field theories of the three nongravitational forces, renormalization cured the infinite quantities that had emerged in various calculations, allowing physicists to generate fantastically accurate predictions. However, when renormalization was brought to bear on the quantum jitters of the gravitational field, it proved ineffective: the method failed to cure infinities that arose in performing quantum calculations involving gravity.

From a more modern vantage point, these infinities are now viewed rather differently. Physicists have come to realize that en route to an ever-deeper understanding of nature's laws, a sensible attitude to take is that any given proposal is provisional, and—if relevant at all—is likely capable of describing physics only down to some particular length scale (or only up to some particular energy scale). Beyond that are phenomena that lie outside the reach of the given proposal. Adopting this perspective, it would be foolhardy to extend the theory to distances smaller than those within its arena of applicability (or to energies above its arena of applicability). And with such inbuilt cutoffs (much as described in the main text), no infinities ever arise. Instead, calculations are undertaken within a theory whose range of applicability is circumscribed from the outset. This means that the ability to make predictions is limited to phenomena that lie within the theory's limits—at very short distances (or at very high energies) the theory offers no insight. The ultimate goal of a complete theory of quantum gravity would be to lift the inbuilt limits, unleashing quantitative, predictive capacities on arbitrary scales.

6. To get a feel for where these particular numbers come from, note that quantum mechanics (discussed in Chapter 8) associates a wave to a particle,

with the heavier the particle the shorter its wavelength (the distance between successive wave crests). Einstein's general relativity also associates a length to any object—the size to which the object would need to be squeezed to become a black hole. The heavier the object, the larger that size. Imagine, then, starting with a particle described by quantum mechanics and then slowly increasing its mass. As you do, the particle's quantum wave gets shorter, while its "black hole size" gets larger. At some mass, the quantum wavelength and the black hole size will be equal—establishing a baseline mass and size at which quantum mechanical and general relativistic considerations are both important. When one makes this thought experiment quantitative, the mass and size are found to be those quoted in the text—the Planck mass and Planck length, respectively. To foreshadow later developments, in Chapter 9 I will discuss the *holographic principle*. This principle uses general relativity and black hole physics to argue for a very particular limit on the number of physical degrees of freedom that can reside in any volume of space (a more refined version of the discussion in Chapter 2 regarding the number of distinct particle arrangements within a volume of space; also mentioned in note 14 of Chapter 2). If this principle is correct, then the conflict between general relativity and quantum mechanics can arise before distances are small and curvatures large. A huge volume containing even a low density gas of particles would be predicted by quantum field theory to have many more degrees of freedom than the holographic principle (which relies on general relativity) would allow.

7. Quantum mechanical spin is a subtle concept. Especially in quantum field theory, where particles are viewed as dots, it is hard to fathom what "spinning" would even mean. What really happens is that experiments show that particles can possess an intrinsic property that behaves much like an immutable quantity of angular momentum. Moreover, quantum theory shows, and experiments confirm, that particles will generally only have angular momentum that is an integer multiple of a fundamental quantity (Planck's constant divided by 2). Since classical spinning objects possess an intrinsic angular momentum (one, however, that is not immutable—it changes as the object's rotational speed changes), theoreticians have borrowed the name "spin" and applied it to this analogous quantum situation. Hence the name "spin angular momentum." While "spinning like a top" provides a reasonable mental image, it's more accurate to imagine that particles are defined not only by their mass, their electric charge, and their nuclear charges, but also by the intrinsic and immutable spin angular momentum they possess. Just as we accept a parti-

cle's electric charge as one of its fundamental defining features, experiments establish that the same is true of its spin angular momentum.

8. Recall that the tension between general relativity and quantum mechanics arises from the powerful quantum jitters of the gravitational field that shake spacetime so violently that the traditional mathematical methods can't cope. Quantum uncertainty tells us that these jitters become ever stronger when space is examined on ever-smaller distances (which is why we don't see these jitters in everyday life). Specifically, the calculations show that it is the wildly energetic jitters over distances shorter than the Planck scale that make the math go haywire (the shorter the distance, the greater the jitters' energy). Since quantum field theory describes particles as points with no spatial extent, the distances these particles probe can be arbitrarily small, and hence the quantum jitters they feel can be arbitrarily energetic. String theory changes this. Strings are *not* points—they have spatial extent. This implies that there is a limit to how small a distance can be accessed, even in principle, since a string can't probe a distance smaller than its own size. In turn, a limit to how small a scale can be probed translates into a limit on how energetic the jitters can become. This limit proves sufficient to tame the unruly mathematics, allowing string theory to merge quantum mechanics and general relativity.

9. If an object were truly one-dimensional, we wouldn't be able to see it directly since it would offer no surface from which photons could reflect and would have no capacity to produce photons of its own through atomic transitions. So, when I say "see" in the text, that's a stand-in for any means of observation or experimentation you might use to seek evidence of an object's spatial extent. The point, then, is that any spatial extent smaller than the resolving power of your experimental procedure will escape your experiment's notice.

10. "What Einstein Never Knew," *NOVA* documentary, 1985.

11. More precisely, the component of the universe most relevant to our existence would be completely different. Since the familiar particles and the objects they compose—stars, planets, people, etc.—amount to less than 5 percent of the mass of the universe, such a disruption would not affect the vast majority of the universe, at least as measured by mass. However, as measured by its effect on life as we know it, the change would be profound.

12. There are some mild restrictions that quantum field theories place on their internal parameters. To avoid certain classes of unacceptable physical behavior (violations of critical conservation laws, violations of certain symmetry transformations, and so on), there can be constraints on the charges (elec-

tric and also nuclear) of the theory's particles. Additionally, to ensure that in all physical processes, probabilities add to 1, there can also be constraints on particle masses. But even with these constraints, there is wide latitude in the allowed values of particle properties.

13. Some researchers will note that even though neither quantum field nor our current understanding of string theory provides an explanation of the particle properties, the issue is more urgent in string theory. The point is a bit involved, but for the technically minded here's the summary. In quantum field theory, the properties of particles—say their masses, to be definite—are controlled by numbers that are inserted into the theory's equations. The fact that quantum field theory's equations allow such numbers to be varied is the mathematical way of saying that quantum field theory does not determine particle masses but instead takes them as input. In string theory, the flexibility in the masses of particles has a similar mathematical origin—the equations allow particular numbers to vary freely—but the manifestation of this flexibility is more significant. The freely varying numbers—numbers, that is, that can be varied with no cost in energy—correspond to the existence of particles with no mass. (Using the language of potential energy curves introduced in Chapter 3, envision a potential energy curve that's completely flat, a horizontal line. Just as walking on a perfectly flat terrain would have no impact on your potential energy, changing the value of such a field would have no cost in energy. Since a particle's mass corresponds to the curvature of its quantum field's potential energy curve around its minimum, the quanta of such fields are massless.) Excessive numbers of massless particles are a particularly awkward feature of any proposed theory since there are tight limits on such particles coming from both accelerator data and cosmological observations. For string theory to be viable it is imperative that these particles acquire mass. In recent years, various discoveries have revealed ways in which this might happen, having to do with fluxes that can thread through holes in the extra-dimensional Calabi-Yau shapes. I will discuss aspects of these developments in Chapter 5.

14. It is not impossible for experiments to provide evidence that would strongly disfavor string theory. The structure of string theory ensures that certain basic principles should be respected by all physical phenomena. Among these are *unitarity* (the sum of all probabilities of all possible outcomes in a given experiment must be 1) and *local Lorentz invariance* (in a small enough domain the laws of special relativity hold), as well as more technical features such as *analyticity* and *crossing symmetry* (the result of particle collisions must

depend on the particles' momentum in a manner that respects a particular collection of mathematical criteria). Should evidence be found—perhaps at the Large Hadron Collider—that any of these principles are violated, it would be a challenge to reconcile those data with string theory. (It would also be a challenge to reconcile those data with the standard model of particle physics, which incorporates these principles too, but the underlying assumption is that the standard model must give way to some kind of new physics at a high enough energy scale since the theory does not incorporate gravity. Data conflicting with any of the principles enumerated would argue that the new physics is not string theory.)

15. It is common to speak of the center of a black hole as if it were a position in space. But it's not. It is a moment in time. When crossing the event horizon of a black hole, time and space (the radial direction) interchange roles. If you fall into a black hole, for example, your radial motion represents progress through time. You are thus pulled toward the black hole's center in the same way you are pulled to the next moment in time. The center of the black hole is, in this sense, akin to a last moment in time.

16. For many reasons, entropy is a key concept in physics. In the case discussed, entropy is being used as a diagnostic tool to determine if string theory is leaving out any essential physics in its description of black holes. If it was, the black hole disorder that the string mathematics is being used to calculate would be inaccurate. The fact that the answer agrees exactly with what Bekenstein and Hawking found using very different considerations is a sign that string theory has successfully captured the fundamental physical description. This is a very encouraging result. For more details, see *The Elegant Universe*, Chapter 13.

17. The first hint of this pairing between Calabi-Yau shapes came from the work of Lance Dixon, as well as independently from Wolfgang Lerche, Nicholas Warner, and Cumrun Vafa. My work with Ronen Plesser found a method for producing the first concrete examples of such pairs, which we named *mirror pairs*, and the relationship between them *mirror symmetry*. Plesser and I also showed that difficult calculations on one member of a mirror pair, involving seemingly impenetrable details such as the number of spheres that can be packed into the shape, could be translated into far more manageable calculations on the mirror shape. This result was picked up by Philip Candelas, Xenia de la Ossa, Paul Green, and Linda Parkes and put into action—they developed techniques for explicitly evaluating the equality

Plesser and I had established between the "difficult" and "easy" formulas. Using the easy formula, they then extracted information about its difficult partner, including the numbers associated with the sphere packing given in the text. In the years since, mirror symmetry has become its own field of research, with a great many important results being established. For a detailed history, see Shing-Tung Yau and Steve Nadis, *The Shape of Inner Space* (New York: Basic Books, 2010).

18. String theory's claim to have successfully melded quantum mechanics and general relativity rests on a wealth of supporting calculations, made yet more convincing by results we will cover in Chapter 9.

Chapter 5: Hovering Universes in Nearby Dimensions

1. Classical Mechanics: $\vec{F} = m\vec{a}$. Electromagnetism: $d^*F = \,^*J; dF = 0$. Quantum mechanics: $H\psi = i\hbar\dfrac{d\psi}{dt}$. General relativity: $R_{\mu\nu} - \dfrac{1}{2}g_{\mu\nu}R = \dfrac{8\pi G}{c^4}\,T_{\mu\nu}$.

2. I am referring here to the *fine structure constant*, $\alpha = e^2/\hbar c$, whose numerical value (at typical energies for electromagnetic processes) is about 1/137, which is roughly .0073.

3. Witten argued that when the Type I string coupling is dialed large, the theory morphs into the Heterotic-O theory with a coupling that's dialed small, and vice versa; the Type IIB at large coupling morphs into *itself*, the Type IIB theory but with small coupling. The cases of the Heterotic-E and Type IIA theories are a little more subtle (see *The Elegant Universe*, Chapter 12, for details), but the overall picture is that all five theories participate in a web of interrelations.

4. For the mathematically inclined reader, the special thing about strings, one-dimensional ingredients, is that the physics describing their motion respects an infinite dimensional symmetry group. That is, as a string moves, it sweeps out a two-dimensional surface, and so the action functional from which its equations of motion are derived is a two-dimensional quantum field theory. Classically, such two-dimensional actions are conformally invariant (invariant under angle-preserving rescalings of the two-dimensional surface), and such symmetry can be preserved quantum mechanically by imposing various restrictions (such as on the number of spacetime dimensions through

which the string moves—the dimension, that is, of spacetime). The conformal group of symmetry transformations is infinite-dimensional, and this proves essential to ensuring that the perturbative quantum analysis of a moving string is mathematically consistent. For example, the infinite number of excitations of a moving string that would otherwise have negative norm (arising from the negative signature of the time component of the spacetime metric) can be systematically "rotated" away using the infinite-dimensional symmetry group. For details, the reader can consult M. Green, J. Schwarz, and E. Witten, *Superstring Theory*, vol. 1 (Cambridge: Cambridge University Press, 1988).

5. As with many major discoveries, credit deserves to be given to those whose insights laid its groundwork as well as to those whose work established its importance. Among those who played such a role for the discovery of branes in string theory are: Michael Duff, Paul Howe, Takeo Inami, Kelley Stelle, Eric Bergshoeff, Ergin Szegin, Paul Townsend, Chris Hull, Chris Pope, John Schwarz, Ashoke Sen, Andrew Strominger, Curtis Callan, Joe Polchinski, Petr Hořava, J. Dai, Robert Leigh, Hermann Nicolai, and Bernard DeWitt.

6. The diligent reader might argue that the Inflationary Multiverse also entwines time in a fundamental way, since, after all, our bubble's boundary marks the beginning of time in our universe; beyond our bubble is thus beyond our time. While true, my point here is meant more generally—the multiverses discussed so far all emerge from analyses that focus fundamentally on processes occurring throughout space. In the multiverse we will now discuss, time is central from the outset.

7. Alexander Friedmann, *The World as Space and Time*, 1923, published in Russian, as referenced by H. Kragh, in "Continual Fascination: The Oscillating Universe in Modern Cosmology," *Science in Context* 22, no. 4 (2009): 587–612.

8. As an interesting point of detail, the authors of the braneworld cyclic model invoke an especially utilitarian application of dark energy (dark energy will be discussed fully in Chapter 6). In the last phase of each cycle, the presence of dark energy in the braneworlds ensures agreement with today's observations of accelerated expansion; this accelerated expansion, in turn, dilutes the entropy density, setting the stage for the next cosmological cycle.

9. Large flux values also tend to destabilize a given Calabi-Yau shape for the extra dimensions. That is, the fluxes tend to push the Calabi-Yau shape to grow large, quickly running into conflict with the criterion that extra dimensions not be visible.

Chapter 6: New Thinking About an Old Constant

1. George Gamow, *My World Line* (New York: Viking Adult, 1970); J. C. Pecker, Letter to the Editor, *Physics Today*, May 1990, p. 117.

2. Albert Einstein, *The Meaning of Relativity* (Princeton: Princeton University Press, 2004), p. 127. Note that Einstein used the term "cosmologic member" for what we now call the "cosmological constant"; for clarity, I have made this substitution in the text.

3. *The Collected Papers of Albert Einstein*, edited by Robert Schulmann et al. (Princeton: Princeton University Press, 1998), p. 316.

4. Of course, some things *do* change. As pointed out in the notes to Chapter 3, galaxies generally have small velocities beyond the spatial swelling. Over the course of cosmological timescales, such additional motion can alter position relationships; such motion can also result in a variety of interesting astrophysical events such as galaxy collisions and mergers. For the purpose of explaining cosmic distances, however, these complications can be safely ignored.

5. There is one complication that does not affect the essential idea I've explained but which does come into play when undertaking the scientific analyses described. As photons travel to us from a given supernova, their number density gets diluted in the manner I've described. However, there is another diminishment to which they are subject. In the next section, I'll describe how the stretching of space causes the wavelength of photons to stretch too, and, correspondingly, their energy to decrease—an effect, as we will see, called *redshift*. As explained there, astronomers use redshift data to learn about the size of the universe when the photons were emitted—an important step toward determining how the expansion of space has varied through time. But the stretching of photons—the diminishment of their energy—has another effect: It accentuates the dimming of a distant source. And so, to properly determine the distance of a supernova by comparing its apparent and intrinsic brightness, astronomers must take account not just of the dilution of photon number density (as I've described in the text), but also the additional diminishment of energy coming from redshift. (More precisely still, this additional dilution factor must be applied twice; the second redshift factor accounts for the rate at which photons arrive being similarly stretched by the cosmic expansion.)

6. Properly interpreted, the second proposed answer for the meaning of the distance being measured may also be construed as correct. In the example

of earth's expanding surface, New York, Austin, and Los Angeles all rush away
from one another, yet each continues to occupy the same location on earth it
always has. The cities separate because the surface swells, not because someone digs them up, puts them on a flatbed, and transports them to a new site.
Similarly, because galaxies separate due to the cosmic swelling, they too
occupy the same location in space they always have. You can think of them as
being stitched to the spatial fabric. When the fabric stretches, the galaxies
move apart, yet each remains tethered to the very same point it has always
occupied. And so, even though the second and third answers appear different—the former focusing on the distance between us and the location a distant galaxy had eons ago, when the supernova emitted the light we now see;
the latter focusing on the distance now between us and that galaxy's current
location—they're not. The distant galaxy is now, and has been for billions of
years, positioned at one and the same spatial location. Only if it moved
through space rather than solely ride the wave of swelling space would its location change. In this sense, the second and third answers are actually the same.

7. For the mathematically inclined reader, here is how you do the calculation of the distance—now, at time t_{now}—that light has traveled since being
emitted at time $t_{emitted}$. We will work in the context of an example in which the
spatial part of spacetime is flat, and so the metric can be written as $ds^2 = c^2dt^2 -
a^2(t)dx^2$, where $a(t)$ is the scale factor of the universe at time t, and c is the
speed of light. The coordinates we are using are called *co-moving*. In the language developed in this chapter, such coordinates can be thought of as labeling points on the static map; the scale factor supplies the information
contained in the map's legend.

The special characteristic of the trajectory followed by light is that $ds^2 = 0$
(equivalent to the speed of light always being c) along the path, which implies
that $|dx| = \dfrac{cdt}{a(t)}$, or, over a finite time interval such as that between $t_{emitted}$
and t_{now}: $\int|dx| = \int_{t_{emitted}}^{t_{now}} \dfrac{cdt}{a(t)}$. The left side of this equation gives the distance light
travels across the static map between emission and now. To turn this into the
distance through real space, we must rescale the formula by today's scale
factor; therefore, the total distance the light traveled equals $a(t_{now})\int_{t_{emitted}}^{t_{now}} \dfrac{cdt}{a(t)}$.
If space were not stretching, the total travel distance would be $\int_{t_{emitted}}^{t_{now}} cdt =
c(t_{now} - t_{emitted})$, as expected. When calculating the distance traveled in an

expanding universe, we thus see that each segment of the light's trajectory is multiplied by the factor $\frac{a(t_{now})}{a(t)}$, which is the amount by which that segment has stretched, since the moment the light traversed it, until today.

8. More precisely, about 7.12×10^{-30} grams per cubic centimeter.

9. The conversion is 7.12×10^{-30} grams/cubic centimeter = $(7.12 \times 10^{-30}$ grams/cubic centimeter) $\times$ $(4.6 \times 10^4$ Planck mass/gram) $\times$ $(1.62 \times 10^{-33}$ centimeter/Planck length$)^3 = 1.38 \times 10^{-123}$ Planck mass/cubic Planck volume.

10. For inflation, the repulsive gravity we considered was intense and brief. This is explained by the enormous energy and negative pressure supplied by the inflaton field. However, by modifying a quantum field's potential energy curve, the amount of energy and negative pressure it supplies can be diminished, thus yielding a mild accelerated expansion. Additionally, a suitable adjustment of the potential energy curve can prolong this period of accelerated expansion. A mild and prolonged period of accelerated expansion is what's required to explain the supernova data. Nevertheless, the small non-zero value for the cosmological constant remains the most convincing explanation to have emerged in the more than ten years since the accelerated expansion was first observed.

11. The mathematically inclined reader should note that each such jitter contributes an energy that's inversely proportional to its wavelength, ensuring that the sum over all possible wavelengths yields an infinite energy.

12. For the mathematically inclined reader, the cancellation occurs because supersymmetry pairs bosons (particles with an integral spin value) and fermions (particles with a half [odd] integral spin value). This results in bosons being described by commuting variables, fermions by anticommuting variables, and that is the source of the relative minus sign in their quantum fluctuations.

13. While the assertion that changes to the physical features of our universe would be inhospitable to life is widely accepted in the scientific community, some have suggested that the range of features compatible with life might be larger than once thought. These issues have been widely written about. See, for example: John Barrow and Frank Tipler, *The Anthropic Cosmological Principle* (New York: Oxford University Press, 1986) for a pioneering discussion of the cosmological constant's impact on galaxy formation and hence on life as we know it; John Barrow, *The Constants of Nature* (New York: Pantheon Books, 2003); Paul Davies, *Cosmic Jackpot* (New York: Houghton Mifflin Harcourt, 2007); and references therein.

14. Based on the material covered in earlier chapters, you might immediately think the answer is a resounding yes. Consider, you say, the Quilted Multiverse, whose infinite spatial expanse contains infinitely many universes. But you need to be careful. Even with infinitely many universes, the list of different cosmological constants represented might not be long. If, for example, the underlying laws don't allow for many different cosmological constant values, then regardless of the number of universes, only the small collection of possible cosmological constants would be realized. So, the question we're asking is whether (a) there are candidate laws of physics that give rise to a multiverse, (b) the multiverse so generated contains far more than 10^{124} different universes, and (c) the laws ensure that the cosmological constant's value varies from universe to universe.

15. These four authors were the first to show fully that by judicious choices of Calabi-Yau shapes, and the fluxes threading their holes, they could realize string models with small, positive cosmological constants, like those found by observations. Together with Juan Maldacena and Liam McAllister, this group subsequently wrote a highly influential paper on how to combine inflationary cosmology with string theory.

16. More precisely, this mountainous terrain would inhabit a roughly 500-dimensional space, whose independent directions—axes—would correspond to different field fluxes. Figure 6.4 is a rough pictorial depiction but gives a feel for the relationships between the various forms for the extra dimensions. Additionally, when speaking of the string landscape, physicists generally envision that the mountainous terrain encompasses, in addition to the possible flux values, all the possible sizes and shapes (the different topologies and geometries) of the extra dimensions. The valleys in the string landscape are locations (specific forms for the extra dimensions and the fluxes they carry) where a bubble universe naturally settles, much as a ball would settle in such a spot in a real mountain terrain. When described mathematically, valleys are (local) minima of the potential energy associated with the extra dimensions. Classically, once a bubble universe acquired an extra dimensional form corresponding to a valley that feature would never change. Quantum mechanically, however, we will see that tunneling events can result in the form of the extra dimensions changing.

17. Quantum tunneling to a higher peak is possible but substantially less likely according to quantum calculations.

Chapter 7: Science and the Multiverse

1. The duration of the bubble's expansion prior to collision determines the impact, and attendant disruption, of the ensuing crash. Such collisions also raise an interesting point to do with time, harking back to the example with Trixie and Norton in Chapter 3. When two bubbles collide, their outer edges—where the inflaton field's energy is high—come into contact. From the perspective of someone within either one of the colliding bubbles, high inflaton energy value corresponds to early moments in time, near that bubble's big bang. And so, bubble collisions happen at the inception of each universe, which is why the ripples created can affect another early universe process, the formation of the microwave background radiation.

2. We will take up quantum mechanics more systematically in Chapter 8. As we will see there, the statement I've made, "slither outside the arena of everyday reality" can be interpreted on a number of levels. What I have in mind here is the conceptually simplest: the equation of quantum mechanics assumes that probability waves generally don't inhabit the spatial dimensions of common experience. Instead, the waves reside in a different environment that takes account not only of the everyday spatial dimensions but also of the *number* of particles being described. It is called *configuration space* and is explained for the mathematically inclined reader in note 4 of Chapter 8.

3. If the accelerated expansion of space that we've observed is not permanent, then at some time in the future the expansion of space will slow down. The slowing would allow light from objects that are now beyond our cosmic horizon to reach us; our cosmic horizon would grow. It would then be yet more peculiar to suggest that realms now beyond our horizon are not real since in the future we would have access to those very realms. (You may recall that toward the end of Chapter 2, I noted that the cosmic horizons illustrated in Figure 2.1 will grow larger as time passes. That's true in a universe in which the pace of spatial expansion is not quickening. However, if the expansion is accelerating, there is distance beyond that we can never see, regardless of how long we wait. In an accelerating universe, the cosmic horizons can't grow larger than a size determined mathematically by the rate of acceleration.)

4. Here is a concrete example of a feature that can be common to all universes in a particular multiverse. In Chapter 2, we noted that current data point strongly toward the curvature of space being zero. Yet, for reasons that are mathematically technical, calculations establish that all bubble universes in

the Inflationary Multiverse have negative curvature. Roughly speaking, the spatial shapes swept out by equal inflaton values—shapes determined by connecting equal numbers in Figure 3.8b—are more like potato chips than like flat tabletops. Even so, the Inflationary Multiverse remains compatible with observation, because as any shape expands its curvature drops; the curvature of a marble is obvious, while that of the earth's surface escaped notice for millennia. If our bubble universe has undergone sufficient expansion, its curvature could be negative yet so exceedingly small that today's measurements can't distinguish it from zero. That gives rise to a potential test. Should more precise observations in the future determine that the curvature of space is very small but *positive* that would provide evidence against our being part of an Inflationary Multiverse as argued by B. Freivogel, M. Kleban, M. Rodríguez Martínez, and L. Susskind, (see "Observational Consequences of a Landscape," *Journal of High Energy Physics* 0603, 039 [2006]), measurement of positive curvature of 1 part in 10^5 would make a strong case against the kind of quantum tunneling transitions (Chapter 6) envisioned to populate the string landscape.

5. The many cosmologists and string theorists who have advanced this subject include Alan Guth, Andrei Linde, Alexander Vilenkin, Jaume Garriga, Don Page, Sergei Winitzki, Richard Easther, Eugene Lim, Matthew Martin, Michael Douglas, Frederik Denef, Raphael Bousso, Ben Freivogel, I-Sheng Yang, Delia Schwartz-Perlov, among many others.

6. An important caveat is that while the impact of modest changes to a few constants can reliably be deduced, more significant changes to a larger number of constants make the task far more difficult. It is at least possible that such significant changes to a variety of nature's constants cancel out one another's effects, or work together in novel ways, and are thus compatible with life as we know it.

7. A little more precisely, if the cosmological constant is negative, but sufficiently tiny, the collapse time would be long enough to allow galaxy formation. For ease, I am glossing over this subtlety.

8. Another point worthy of note is that the calculations I've described were undertaken without making a specific choice for the multiverse. Instead, Weinberg and his collaborators proceeded by positing a multiverse in which features could vary and calculated the abundance of galaxies in each of their constituent universes. The more galaxies a universe had, the more weight Weinberg and collaborators gave to its properties in their calculation of the average features a typical observer would encounter. But because they didn't

commit to an underlying multiverse theory, the calculations necessarily failed to account for the probability that a universe with this or that property would actually be found in the multiverse (the probabilities, that is, that we discussed in the previous section). Universes with cosmological constants and primordial fluctuations in certain ranges might be ripe for galaxy formation, but if such universes are rarely created in a given multiverse, it would nevertheless be highly unlikely for us to find ourselves in one of them.

To make the calculations manageable, Weinberg and collaborators argued that since the range of cosmological constant values they were considering was so narrow (between 0 and about 10^{-120}), the intrinsic probabilities that such universes would exist in a given multiverse were not likely to vary wildly, much as the probabilities that you'll encounter a 59.99997-pound dog or one weighing 59.99999 pounds also don't differ substantially. They thus assumed that every value for the cosmological constant in the small range consistent with the formation of galaxies is as intrinsically probable as any other. With our rudimentary understanding of multiverse formation, this might seem like a reasonable first pass. But subsequent work has questioned the validity of this assumption, emphasizing that a full calculation needs to go further: committing to a definite multiverse proposal and determining the actual distribution of universes with various properties. A self-contained anthropic calculation that relies on a bare minimum of assumptions is the only way to judge whether this approach will ultimately bear explanatory fruit.

9. The very meaning of "typical" is also burdened, as it depends on how it's defined and measured. If we use numbers of kids and cars as our delimeter, we arrive at one kind of "typical" American family. If we use different scales such as interest in physics, love of opera, or immersion in politics, the characterization of a "typical" family will change. And what's true for the "typical" American family is likely true for "typical" observers in the multiverse: consideration of features beyond just population size would yield a different notion of who is "typical." In turn, this would affect the predictions for how likely it is that we will see this or that property in our universe. For an anthropic calculation to be truly convincing, it would have to address this issue. Alternatively, as indicated in the text, the distributions would need to be so sharply peaked that there would be minimal variation from one life-supporting universe to another.

10. The mathematical study of sets with an infinite number of members is rich and well developed. The mathematically inclined reader may be familiar with the fact that research going back to the nineteenth century established

there are different "sizes" or, more commonly, "levels" of infinity. That is, one infinite quantity can be larger than another. The level of infinity that gives the size of the set containing all the whole numbers is called $\aleph_0$. This infinity was shown by Georg Cantor to be smaller than that giving the number of members contained in the set of real numbers. Roughly speaking, if you try to match up whole numbers and real numbers, you necessarily exhaust the former before the latter. And if you consider the set of all subsets of real numbers, the level of infinity grows larger still.

Now, in all of the examples we discuss in the main text, the relevant infinity is $\aleph_0$, since we are dealing with infinite collections of discrete, or "countable," objects—various collections, that is, of whole numbers. In the mathematical sense, then, all of the examples have the same size; their total membership is described by the same level of infinity. However, for physics, as we will shortly see, a conclusion of this sort would not be particularly useful. The goal instead is to find a physically motivated scheme for comparing infinite collections of universes that would yield a more refined hierarchy, one that reflects the relative abundance across the multiverse of one set of physical features compared with another. A typical physics approach to a challenge of this sort is to first make comparisons between finite subsets of the relevant infinite collections (since in the finite case, all of the puzzling issues evaporate), and then allow the subsets to include ever more members, ultimately embracing the full infinite collections. The hurdle is finding a physically justifiable way of picking out the finite subsets for comparison, and then also establishing that comparisons remain sensible as the subsets grow larger.

11. Inflation is credited with other successes too, including the solution to the *magnetic monopole problem*. In attempts to meld the three nongravitational forces into a unified theoretical structure (known as *grand unification*) researchers found that the resulting mathematics implied that just after the big bang a great many magnetic monopoles would have been formed. These particles would be, in effect, the north pole of a bar magnet without the usual pairing with a south pole (or vice versa). But no such particles have ever been found. Inflationary cosmology explains the absence of monopoles by noting that the brief but stupendous expansion of space just after the big bang would have diluted their presence in our universe to nearly zero.

12. Currently, there are differing views on how great a challenge this presents. Some view the measure problem as a knotty technical issue that once solved will provide inflationary cosmology with an important additional detail.

Others (for example, Paul Steinhardt) have expressed the belief that solving the measure problem will require stepping so far outside the mathematical formulation of inflationary cosmology that the resulting framework will need to be interpreted as a completely new cosmological theory. My view, one held by a small but growing number of researchers, is that the measure problem is tapping into a deep problem at the very root of physics, one that may require a substantial overhaul of foundational ideas.

Chapter 8: The Many Worlds of Quantum Measurement

1. Both Everett's original 1956 thesis and the shortened 1957 version can be found in *The Many-Worlds Interpretation of Quantum Mechanics*, edited by Bryce S. DeWitt and Neill Graham (Princeton: Princeton University Press, 1973).

2. On January 27, 1998, I had a conversation with John Wheeler to discuss aspects of quantum mechanics and general relativity that I would be writing about in *The Elegant Universe*. Before getting into the science proper, Wheeler noted how important it was, especially for young theoreticians, to find the right language for expressing their results. At the time, I took this as nothing more than sage advice, perhaps inspired by his speaking with me, a "young theoretician" who'd expressed interest in using ordinary language to describe mathematical insights. On reading the illuminating history laid out in *The Many Worlds of Hugh Everett III* by Peter Byrne (New York: Oxford University Press, 2010), I was struck by Wheeler's emphasis of the same theme some forty years earlier in his dealings with Everett, but in a context whose stakes were far higher. In response to Everett's first draft of his thesis, Wheeler told Everett that he needed to "get the bugs out of the words, not the formalism" and warned him of "the difficulty of expressing in everyday words the goings-on in a mathematical scheme that is about as far removed as it could be from the everyday description; the contradictions and misunderstandings that will arise; the very very heavy burden and responsibility you have to state everything in such a way that these misunderstandings can't arise." Byrne makes a compelling case that Wheeler was walking a delicate line between his admiration for Everett's work and his respect for the quantum mechanical framework that Bohr and many other renowned physicists had labored to build. On the one hand, he didn't want Everett's insights to be summarily dismissed by the

old guard because the presentation was deemed overreaching, or because of hot-button words (like universes "splitting") that could appear fanciful. On the other hand, Wheeler didn't want the established community of physicists to conclude that he was abandoning the demonstrably successful quantum formalism by spearheading an unjustified assault. The compromise Wheeler was imposing on Everett and his dissertation was to keep the mathematics he'd developed but frame its meaning and utility in a softer, more conciliatory tone. At the same time, Wheeler strongly encouraged Everett to visit Bohr and make his case in person, at a blackboard. In 1959 Everett did just that, but what Everett thought would be a two-week showdown amounted to a few unproductive conversations. No minds changed; no positions altered.

3. Let me clarify one imprecision. Schrödinger's equation shows that the values attained by a quantum wave (or, in the language of the field, the wavefunction) can be positive or negative; more generally, the values can be complex numbers. Such values cannot be interpreted directly as probabilities—what would a negative or complex probability mean? Instead, probabilities are associated with the *squared magnitude* of the quantum wave at a given location. Mathematically, this means that to determine the probability that a particle will be found at a given location, we take the *product of wave's value at that point and its complex conjugate*. This clarification also addresses an important related issue. Cancellations between overlapping waves are vital to creating an interference pattern. But if the waves themselves were properly described as probability waves, such cancellations couldn't happen because probabilities are positive numbers. As we now see, however, quantum waves do not only have positive values; this allows cancellations to take place between positive and negative numbers, as well as, more generally, between complex numbers. Because we will only need qualitative features of such waves, for ease of discussion in the main text I will not distinguish between a quantum wave and the associated probability wave (derived from its squared magnitude).

4. For the mathematically inclined reader, note that the quantum wave (*wavefunction*) for a single particle with large mass would conform to the description I've given in the text. However, very massive objects are generally composed of many particles, not one. In such a situation, the quantum mechanical description is more involved. In particular, you might have thought that all of the particles could be described by a quantum wave defined on the same coordinate grid we employ for a single particle—using the same three spatial axes. But that's not right. The probability wave takes as input the

possible position of each particle and produces the probability that the particles occupy those positions. Consequently, the probability wave lives in a space with three axes for each particle — that is, in total three times as many axes as there are particles (or ten times as many, if you embrace string theory's extra spatial dimensions). This means that the wavefunction for a composite system made of n fundamental particles is a complex-valued function whose domain is not ordinary three-dimensional space but rather $3n$-dimensional space; if the number of spatial dimensions is not 3 but rather m, the number 3 in these expressions would be replaced by m. This space is called *configuration space*. That is, in the general setting, the wavefunction would be a map $\psi : \mathfrak{R}^{mn} \to C$. When we speak of such a wavefunction as being sharply peaked, we mean that this map would have support in a small mn-dimensional ball within its domain. Note, in particular, that wavefunctions don't generally reside in the spatial dimensions of common experience. It is only in the idealized case of the wavefunction for a completely isolated single particle that its configuration space coincides with the familiar spatial environment. Note as well that when I say that the quantum laws show that the sharply peaked wavefunction for a massive object traces the same trajectory that Newton's equations imply for the object itself, you can think of the wavefunction describing the object's center of mass motion.

5. From this description, you might conclude that there are infinitely many locations that the electron could be found: to properly fill out the gradually varying quantum wave you would need an infinite number of spiked shapes, each associated with a possible position of the electron. How does this relate to Chapter 2 in which we discussed there being finitely many distinct configurations for particles? To avoid constant qualifications that would be of minimal relevance to the major points I am explaining in this chapter, I have not emphasized the fact, encountered in Chapter 2, that to pinpoint the electron's location with ever-greater accuracy your device would need to exert ever-greater energy. As physically realistic situations have access to finite energy, resolution is thus imperfect. For the spiked quantum waves, this means that in any finite energy context, the spikes have nonzero width. In turn, this implies that in any bounded domain (such as a cosmic horizon) there are finitely many measurably distinct electron locations. Moreover, the thinner the spikes are (the more refined the resolution of the particle's position) the wider are the quantum waves describing the particle's energy, illustrating the trade-off necessitated by the uncertainty principle.

6. For the philosophically inclined reader, I'll note that the two-tiered story

for scientific explanation which I've outlined has been the subject of philosophical discussion and debate. For related ideas and discussions see Frederick Suppe, *The Semantic Conception of Theories and Scientific Realism* (Chicago: University of Illinois Press, 1989); James Ladyman, Don Ross, David Spurrett, and John Collier, *Every Thing Must Go* (Oxford: Oxford University Press, 2007).

7. Physicists often speak loosely of there being infinitely many universes associated with the Many Worlds approach to quantum mechanics. Certainly, there are infinitely many possible probability wave shapes. Even at a single location in space you can continuously vary the value of a probability wave, and so there are infinitely many different values it can have. However, probability waves are not the physical attribute of a system to which we have direct access. Instead, probability waves contain information about the possible distinct outcomes in a given situation, and these need not have infinite variety. Specifically, the mathematically inclined reader will note that a quantum wave (a wavefunction) lies in a Hilbert space. If that Hilbert space is finite-dimensional, then there are finitely many distinct possible outcomes for measurements on the physical system described by that wavefunction (that is, any Hermitian operator has finitely many distinct eigenvalues). This would entail finitely many worlds for a finite number of observations or measurements. It is believed that the Hilbert space associated with physics taking place within any finite volume of space, and limited to having a finite amount of energy, is necessarily finite dimensional (a point we will take up more generally in Chapter 9), which suggests that the number of worlds would similarly be finite.

8. See Peter Byrne, *The Many Worlds of Hugh Everett III* (New York: Oxford University Press, 2010), p. 177.

9. Over the years, a number of researchers including Neill Graham; Bryce DeWitt; James Hartle; Edward Farhi, Jeffrey Goldstone, and Sam Gutmann; David Deutsch; Sidney Coleman; David Albert; and others, including me, have independently come upon a striking mathematical fact that seems central to understanding the nature of probability in quantum mechanics. For the mathematically inclined reader, here's what it says: Let $|\psi\rangle$ be the wavefunction for a quantum mechanical system, a vector that's an element of the Hilbert space H. The wavefunction for n-identical copies of the system is thus $|\psi\rangle^{\otimes n}$. Let A be any Hermitian operator with eigenvalues α_k, and eigenfunctions $|\lambda_k\rangle$. Let $F_k(A)$ be the "frequency" operator that counts the number of times $|\lambda_k\rangle$ appears in a given state lying in $H^{\otimes n}$. The mathematical result is that

$\lim_{n \to \infty}[F_k(A)|\psi\rangle^{\otimes n}] = |\langle\psi|\lambda_k\rangle|^2|\psi\rangle^{\otimes n}$. That is, as the number of identical copies of the system grows without bound, the wavefunction of the composite system approaches an eigenfunction of the frequency operator, with eigenvalue $|\langle\psi|\lambda_k\rangle|^2$. This is a remarkable result. Being an eigenfunction of the frequency operator means that, in the stated limit, the fractional number of times an observer measuring A will find α_k is $|\langle\psi|\lambda_k\rangle|^2$—which looks like the most straightforward derivation of the famous Born rule for quantum mechanical probability. From the Many Worlds perspective, it suggests that those worlds in which the fractional number of times that α_k is observed fails to agree with the Born rule have zero Hilbert space norm in the limit of arbitrarily large n. In this sense, it seems as though quantum mechanical probability has a direct interpretation in the Many Worlds approach. All observers in the Many Worlds will see results with frequencies that match those of standard quantum mechanics, except for a set of observers whose Hilbert space norm becomes vanishingly small as n goes to infinity. As promising as this seems, on reflection it is less convincing. In what sense can we say that an observer with a small Hilbert space norm, or a norm that goes to zero as n goes to infinity, is unimportant or doesn't exist? We want to say that such observers are anomalous or "unlikely," but how do we draw a link between a vector's Hilbert space norm and these characterizations? An example makes the issue manifest. In a two-dimensional Hilbert space, say with states spin-up $|\uparrow\rangle$, and spin-down $|\downarrow\rangle$, consider a state $|\psi\rangle = .99|\uparrow\rangle + .14|\downarrow\rangle$. This state yields the probability for measuring spin-up of about .98 and for measuring spin-down to be about .02. If we consider n copies of this spin system, $|\psi\rangle^{\otimes n}$, then as n goes to infinity, the vast majority of terms in the expansion of this vector have roughly equal numbers of spin-up and spin-down states. So from the standpoint of observers (copies of the experimenter) the vast majority would see spin-ups and spin-downs in a ratio that does not agree with the quantum mechanical predictions. Only the very few terms in the expansion of $|\psi\rangle^{\otimes n}$ that have 98 percent spin-ups and 2 percent spin-downs are consistent with the quantum mechanical expectation; the result above tells us that these states are the only ones with nonzero Hilbert space norm as n goes to infinity. In some sense, then, the vast majority of terms in the expansion of $|\psi\rangle^{\otimes n}$ (the vast majority of copies of the experimenter) need to be considered as "nonexistent." The challenge lies in understanding what, if anything, that means.

I also independently found the mathematical result described above, while preparing lectures for a course on quantum mechanics I was teaching. It was a notable thrill to have the probabilistic interpretation of quantum mechanics seemingly fall out directly from the mathematical formalism—I would imagine the list of physicists (on page 393) who found this result before me had the same experience. I'm surprised at how little known the result is among mainstream physics. For instance, I don't know of any standard quantum physics textbook that includes it. My take on the result is that it is best thought of as (1) a strong mathematical motivation for the Born probability interpretation of the wavefunction—had Born not "guessed" this interpretation, the math would have led someone there eventually; (2) a consistency check on the probability interpretation—had this mathematical result *not* held, it would have challenged the internal sensibility of the probability interpretation of the wavefunction.

10. I've been using the phrase "Zaxtarian-type reasoning" to denote a framework in which probability enters through the ignorance of each inhabitant of the Many Worlds as to which particular world he or she inhabits. Lev Vaidman has suggested taking more of the particulars of the Zaxtarian scenario to heart. He argues that probability enters the Many Worlds approach in the temporal window between an experimenter completing a measurement and reading the result. But, skeptics counter, this is too late in the game: it's incumbent on quantum mechanics, and science more generally, to make predictions about what *will* happen in an experiment, not what *did* happen. What's more, it seems precarious for the bedrock of quantum probability to rely on what seems to be an avoidable time delay: if a scientist gains immediate access to the result of his or her experiment, quantum probability seems in danger of being squeezed out of the picture. (For a detailed discussion see David Albert, "Probability in the Everett Picture" in *Many Worlds: Everett, Quantum Theory, and Reality*, eds. Simon Saunders, Jonathan Barrett, Adrian Kent, and David Wallace (Oxford: Oxford University Press, 2010) and "Uncertainty and Probability for Branching Selves," Peter Lewis, philsciarchive.pitt.edu/archive/00002636.) A final issue of relevance to Vaidman's suggestion and also to this type of ignorance probability is this: when I flip a fair coin in the familiar context of a single universe, the reason I say there's a 50 percent chance the coin will land heads is that while I'll experience only one outcome, there are two outcomes that I *could* have experienced. But let me now close my eyes and imagine I've just measured the position of the somber electron. I know that my detector display says either Strawberry Fields or Grant's Tomb, but I don't know which. You then confront me. "Brian," you say, "what's the probability that your

screen says Grant's Tomb?" To answer, I think back on the coin toss, and just as I'm about to follow the same reasoning, I hesitate. "Hmmm," I think. "Are there really two outcomes that *I* could have experienced? The *only* detail that differentiates me from the other Brian is the reading on my screen. To imagine that my screen could have returned a different reading is to imagine that I'm not me. It's to imagine I'm the other Brian." So even though I don't know what my screen says, I—*this guy talking in my head right now*—couldn't have experienced any other outcome; that suggests that my ignorance doesn't lend itself to probabilistic thinking.

11. Scientists are meant to be objective in their judgments. But I feel comfortable admitting that because of its mathematical economy and far-reaching implications for reality, I'd like the Many Worlds approach to be right. At the same time, I maintain a healthy skepticism, fueled by the difficulties of integrating probability into the framework, so I'm fully open to alternative lines of attack. Two of these provide good bookends for the discussion in the text. One tries to develop the incomplete Copenhagen approach into a full theory; the other can be viewed as Many Worlds without the many worlds.

The first direction, spearheaded by Giancarlo Ghirardi, Alberto Rimini, and Tullio Weber, tries to make sense of the Copenhagen scheme by changing Schrödinger's math so that it *does* allow probability waves to collapse. This is easier said than done. The modified math should barely affect the probability waves for small things like individual particles or atoms, since we don't want to change the theory's successful descriptions in this domain. But the modifications must kick in with a vengeance when a large object like a piece of laboratory equipment comes into play, causing the commingled probability wave to collapse. Ghirardi, Rimini, and Weber developed math that does just that. The upshot is that with their modified equations, measuring does indeed make a probability wave collapse; it sets in motion the evolution pictured in Figure 8.6.

The second approach, initially developed by Prince Louis de Broglie back in the 1920s, and then more fully decades later by David Bohm, starts from a mathematical premise that resonates with Everett. Schrödinger's equation should always, in every circumstance, govern the evolution of quantum waves. So, in the de Broglie–Bohm theory, probability waves evolve just as they do in the Many Worlds approach. The de Broglie–Bohm theory goes on, however, to propose the very idea I emphasized earlier as being wrongheaded: in the de Broglie–Bohm approach, all but one of the many worlds encapsulated in a probability wave are merely *possible* worlds; only one world is singled out as real.

To accomplish this, the approach jettisons the traditional quantum haiku of wave *or* particle (an electron is a wave until it's measured, whereupon it reverts to being a particle) and instead advocates a picture that embraces waves *and* particles. Contrary to the standard quantum view, de Broglie and Bohm envision particles as tiny, localized entities that travel along definite trajectories, yielding an ordinary, unambiguous reality, much as in the classical tradition. The only "real" world is the one in which the particles inhabit their unique, definite positions. Quantum waves then play a very different role. Rather than embodying a multitude of realities, a quantum wave acts to *guide the motion of particles.* The quantum wave pushes particles toward locations where the wave is large, making it likely that particles will be found at such locations, and away from locations where the wave is small, making it unlikely that particles will be found at those. To account for the process, de Broglie and Bohm needed an additional equation describing the effect of a quantum wave on a particle, so in their approach, Schrödinger's equation, while not superseded, shares the stage with another mathematical player. (The mathematically inclined reader can see these equations below.)

For many years, the word on the street was that the de Broglie–Bohm approach was not worth considering, laden as it was with unnecessary baggage—not only a second equation but also, since it involves both particles and waves, a doubly long list of ingredients. More recently, there has been a growing recognition that these criticisms need context. As the Ghirardi-Rimini-Weber work makes explicit, even a sensible version of the standard-bearer Copenhagen approach requires a second equation. Additionally, the inclusion of both waves and particles yields an enormous benefit: it restores the notion of objects moving from here to there along definite trajectories, a return to a basic and familiar feature of reality that the Copenhagenists may have persuaded everyone to relinquish a little too quickly. More technical criticisms are that the approach is *nonlocal* (the new equation shows that influences exerted at one location appear to instantaneously affect distant locations) and that it is difficult to reconcile the approach with special relativity . The potency of the former criticism is diminished by the recognition that even the Copenhagen approach has non-local features that, moreover, have been confirmed experimentally. The latter point regarding relativity, though, is certainly an important one that has yet to be fully resolved.

Part of the resistance to the de Broglie–Bohm theory arose because the theory's mathematical formalism has not always been presented in its most straightforward form. Here, for the mathematically inclined reader, is the most direct derivation of the theory.

Begin with Schrödinger's equation for the wavefunction of a particle: $H\psi = i\hbar\frac{\partial\psi}{\partial t}$, where the probability density for the particle to be at position x, $\rho(x)$, is given by the standard equation $\rho(x) = |\psi(x)|^2$. Then, imagine assigning a definite trajectory to the particle, with velocity at x given by a function $v(x)$. What physical condition should this velocity function satisfy? Certainly, it should ensure conservation of probability: if the particle is moving with velocity $v(x)$ from one region into another, the probability density should adjust accordingly: $\frac{\partial\rho}{\partial t} + \frac{\partial(\rho v)}{\partial x} = 0$. It is now straightforward to solve for $v(x)$ and find

$$v(x, t) = \frac{-1}{\rho(x, t)}\int\frac{\partial\rho}{\partial t} = \frac{\hbar}{m}\,\mathrm{Im}(\frac{\psi^*\frac{\partial\psi}{\partial x}}{\psi^*\psi}),$$ where m is the particle's mass.

Together with Schrödinger's equation, this latter equation defines the de Broglie–Bohm theory. Note that this latter equation is nonlinear, but this has no bearing on Schrödinger's equation, which retains its full linearity. The proper interpretation, then, is that this approach to filling in the gaps left by the Copenhagen approach adds a new equation, which depends nonlinearly on the wavefunction. All of the power and beauty of the underlying wave equation, that of Schrödinger, is fully preserved.

I might also add that the generalization to many particles is immediate: on the right-hand side of the new equation, we substitute the wavefunction of the multiparticle system: $\psi(x_1, x_2, x_3, \ldots x_n)$, and in calculating the velocity of the kth particle, we take the derivative with respect to the k-th coordinate (working, for ease, in a one-dimensional space; for higher dimensions, we suitably increase the number of coordinates). This generalized equation manifests the nonlocality of this approach: the velocity of the kth particle depends, instantaneously, on the positions of all other particles (as the particles' coordinate locations are the arguments of the wavefunction).

12. Here is a concrete in-principle experiment for distinguishing the Copenhagen and Many Worlds approaches. An electron, like all other elementary particles, has a property known as *spin*. Somewhat as a top can spin about an axis, an electron can too, with one significant difference being that the rate of this spin—regardless of the direction of the axis—is always the same.

It is an intrinsic property of the electron, like its mass or its electrical charge. The only variable is whether the spin is clockwise or counterclockwise about a given axis. If it is counterclockwise, we say the electron's spin about that axis is *up*; if it is clockwise, we say the electron's spin is *down*. Because of quantum mechanical uncertainty, if the electron's spin about a given axis is definite— say, with 100 percent certainty its spin is up about the z-axis—then its spin about the x- or y-axis is uncertain: about the x-axis the spin would be 50 percent up and 50 percent down; and similarly for the y-axis.

Imagine, then, starting with an electron whose spin about the z-axis is 100 percent up and then measuring its spin about the x-axis. According to the Copenhagen approach, if you find spin-down, that means the probability wave for the electron's spin has collapsed: the spin-up possibility has been erased from reality, leaving the sole spike at spin-down. In the Many Worlds approach, by contrast, both the spin-up and spin-down outcomes occur, so, in particular, the spin-up possibility survives fully intact.

To adjudicate between these two pictures, imagine the following. After you measure the electron's spin about the x-axis, have someone fully *reverse* the physical evolution. (The fundamental equations of physics, including that of Schrödinger, are time-reversal invariant, which means, in particular, that, at least in principle, any evolution can be undone. See *The Fabric of the Cosmos* for an in-depth discussion of this point.) Such reversal would be applied to everything: the electron, the equipment, and anything else that's part of the experiment. Now, if the Many Worlds approach is correct, a subsequent measurement of the electron's spin about the z-axis should yield, with 100 percent certainty, the value with which we began: spin-up. However, if the Copenhagen approach is correct (by which I mean a mathematically coherent version of it, such as the Ghirardi-Rimini-Weber formulation), we would find a different answer. Copenhagen says that upon measurement of the electron's spin about the x-axis, in which we found spin-down, the spin-up possibility was annihilated. It was wiped off reality's ledger. And so, upon reversing the measurement we don't get back to our starting point because we've permanently lost part of the probability wave. Upon subsequent measurement of the electron's spin about the z-axis, then, there is not 100 percent certainty that we will get the same answer we started with. Instead, it turns out that there's a 50 percent chance that we will and a 50 percent chance that we won't. If you were to undertake this experiment repeatedly, and if the Copenhagen approach is correct, on average, half the time you would not recover the same answer you initially did for the electron's spin about the *z*-axis. The challenge, of course, is in carrying out the full reversal of a physical evolution. But, in principle, this is an experiment that would provide insight into which of the two theories is correct.

Chapter 9: Black Holes and Holograms

1. Einstein undertook calculations within general relativity to prove mathematically that Schwarzschild's extreme configurations—what we would now call a black hole—could not exist. The mathematics underlying his calculations was invariably correct. But he made additional assumptions that, given the intense folding of space and time that would be caused by a black hole, turn out to be too restrictive; in essence, the assumption left out the possibility of matter imploding. The assumptions meant that Einstein's mathematical formulation did not have the latitude to reveal black holes as possibly real. But this was an artifact of Einstein's approach, not an indication of whether black holes might actually form. The modern understanding makes clear that general relativity allows for black hole solutions.

2. Once a system reaches a maximal entropy configuration (such as steam, at a fixed temperature, that is uniformly spread throughout a vat), it will have exhausted its capacity for yet further entropic increase. So, the more precise statement is that entropy tends to increase, until it reaches the largest value the system can support.

3. In 1972, James Bardeen, Brandon Carter, and Stephen Hawking worked out the mathematical laws underlying the evolution of black holes, and found that the equations looked just like those of thermodynamics. To translate between the two sets of laws, all one needed to do was substitute "area of black hole's horizon" for "entropy" (and vice versa), and "gravity at the surface of the black hole" for "temperature." So, for Bekenstein's idea to hold—for this similarity to not just be a coincidence, but to reflect the fact that black holes have entropy—black holes would also need to have a nonzero temperature.

4. The reason for the apparent change in energy is far from obvious; it relies on an intimate connection between energy and time. You can think of a particle's energy as the vibrational speed of its quantum field. Noting that the very meaning of speed invokes the concept of time, a relationship between energy and time becomes apparent. Now, black holes have a profound effect

on time. From a distant vantage point, time appears to slow for an object approaching the horizon of a black hole, and comes to a stop at the horizon itself. Upon crossing the horizon, time and space interchange roles — inside the black hole, the radial direction becomes the time direction. This implies that within the black hole, the notion of positive energy coincides with motion in the radial direction toward the black hole's singularity. When the negative energy member of a particle pair crosses the horizon, it does indeed fall toward the black hole's center. Thus the negative energy it had from the perspective of someone watching from afar becomes positive energy from the perspective of someone situated within the black hole itself. This makes the interior of the black hole a place where such particles can exist.

5. When a black hole shrinks, the surface area of its event horizon shrinks too, conflicting with Hawking's pronouncement that total surface area increases. Remember, however, that Hawking's area theorem is based on classical general relativity. We are now taking account of quantum processes and coming to a more refined conclusion.

6. To be a little more precise, it's the minimum number of yes-no questions whose answers uniquely specify the microscopic details of the system.

7. Hawking found that the entropy is the area of the event horizon in Planck units, divided by four.

8. For all the insights that will be described as this chapter unfolds, the issue of a black hole's microscopic makeup has yet to be fully resolved. As I mentioned in Chapter 4, in 1996, Andrew Strominger and Cumrun Vafa discovered that if one (mathematically) gradually turns down the strength of gravity, then certain black holes morph into particular collections of strings and branes. By counting the possible rearrangements of these ingredients, Strominger and Vafa recovered, in the most explicit manner ever achieved, Hawking's famous black hole entropy formula. Even so, they were not able to describe these ingredients at stronger gravitational strength, i.e., when the black hole actually forms. Other authors, such as Samir Mathur and various of his collaborators, have put forward other ideas, such as the possibility that black holes are what they call "fuzz balls," accumulations of vibrating strings strewn throughout the black hole's interior. These ideas remain tentative. The results we discuss later in this chapter (in the section "String Theory and Holography") provide some of the sharpest insight into this question.

9. More precisely, gravity can be canceled in a region of space by going into a freely falling state of motion. The size of the region depends on the

scales over which the gravitational field varies. If the gravitational field varies only over large scales (that is, if the gravitational field is uniform, or nearly so), your free-fall motion will cancel gravity over a large region of space. But if the gravitational field varies over short-distance scales—the scales of your body, say—then you might cancel gravity at your feet and yet still feel it at your head. This becomes particularly relevant later in your fall because the gravitational field gets ever stronger ever closer to the black hole's singularity; its strength rises sharply as your distance from the singularity decreases. The rapid variation means there is no way to cancel the effects of the singularity, which will ultimately stretch your body to its breaking point since the gravitational pull on your feet, if you jump in feetfirst, will be ever stronger than the pull on your head.

10. This discussion exemplifies the discovery, made in 1976 by William Unruh, that links one's motion and the particles one encounters. Unruh found that if you accelerate through otherwise empty space, you will encounter a bath of particles at a temperature determined by your motion. General relativity instructs us to determine one's rate of acceleration by comparing with the benchmark set by free-fall observers (see *Fabric of the Cosmos*, Chapter 3). A distant, non-free-fall observer thereby sees radiation emerging from a black hole; a free-fall observer does not.

11. A black hole forms if the mass M within a sphere of radius R exceeds $c^2R/2G$, where c is the speed of light and G is Newton's constant.

12. In actuality, as the matter collapsed under its own weight and a black hole formed, the event horizon would generally be located within the boundary of the region we've been discussing. This means that we would not have so far maxed out the entropy that the region itself could contain. This is easily remedied. Throw more material into the black hole, causing the event horizon to swell out to the region's original boundary. Since entropy would again increase throughout this somewhat more elaborate process, the entropy of the material we put within the region would be less than that of the black hole that fills the region, i.e., the surface area of the region in Planck units.

13. G. 't Hooft, "Dimensional Reduction in Quantum Gravity." In *Salam Festschrift*, edited by A. Ali, J. Ellis, and S. Randjbar-Daemi (River Edge, N.J.: World Scientific, 1993), pp. 284–96 (QCD161:C512:1993).

14. We've discussed that "tired" or "exhausted" light is light whose wavelength is stretched (redshifted) and vibrational frequency reduced by virtue of its having expended energy climbing away from a black hole (or climbing

away from any source of gravity). Like more familiar cyclical processes (the earth's orbit around the sun; the earth's rotation on its axis, etc.), the vibrations of light can be used to define elapsed time. In fact, the vibrations of light emitted by excited Cesium-133 atoms are now used by scientists to *define* the second. The tired light's slower vibrational frequency thus implies that the passage of time near the black hole—as viewed by the faraway observer—is slower too.

15. With most important discoveries in science, the pinnacle result relies on a collection of earlier works. Such is the case here. In addition to 't Hooft, Susskind, and Maldacena, the researchers who helped blaze the trail to this result and develop its consequences include Steve Gubser, Joe Polchinski, Alexander Polyakov, Ashoke Sen, Andy Strominger, Cumrun Vafa, Edward Witten, and many others.

For the mathematically inclined reader, the more precise statement of Maldacena's result is the following. Let N be the number of three-branes in the brane stack, and let g be the value of the coupling constant in the Type IIB string theory. When gN is a small number, much less than one, the physics is well described by low-energy strings moving on the brane stack. In turn, such strings are well described by a particular four-dimensional supersymmetric conformally invariant quantum field theory. But when gN is a large number, this field theory is strongly coupled, making its analytical treatment difficult. However, in this regime, Maldacena's result is that we can use the description of strings moving on the near horizon geometry of the brane stack, which is $AdS_5 \times S^5$ (anti-de Sitter five-space times the five sphere). The radius of these spaces is controlled by gN (specifically, the radius is proportional to $(gN)^{1/4}$), and thus for large gN, the curvature of $AdS_5 \times S^5$ is small, ensuring that string theory calculations are tractable (in particular, they are well approximated by calculations in a particular modification of Einsteinian gravity). Therefore, as the value of gN varies from small to large values, the physics morphs from being described by four-dimensional supersymmetric conformally invariant quantum field theory to being described by ten-dimensional string theory on $AdS_5 \times S^5$. This is the so-called AdS/CFT (anti-de Sitter space/conformal field theory) correspondence.

16. Although a full proof of Maldacena's argument remains beyond reach, in recent years the link between the bulk and boundary descriptions has become increasingly well understood. For example, a class of calculations has been identified whose results are accurate for any value of the coupling con-

stant. The results can therefore be explicitly tracked from small to large values. This provides a window onto the "morphing" process by which a description of physics from the bulk perspective transforms into a description in the boundary perspective, and vice versa. Such calculations have shown, for instance, how chains of interacting particles from the boundary perspective can transform into strings in the bulk perspective—a particularly convincing interpolation between the two descriptions.

17. More precisely, this is a variation on Maldacena's result, modified so that the quantum field theory on the boundary is not the one that originally arose in his investigations, but instead closely approximates quantum chromodynamics. This variation also entails parallel modifications to the bulk theory. Specifically, following the work of Witten, the high temperature of the boundary theory translates into a black hole in the interior description. In turn, the dictionary between the two descriptions shows that the difficult viscosity calculations of the quark-gluon plasma translate into the response of the black hole's event horizon to particular deformations—a technical but tractable calculation.

18. Another approach to providing a full definition of string theory emerged from earlier work in an area called Matrix theory (another possible meaning of the "M" in M-theory), developed by Tom Banks, Willy Fischler, Steve Shenker, and Leonard Susskind.

Chapter 10: Universes, Computers, and Mathematical Reality

1. The number I quoted, 10^{55} grams, accounts for the contents of the observable universe today, but at ever-earlier times, the temperature of these constituents would be larger and so they would contain higher energy. The number 10^{65} grams is a better estimate of what you'd need to gather into a tiny speck to recapitulate the evolution of our universe from when it was roughly one second old.

2. You might think that because your speed is constrained to be less than the speed of light, your kinetic energy will also be limited. But that's not the case. As your speed gets ever closer to that of light, your energy grows ever larger; according to special relativity, it has no bounds. Mathematically, the formula for your energy is: $E = mc^2/\sqrt{1 - \frac{v^2}{c^2}}$, where c is the speed of light and v is your speed. As you can see, as v approaches c, E grows arbitrarily large.

Note too that the discussion is from the perspective of someone watching you fall, say someone stationary on the surface of the earth. From your perspective, while you are in free fall, you are stationary and all the surrounding matter is acquiring increasing speed.

3. With our current level of understanding, there is significant flexibility in such estimates. The number "10 grams" comes from the following consideration: the energy scale at which inflation takes place is thought to be about 10^{-5} or so times the Planck energy scale, where the latter is about 10^{19} times the energy equivalent of the mass of a proton. (If inflation happened at a higher energy scale, models suggest that evidence for gravitational waves produced in the early universe should already have been seen.) In more conventional units, the Planck scale is about 10^{-5} grams (small by everyday standards, but enormous by the scales of elementary particle physics, where such energies would be carried by individual particles). The energy density of an inflaton field would therefore have been about 10^{-5} grams packed in every cubic volume whose linear dimension is set by roughly 10^5 times the Planck length (recall, from quantum uncertainty, that energies and lengths scale inversely proportional to each other), which is about 10^{-28} centimeters. The total mass-energy carried by such an inflaton field in a volume that is 10^{-26} centimeters on a side is thus: 10^{-5} grams/(10^{-28} centimeters)^3 × (10^{-26} centimeters)^3, which is about 10 grams. Readers of *The Fabric of the Cosmos* may recall that there I used a slightly different value. The difference came from the assumption that the energy scale of the inflaton was slightly higher.

4. Hans Moravec, *Robot: Mere Machine to Transcendent Mind* (New York: Oxford University Press, 2000). See also Ray Kurzweil, *The Singularity Is Near: When Humans Transcend Biology* (New York: Penguin, 2006).

5. See, for example, Robin Hanson, "How to Live in a Simulation," *Journal of Evolution and Technology* 7, no. 1 (2001).

6. The Church-Turing thesis argues that any computer of the so-called universal Turing type can simulate the actions of another, and so it's perfectly reasonable for a computer that's within the simulation—and hence is itself simulated by the parent computer running the whole simulated world—to perform particular tasks equivalent to those undertaken by the parent computer.

7. Philosopher David Lewis developed a similar idea through what he called Modal Realism. See *On the Plurality of Worlds* (Malden, Mass.: Wiley-Blackwell, 2001). However, Lewis's motivation in introducing all possible universes differs from Nozick's. Lewis wanted a context where, for example,

counterfactual statements (such as, "If Hitler had won the war, the world today would be very different") would be instantiated.

8. John Barrow has made a similar point in *Pi in the Sky* (New York: Little, Brown, 1992).

9. As explained in endnote 10 of Chapter 7, the size of this infinity exceeds that of the infinite collection of whole numbers 1, 2, 3, . . . and so on.

10. This is a variation on the famous Barber of Seville paradox, in which a barber shaves all those who don't shave themselves. The question then is: Who shaves the barber? (The barber is usually stipulated to be male, to avoid the easy answer—the barber is a woman and so doesn't need to shave.) Let me stress that this example is meant to provide an intuitive sense of how a computer can get "stuck," but should not be thought of as a literal example of a noncomputable function. See also note 12.

11. Schmidhuber notes that an efficient strategy would be to have the computer evolve each simulated universe forward in time in a "dovetailed" manner: the first universe would be updated on every other time-step of the computer, the second universe would be updated on every other of the remaining time-steps, the third universe would be updated on every other time-step not already devoted to the first two universes, and so on. In due course, every computable universe would be evolved forward by an arbitrarily large number of time-steps.

12. A more refined discussion of computable and noncomputable functions would also include *limit computable functions*. These are functions for which there is a finite algorithm that evaluates them to ever greater precision. For instance, such is the case for producing the digits of π: a computer can produce each successive digit of π, even though it will never reach the end of the computation. So, while π is strictly speaking noncomputable, it is limit computable. Most real numbers, however, are not like π. They are not just noncomputable, they are also not limit computable.

When we consider "successful" simulations, we should include those based on limit computable functions. In principle, a convincing reality could be generated by the partial output of a computer evaluating limit computable functions.

For the laws of physics to be computable, or even limit computable, the traditional reliance on real numbers would have to be abandoned. This would apply not just to space and time, usually described using coordinates whose values can range over the real numbers, but also for all other mathematical ingredients the laws use. The strength of an electromagnetic field, for example, could not vary over real numbers, but only over a discrete set of values. Similarly for the probability that an electron is here or there. Schmidhuber has emphasized that all calculations that physicists have ever carried out have involved the manipulation of discrete symbols (written on paper, on a blackboard, or input to a computer). And so, even though this body of scientific work has always been viewed as invoking the real numbers, in practice it doesn't. Similarly for all quantities ever measured. No device has infinite accuracy and so our measurements always involve discrete numerical outputs. In that sense, all the successes of physics can be read as successes for a digital paradigm. Perhaps, then, the true laws themselves are, in fact, computable (or limit computable).

There are many different perspectives on the possibility of "digital physics." See, for example, Stephen Wolfram's *A New Kind of Science* (Champaign, Ill.: Wolfram Media, 2002) and Seth Lloyd's *Programming the Universe* (New York: Alfred A. Knopf, 2006). The mathematician Roger Penrose believes that the human mind is based on noncomputable processes and hence the universe we inhabit must involve noncomputable mathematical functions. From this perspective, our universe does not fall into the digital paradigm. See, for instance, *The Emperor's New Mind* (New York: Oxford University Press, 1989) and *Shadows of the Mind* (New York: Oxford University Press, 1994).

Chapter 11: The Limits of Inquiry

1. Steven Weinberg, *The First Three Minutes* (New York: Basic Books, 1993), p. 131.

Suggestions for Further Reading

The subject of parallel universes draws on a broad range of scientific material. There is a growing literature that focuses on various aspects of such material, mostly intended for the nonexpert, but often well-suited for those with more background. In addition to the references called out in the notes, here is a collection of books, from the many wonderful ones that have been written, through which the reader can continue exploring topics discussed in *The Hidden Reality*.

Albert, David. *Quantum Mechanics and Experience*. Cambridge, Mass.: Harvard University Press, 1994.

Alexander, H. G. *The Leibniz-Clarke Correspondence*. Manchester: Manchester University Press, 1956.

Barrow, John. *Pi in the Sky*. Boston: Little, Brown, 1992.

———. *The World Within the World*. Oxford: Clarendon Press, 1988.

Barrow, John, and Frank Tipler. *The Anthropic Cosmological Principle*. Oxford: Oxford University Press, 1986.

Bartusiak, Marcia. *The Day We Found the Universe*. New York: Vintage, 2010.

Bell, John. *Speakable and Unspeakable in Quantum Mechanics*. Cambridge, Eng.: Cambridge University Press, 1993.

Bronowski, Jacob. *The Ascent of Man*. Boston: Little, Brown, 1973.

Byrne, Peter. *The Many Worlds of Hugh Everett III*. New York: Oxford University Press, 2010.

Callender, Craig, and Nick Huggett. *Physics Meets Philosophy at the Planck Scale*. Cambridge, Eng.: Cambridge University Press, 2001.

Carroll, Sean. *From Eternity to Here.* New York: Dutton, 2010.

Clark, Ronald. *Einstein: The Life and Times.* New York: Avon, 1984.

Cole, K. C. *The Hole in the Universe.* New York: Harcourt, 2001.

Crease, Robert P., and Charles C. Mann. *The Second Creation.* New Brunswick, N.J.: Rutgers University Press, 1996.

Davies, Paul. *Cosmic Jackpot.* Boston: Houghton Mifflin, 2007.

Deutsch, David. *The Fabric of Reality.* New York: Allen Lane, 1997.

DeWitt, Bryce, and Neill Graham, eds. *The Many-Worlds Interpretation of Quantum Mechanics.* Princeton: Princeton University Press, 1973.

Einstein, Albert. *The Meaning of Relativity.* Princeton: Princeton University Press, 1988.

———. *Relativity.* New York: Crown, 1961.

Ferris, Timothy. *Coming of Age in the Milky Way.* New York: Anchor, 1989.

———. *The Whole Shebang.* New York: Simon & Schuster, 1997.

Feynman, Richard. *The Character of Physical Law.* Cambridge, Mass.: MIT Press, 1995.

———. *QED.* Princeton: Princeton University Press, 1986.

Gamow, George. *Mr. Tompkins in Paperback.* Cambridge, Eng.: Cambridge University Press, 1993.

Gleick, James. *Isaac Newton.* New York: Pantheon, 2003.

Gribbin, John. *In Search of the Multiverse.* Hoboken, N.J.: Wiley, 2010.

———. *Schrödinger's Kittens and the Search for Reality.* Boston: Little, Brown, 1995.

Guth, Alan H. *The Inflationary Universe.* Reading, Mass.: Addison-Wesley, 1997.

Hawking, Stephen. *A Brief History of Time.* New York: Bantam Books, 1988.

———. *The Universe in a Nutshell.* New York: Bantam Books, 2001.

Isaacson, Walter. *Einstein.* New York: Simon & Schuster, 2007.

Kaku, Michio. *Parallel Worlds.* New York: Anchor, 2006.

Kirschner, Robert. *The Extravagant Universe.* Princeton: Princeton University Press, 2002.

Krauss, Lawrence. *Quintessence.* New York: Perseus, 2000.

Kurzweil, Ray. *The Age of Spiritual Machines.* New York: Viking, 1999.

———. *The Singularity Is Near.* New York: Viking, 2005.

Lederman, Leon, and Christopher Hill. *Symmetry and the Beautiful Universe.* Amherst, N.Y.: Prometheus Books, 2004.

Livio, Mario. *The Accelerating Universe.* New York: Wiley, 2000.

Lloyd, Seth. *Programming the Universe*. New York: Knopf, 2006.

Moravec, Hans. *Robot*. New York: Oxford University Press, 1998.

Pais, Abraham. *Subtle Is the Lord*. Oxford: Oxford University Press, 1982.

Penrose, Roger. *The Emperor's New Mind*. New York: Oxford University Press, 1989.

————. *Shadows of the Mind*. New York: Oxford University Press, 1994.

Randall, Lisa. *Warped Passages*. New York: Ecco, 2005.

Rees, Martin. *Before the Beginning*. Reading, Mass.: Addison-Wesley, 1997.

————. *Just Six Numbers*. New York: Basic Books, 2001.

Schrödinger, Erwin. *What Is Life?* Cambridge, Eng.: Canto, 2000.

Siegfried, Tom. *The Bit and the Pendulum*. New York: John Wiley & Sons, 2000.

Singh, Simon. *Big Bang*. New York: Fourth Estate, 2004.

Stenger, Victor. *Has Science Found God?* Amherst, N.Y.: Prometheus Books, 2003.

Susskind, Leonard. *The Black Hole War*. New York: Little, Brown, 2008.

————. *The Cosmic Landscape*. New York: Little, Brown, 2005.

Thorne, Kip. *Black Holes and Time Warps*. New York: W. W. Norton, 1994.

Tyson, Neil deGrasse. *Death by Black Hole*. New York: W. W. Norton, 2007.

Vilenkin, Alexander. *Many Worlds in One*. New York: Hill and Wang, 2006.

von Weizsäcker, Carl Friedrich. *The Unity of Nature*. New York: Farrar, Straus and Giroux, 1980.

Weinberg, Steven. *Dreams of a Final Theory*. New York: Pantheon, 1992.

————. *The First Three Minutes*. New York: Basic Books, 1993.

Wheeler, John. *A Journey into Gravity and Spacetime*. New York: Scientific American Library, 1990.

Wilczek, Frank. *The Lightness of Being*. New York: Basic Books, 2008.

Wilczek, Frank, and Betsy Devine. *Longing for the Harmonies*. New York: W. W. Norton, 1988.

Yau, Shing-Tung, and Steve Nadis. *The Shape of Inner Space*. New York: Basic Books, 2010.

Index

Page references in *italic* refer to illustrations.

ALSO BY BRIAN GREENE

"Send[s] the reader's imagination hurtling through the universe on an astonishing ride. . . . He is both a skilled and kindly explicator. His excitement for science on the threshold of vital breakthroughs is extremely contagious."
—Janet Maslin, *The New York Times*

THE FABRIC OF THE COSMOS
Space, Time, and the Texture of Reality

Space and time form the very fabric of the cosmos. Yet they remain among the most mysterious of concepts. Is space an entity? Why does time have a direction? Could the universe exist without space and time? Can we travel to the past? From Newton's unchanging realm in which space and time are absolute, to Einstein's fluid conception of spacetime, to quantum mechanics' entangled arena where vastly different objects can instantaneously coordinate their behavior, Greene takes us all, regardless of our scientific backgrounds, on an irresistible and revelatory journey to the new layers of reality that modern physics has discovered lying just beneath the surface of our everyday world.

Science

VINTAGE BOOKS
Available wherever books are sold.
www.randomhouse.com